# TEMPLATE MATCHING TECHNIQUES IN COMPUTER VISION

# TEMPLATE MATCHING TECHNIQUES IN COMPUTER VISION

## THEORY AND PRACTICE

**Roberto Brunelli**

*Fondazione Bruno Kessler, Italy*

WILEY

A John Wiley and Sons, Ltd, Publication

This edition first published 2009


*Registered office*
John Wiley & Sons Ltd, The Atrium, Southern Gate, Chichester, West Sussex, PO19 8SQ, United Kingdom

For details of our global editorial offices, for customer services and for information about how to apply for permission to reuse the copyright material in this book please see our website at www.wiley.com.



*Library of Congress Cataloging-in-Publication Data*

Brunelli, Roberto, 1961-
Template matching techniques in computer vision : theory and practice / Roberto Brunelli.
p. cm.
Includes bibliographical references and index.
ISBN 978-0-470-51706-2 (cloth)
1. Computer vision. 2. Template matching (Digital image processing) 3. Image processing–Digital techniques. 4. Optical pattern recognition. 5. Image analysis. I. Title.
TA1634.B78 2009
006.3'7–dc22
2008052237

A catalogue record for this book is available from the British Library.

ISBN 978-0-470-51706-2

Set in 10/12pt Times by Sunrise Setting Ltd, Torquay, UK.

MIX
Paper from responsible sources
FSC® C013604

# CONTENTS

# PREFACE

Detection and recognition of objects from their images, irrespective of their orientation, scale, and view, is a very important research subject in computer vision, if not computer vision itself. This book focuses on a subset of the object recognition techniques proposed so far by the computer vision community, those employing the idea of projection to match image patterns, and on a specific class of target objects, faces, to illustrate general object recognition approaches. Face recognition and interpretation is a critical task for people, and one at which they excel. Over the last two decades it has received increasing attention from the computer vision community for a variety of reasons, ranging from the possibility of creating computational models of interesting human recognition tasks, to the development of practical biometric systems and interactive, emotion-aware, and capable human–machine interfaces.

The topics covered in this book have been investigated over a period of about 30 years by the image processing community, providing increasingly better computer vision solutions to the problem of automatic object location and recognition. While many books on computer vision are currently available that touch upon some of the topics addressed in the present book, none of them, to the best of the author's knowledge, provides a coherent, in-depth coverage of template matching, presenting a varied set of techniques from a common perspective. The methods considered present both theoretical and practical interest by themselves as well as enabling techniques for more complex vision systems (stereo vision, robot navigation, image registration, multimedia retrieval, target tracking, landmark detection, just to mention a few). The book contains many photographs and diagrams that help the user grasp qualitative and quantitative aspects of the material presented. The software available on the book's web site provides a high-level image processing environment and image datasets to explore the techniques presented in the book.

Knowledge of basic calculus, statistics, and probability theory is a prerequisite for the reader. The material covered in the book is at the level of (advanced) undergraduate students or introductory Ph.D. courses and will prove useful to researchers and developers of computer vision systems addressing a variety of tasks, from robotic vision to quality control and biometric systems. It may be used for a special topics course on image analysis at the graduate level. Another expected use is as a supporting textbook for an intensive short course on template matching, with the possibility of choosing between a theoretical and an application-oriented bias. The techniques are discussed at a level that makes them useful also for the experienced researcher and make the book an essential learning kit for practitioners in academia and industry.

Rarely, if ever, does a book owe its existence to the sole author, and this one certainly does not. First a tribute to the open source software community, for providing the many tools necessary to describe ideas and making them operational. To Jaime Vives Piqueres and to Matthias Baas, my gratitude for providing me with technical help on the rendering of the three-dimensional models appearing in the book. To Andrew Beatty at Singular Inversions, appreciation for providing me with a free copy of their programs for the generation of three-dimensional head models. A blossomy 'whoa' to Filippo Brunelli, for using these programs to generate the many virtual heads popping up in the figures of the book and feeding some of the algorithms described. To Carla Maria Modena, a lot of thanks for helping in the revision of the manuscript. And, finally, very huge thanks indeed to Tomaso Poggio, the best colleague I ever had, and the main culprit for the appearance of this book, as the first epigraph in the book tells you.

Roberto Brunelli
Trento, Italy

# 1 INTRODUCTION

Somewhere, somewhen,
a two headed strategic meeting
on face recognition and matters alike:
t: What about using template matching?
r: Template matching?
t: Yes, a simple technique to compare patterns . . .
r: I'll have a look.

*Faces' faces – r's virtual autobiography*
ROBERTO BRUNELLI

Go thither; and, with unattainted eye,
Compare her face with some that I shall show,
And I will make thee think thy swan a crow.

*Romeo and Juliet*
WILLIAM SHAKESPEARE

Computer vision is a wide research field that aims at creating machines that see, not in the limited meaning that they are able to sense the world by optical means, but in the more general meaning that they are able to understand its perceivable structure. Template matching techniques, as now available, have proven to be a very useful tool for this intelligent perception process and have led machines to superhuman performance in tasks such as face recognition. This introductory chapter sets the stage for the rest of the book, where template matching techniques for monochromatic images are discussed.

## 1.1. Template Matching and Computer Vision

The whole book is dedicated to the problem of template matching in computer vision. While template matching is often considered to be a very basic, limited approach to the most interesting problems of computer vision, it touches upon many old and new techniques in the field.

The two terms *template* and *matching* are used in everyday language, but recalling the definitions more closely related to their technical meaning is useful:

**template/pattern**

1. Anything fashioned, shaped, or designed to serve as a model from which something is to be made: a model, design, plan, outline.

*Template Matching Techniques in Computer Vision: Theory and Practice* Roberto Brunelli

2. Something formed after a model or prototype, a copy; a likeness, a similitude.
3. An example, an instance; esp. a typical model or a representative instance.

**matching**

1. Comparing in respect of similarity; to examine the likeness or difference of.

A template may additionally exhibit some variability: not all of its instances are exactly equal (see Figure 1.1). A simple example of template variability is related to its being corrupted by additive noise. Another important example of variability is due to the different viewpoints from which a single object might be observed. Changes in illumination, imaging sensor, or sensor configuration may also cause significant variations. Yet another form of variability derives from intrinsic variability across physical object instances that causes variability of the corresponding image patterns: consider the many variations of faces, all of them sharing a basic structure, but also exhibiting marked differences. Another important source of variability stems from the temporal evolution of a single object, an interesting example being the mouth during speech. Many tasks of our everyday life require that we identify classes of objects in order to take appropriate actions in spite of the significant variations that these objects may exhibit. The purpose of this book is to present a set of techniques by which a computer can perform some of these identifications. The techniques presented share two common features:

- all of them rely on explicit templates, or on representations by which explicit templates can be generated;
- recognition is performed by matching: images, or image regions, are set in comparison to the stored representative templates and are compared in such a way that their appearance (their image representation) plays an explicit and fundamental role.

The simplest template matching technique used in computer vision is illustrated in Figure 1.2. A planar distribution of light intensity values is transformed into a vector $\boldsymbol{x}$ which can be compared, in a coordinate-wise fashion, to a spatially congruent light distribution similarly represented by vector $\boldsymbol{y}$:

$$d(\boldsymbol{x}, \boldsymbol{y}) = \frac{1}{N} \sum_{i=1}^{N} (x_i - y_i)^2 = \frac{1}{N} \|\boldsymbol{x} - \boldsymbol{y}\|_2^2 \tag{1.1}$$

$$s(\boldsymbol{x}, \boldsymbol{y}) = \frac{1}{1 + d(\boldsymbol{x}, \boldsymbol{y})}. \tag{1.2}$$

A small value of $d(\boldsymbol{x}, \boldsymbol{y})$ or a high value of $s(\boldsymbol{x}, \boldsymbol{y})$ is indicative of pattern similarity. A simple variation is obtained by substituting the $L_2$ norm with the $L_p$ norm:

$$d_p(\boldsymbol{x}, \boldsymbol{y}) = \frac{1}{N} \sum_{i=1}^{N} (x_i - y_i)^p = \frac{1}{N} \|\boldsymbol{x} - \boldsymbol{y}\|_p^p. \tag{1.3}$$

If $\boldsymbol{x}$ is representative of our template, we search for other instances of it by superposing it on other images, or portions thereof, searching for the locations of lowest distance $d(\boldsymbol{x}, \boldsymbol{y})$ (or highest similarity $s(\boldsymbol{x}, \boldsymbol{y})$).

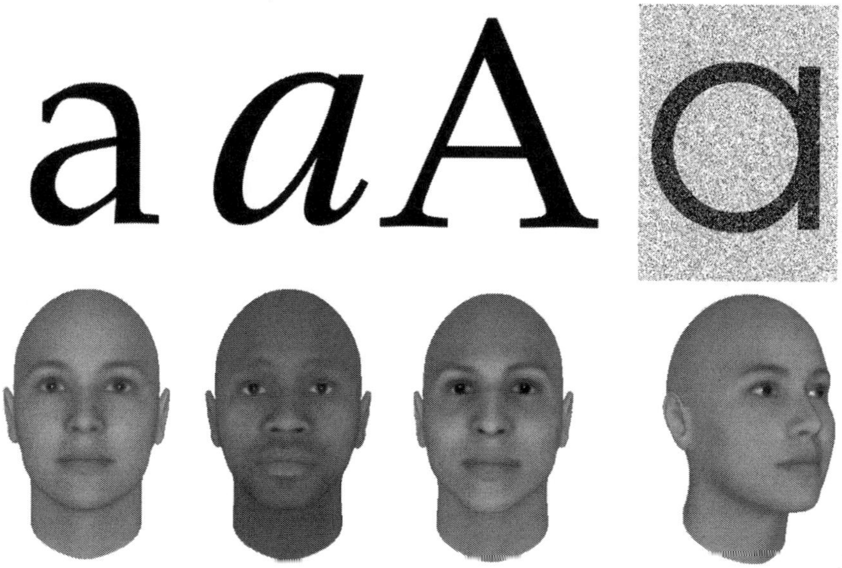

**Figure 1.1.** Templates from two very common classes: characters and 'characters', i.e. faces. Both classes exhibit intrinsic variability and can appear corrupted by noise.

The book shows how this simple template matching technique can be extended to become a flexible and powerful tool supporting the development of sophisticated computer vision systems, such as face recognition systems.

While not a face recognition book, its many examples are related to automated face perception. The main reason for the bias is certainly the background of the author, but there are at least three valid reasons for which face recognition is a valid test bed for template matching techniques. The first one is the widespread interest in the development of high-performing face recognition systems for security applications and for the development of novel services. The second, related reason is that, over the last 20 years, the task has become very popular and it has seen a significant research effort. This has resulted in the development of many algorithms, most of them of the template matching type, providing material for the book. The third reason is that face recognition and facial expression interpretation are two tasks where human performance is considered to be flawless and key to social human behavior. Psychophysical experiments and the evolution of matching techniques have shown that human performance is not flawless and that machines can, sometimes, achieve super human performance.

## 1.2. The Book

A modern approach to template matching in computer vision touches upon many aspects, from imaging, the very first step in getting the templates, to learning techniques that are key to the possibility of developing new systems with minimal human intervention. The chapters present a balanced description of all necessary concepts and techniques, illustrating them with examples taken from face processing tasks.

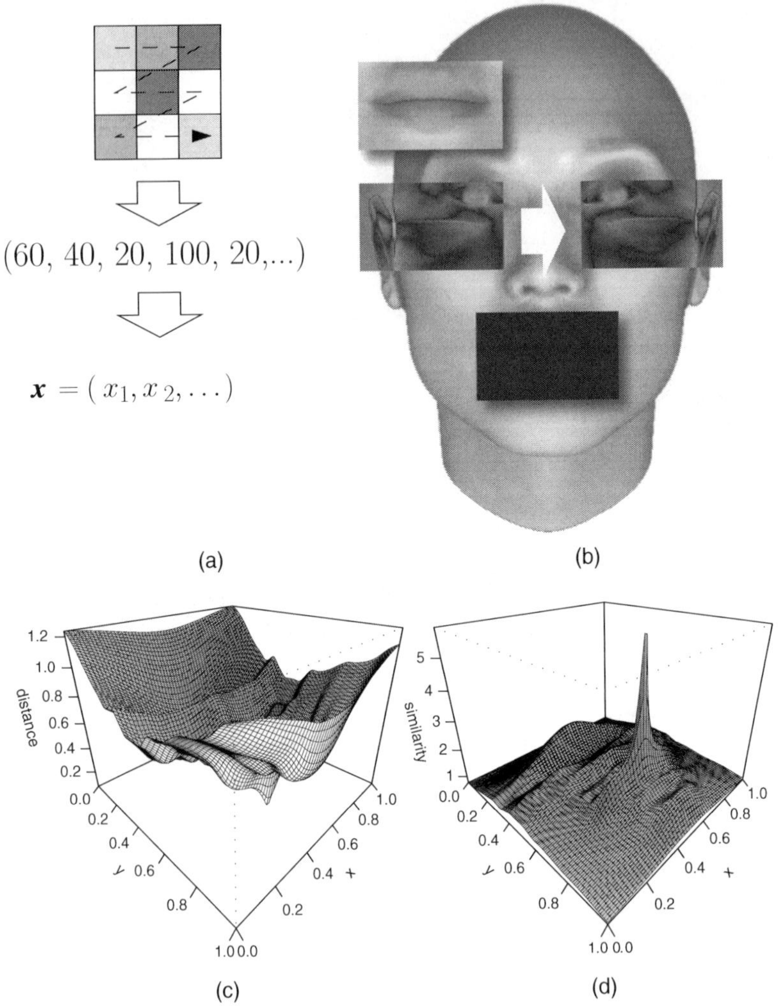


**Figure 1.2.** The simplest template matching technique: templates are represented as vectors (a) and they are matched by computing their distance in the associated vector space. The template is moved over the image, as a sliding window (b), and the difference between the template and the image is quantified using Equation 1.1, searching for the minimum value (c), or Equation 1.2, searching for the maximum value (d).

A complete description of the imaging process, be it in the case of humans, animals, or computers, would require a (very large) book by itself and we will not attempt it. Chapter 2 discusses some aspects of it that turn out to be critical in the design of artificial vision systems. The basics of how images are created using electromagnetic stimuli and imaging devices are considered. Simple concepts from optics are introduced (including distortion, depth of field, aperture, telecentric lens design) and eyes and digital imaging sensors briefly described. The sampling theorem is presented and its impact on image representation and common image processing operations such as resizing is discussed.

Chapter 3 formally introduces template matching as an hypothesis testing problem. The Bayesian and frequentist approaches are considered with particular emphasis on the Neyman–Pearson paradigm. Matched filters are introduced from a signal processing perspective and simple pattern variability is addressed with the normalized Pearson correlation coefficient. Hypothesis testing often requires the statistical estimation of the parameters characterizing the associated decision function; some subtleties in the estimation of covariance matrices are discussed.

A major issue in template matching is the stability of similarity scores with respect to noise extended to include unmodelled phenomena. Many commonly used estimators suffer from a lack of robustness: small perturbations in the data can drive them towards uninformative values. Chapter 4 addresses the concept of estimator robustness in a technical way, presenting applications of robust statistics to the problem of pattern matching.

Linear correspondence measures like correlation and the sum of squared differences between intensity distributions are fragile. Chapter 5 introduces similarity measures based on the relative ordering of intensity values. These measures have demonstrable robustness both to monotonic image mappings and to the presence of outliers.

While finding a single, well-defined shape is useful, finding instances of a class of shapes can be even more useful. Intraclass variability poses new problems for template matching and several interesting solutions are available. Chapter 6 focuses on the use of projection operators on a one-dimensional space to solve the task. The use of projection operators on multidimensional spaces is covered in Chapter 8.

Finding simple shapes, such as lines and circles, in images may look like a simple task but computational issues coupled with noise and occlusions require some not so naive solutions. In spite of the apparent diversity of lines and areas, it turns out that common approaches to the detection of linear structures can be seen as an efficient implementation of matched filters. Chapter 7 describes how to compute salient image discontinuities and how simple shapes embedded in the resulting map can be located with the Radon/Hough transform.

The representation of images of even moderate resolution requires a significant amount of numeric data, usually one (three) values per pixel if the typical array-based method is adopted. Chapter 8 investigates the possibility of alternative ways of representing iconic data so that a large variety of images can be represented using vectors of reduced dimensionality. Besides significant storage savings, these approaches provide significant benefits to template detection and recognition algorithms, improving their efficiency and effectiveness.

Chapter 9 addresses a couple of cases that are not easily reduced to pattern detection and classification. One such case is the detailed estimation of the parameters of a parametric curve: while Hough/Radon techniques may be sufficient, accurate estimation may benefit from specific approaches. Another important case is the comparison of anatomical structures, such as brain sections, across different individuals or, for the same person, over time. Instead of modeling the variability of the patterns within a class as a static multidimensional manifold, we may focus on the constrained deformation of a parameterized model and measure similarity by the deformation stress.

The drawback of template matching is its high computational cost which has two distinct origins. The first source of complexity is the necessity of using multiple templates to accommodate the variability exhibited by the appearance of complex objects. The second source of complexity is related to the representation of the templates: the higher the resolution, i.e. the number of pixels, the heavier the computational requirements. Besides some computational tricks, Chapter 10 presents more organized, structural ways to improve the speed at which template matching can be performed.

Matching sets of points using techniques targeted at area matching is far from optimal, with regard to both efficiency and effectiveness. Chapter 11 shows how to compare sparse templates, composed by points with no textural properties, using an appropriate distance. Robustness to noise and template deformation as well as computational efficiency are analyzed.

When the probability distribution of the templates is unknown, the design of a classifier becomes more complex and many critical estimation issues surface. Chapter 12 presents basic results upon which two interrelated, powerful classifier design paradigms stand: regularization networks and support vector machines.

Many applications in image processing rely on robust detection of image features and accurate estimation of their parameters. Features may be too numerous to justify the process of deriving a new detector for each one. Chapter 13 exploits the results presented in Chapter 8 to build a single, flexible, and efficient detection mechanism. The complementary aspect of detecting templates considered as a set of separate features will also be addressed and an efficient architecture presented.

Template matching techniques are a key ingredient of many computer vision systems, ranging from quality control to object recognition systems among which biometric identification systems have today a prominent position. Among biometric systems, those based on face recognition have been the subject of extensive research. This popularity is due to many factors, from the non-invasiveness of the technique, to the high expectations due to the widely held belief that human face recognition mechanisms perform flawlessly. Building a face recognition system from the ground up is a complex task and Chapter 14 addresses all the required practical steps: preprocessing issues, feature scoring, the integration of multiple features and modalities, and the final classification stage.

The process of developing a computer vision system for a specific task often requires the interactive exploration of several alternative approaches and variants, preliminary parameter tuning, and more. Appendix A introduces AnImAl, an image processing package written for the R statistical software system. AnImAl, which relies on an algebraic formalization of the concept of image, supports interactive image processing by adding to images a self-documenting capability based on a history mechanism. The documentation facilities of the resulting interactive environment support a practical approach to reproducible research.

A key need in the development of algorithms in computer vision (as in many other fields) is the availability of large datasets for training and testing them. Ideally, datasets should cover the expected variability range of data and be supported by high-quality annotations describing what they represent so that the response of an algorithm can be compared to reality. Gathering large, high-quality datasets is, however, time consuming. An alternative is available for computer vision research: computer graphics systems can be used to generate photorealistic images of complex environments together with supporting ground truth information. Appendix B shows how these systems can be exploited to generate a flexible (and cheap) evaluation environment.

Evaluation of algorithms and systems is a complex task. Appendix C addresses four related questions that are important from a practical and methodological point of view: what is a good response of a template matching system, how can we exploit data to train and at the same time evaluate a classification system, how can we describe in a compact but informative way the performance of a classification system, and, finally, how can we compare multiple classification systems for the same task in order to assess the state of the art of a technology?

The exposition of the main chapter topics is complemented by several intermezzos which provide ancillary material or refresh the memory of useful results. The arguments presented are illustrated with several examples from a very specific computer vision research topic: face detection, recognition, and analysis. There are three main reasons for the very biased choice: the research background of the author, the relevance of the task in the development of biometrics systems, and the possibility that a computational solution to these problems helps understanding (and benefits from the understanding of) the way people do it. Some of the images appearing in the book are generated using the computer graphics techniques described in Appendix B and the packages POV-ray (The Povray Team 2008), POVMan (Krouverk 2005), Aqsis (The Aqsis Team 2007), and FaceGen (Singular Inversions 2008).

**Intermezzo 1.1.** The definition of intermezzo

**intermezzo**

1. A brief entertainment between two acts of a play; an entr'acte.
2. A short movement separating the major sections of a lengthy composition or work.

References to relevant literature are not inserted throughout chapter text but are postponed to a final chapter section. Their order of presentation follows the structure of the chapter. All papers on which the chapter is based are listed and pointers to additional material are also provided. References are not meant to be exhaustive, but the interested reader can find additional literature coverage in the cited papers.

## 1.3. Bibliographical Remarks

This book, while addressing a very specific technique of computer vision, touches upon concepts and methods typical of other fields, from optics to machine learning and its comprehension benefits from readings in these fields.

Among the many books on computer vision, the one by Marr (1982) is perhaps the most fascinating. The book by Horn (1986) still provides an excellent introduction to the fundamental aspects of computer vision, with a careful treatment of image formation. A more recent book is that by Forsyth and Ponce (2002).

Basic notions of probability and statistics can be found in Papoulis (1965). Pattern classification is considered in detail in the books by Fukunaga (1990) and Duda *et al.* (2000). Other important reference books are Moon and Stirling (2000) and Bishop (2007).

A very good, albeit concise, reference for basic mathematical concepts and results is the *Encyclopedic Dictionary of Mathematics* (Mathematical Society of Japan 1993). A wide coverage of numerical techniques is provided by Press *et al.* (2007).

Two interesting papers on computer and human face recognition are those by Sinha *et al.* (2006) and O'Toole *et al.* (2007). The former presents several results on human face analysis processes that may provide guidance for the development of computer vision algorithms. The latter presents some results showing that, at least in some situations, computer vision efforts resulted in algorithms capable of superhuman performance.

## References

Bishop C 2007 *Pattern Recognition and Machine Learning*. Springer.

Duda R, Hart P and Stork D 2000 *Pattern Classification* 2nd edn. John Wiley & Sons, Ltd.

Forsyth D and Ponce J 2002 *Computer Vision: A Modern Approach*. Prentice Hall.

Fukunaga K 1990 *Statistical Pattern Recognition* 2nd edn. Academic Press.

Horn B 1986 *Robot Vision*. MIT Press.

Krouverk V 2005 POVMan v1.2. http://www.aetec.ee/fv/vkhomep.nsf/pages/povman2.

Marr D 1982 *Vision*. W.H. Freeman.

Mathematical Society of Japan 1993 *Encyclopedic Dictionary of Mathematics* 2 edn. MIT Press.

Moon T and Stirling W 2000 *Mathematical Methods and Algorithms for Signal Processing*. Prentice Hall.

O'Toole A, Phillips P, Fang J, Ayyad J, Penard N and Abdi H 2007 Face recognition algorithms surpass humans matching faces over changes in illumination. *IEEE Transactions on Pattern Analysis and Machine Intelligence* **29**(9), 1642–1646.

Papoulis A 1965 *Probability, Random Variables and Stochastic Processes*. McGraw-Hill.

Press W, Teukolsky S, Vetterling W and Flannery B 2007 *Numerical Recipes* 3rd edn. Cambridge University Press.

Singular Inversions 2008 FaceGen Modeller 3.2. http://www.facegen.com.

Sinha P, Balas B, Ostrovsky Y and Russell R 2006 Face recognition by humans: nineteen results all computer vision researchers should know about. *Proceedings of the IEEE* **94**, 1948–1962. Face Recognition.

The Aqsis Team 2007 Aqsis v1.2. http://www.aqsis.org/.

The Povray Team 2008 The Persistence of Vision Raytracer v3.6. http://www.povray.org/.

# 2 THE IMAGING PROCESS

I have a good eye, uncle; I can see a church by daylight.

*Much Ado About Nothing*
WILLIAM SHAKESPEARE

A complete description of the imaging process, be it in the case of humans, animals, or computers, would require a (very large) book by itself and we will not attempt it. Rather, we discuss some aspects of it that turn out to be critical in the design of artificial vision systems. The basics of how images are created using electromagnetic stimuli and imaging devices will be considered. Simple concepts from optics will be introduced (including distortion, depth of field, aperture, telecentric lens design). Eyes and digital imaging sensors will be briefly considered. The sampling theorem is presented and its impact on image representation and common image processing operations such as resizing is discussed.

## 2.1. Image Creation

Computer vision can be considered as the science and technology of machines that see, obtaining information on the real world from images of it. Images are created by the interaction of light with objects and they are captured by optical devices whose nature may differ significantly.

### 2.1.1. LIGHT

What is commonly understood by light is actually a propagating oscillatory disturbance in the electromagnetic field which describes the interaction of charged particles, such as electrons. The field derives from the close interaction of varying electric and magnetic fields whose coupling is described by a set of partial differential equations known as Maxwell's equations. An important consequence of them is the second-order partial differential equation that describes the propagation of electromagnetic waves through a medium or in vacuum. The free space version of the electromagnetic wave equation is

$$\left(\nabla^2 - \frac{1}{c}\frac{\partial^2}{\partial t^2}\right)\boldsymbol{E} = 0 \tag{2.1}$$

where $\boldsymbol{E}$ is the electric field (and similarly for the magnetic field $\boldsymbol{B}$): light is a visible solution of this equation. From the theory of Fourier decomposition, the finite spatio-temporal extent of real physical waves results in the fact that they can be described as a superposition of an infinite set of sinusoidal frequencies. In many cases we may then limit our analysis to pure

**Intermezzo 2.1.** Convolution and its properties

As convolution, and the closely related cross-correlation, play a significant role in our analysis of template matching techniques, it is useful to recall their definitions and their most important properties. Both operations feature a continuous and a discrete definition. The convolution of two real continuous functions $f$ and $g$ is a new function defined as

$$(f * g)(x) = \int f(y)g(x-y)\,dy \tag{2.2}$$

where integration is extended to the domain over which the two functions are defined. The cross-correlation is a new function defined as

$$(f \otimes g)(x) = \int f(y)g(x+y)\,dy. \tag{2.3}$$

The discrete versions are

$$(f * g)(i) = \sum_j f(j)g(i-j) \tag{2.4}$$

$$(f \otimes g)(i) = \sum_j f(j)g(i+j). \tag{2.5}$$

The two operations provide the same result when one of the two argument functions is even. The main properties of convolution are

$$f * g = g * f \tag{2.6}$$

$$f * (g + h) = (f * g) + (f * h) \tag{2.7}$$

$$f * (g * h) = (f * g) * h \tag{2.8}$$

$$(f * g)(x - a) = f(x - a) * g(x) \tag{2.9}$$

$$\frac{d}{dx}(f * g)(x) = \left(\frac{d}{dx} f\right) * g \tag{2.10}$$

$$\text{support}(f * g) = \text{support}(f) \cup \text{support}(g). \tag{2.11}$$

Convolution and correlation can be extended in a straightforward way to the multidimensional case.

sinusoidal components that we can write conveniently in complex form, remembering that we should finally get the real or imaginary part of the complex results:

$$\boldsymbol{E}(\boldsymbol{x}, t) = \boldsymbol{E}_0 e^{\boldsymbol{k}\cdot\boldsymbol{x} - \omega t} \tag{2.12}$$

where $\boldsymbol{k}$ is the wave vector representing the propagating direction and the angular frequency $\omega$ is related to frequency $f$ by $\omega = 2\pi f$. The electric and magnetic fields for the plane wave represented by Equation 2.12 are perpendicular to each other and to the direction of propagation of the wave. The velocity of the wave $c$, the wavelength $\lambda$, the angular frequency $\omega$, and the size of the wave vector $\boldsymbol{k}$ are related by

$$c = \frac{\omega}{k} = \frac{\omega\lambda}{2\pi} = f\lambda. \tag{2.13}$$

Besides wavelength, two other properties of light are of practical importance: its polarization and its intensity. Polarization and its usefulness in imaging are considered in Intermezzo 2.2. Even if we described light using the wave equation, there are no sensors to detect directly its amplitude and phase. The only quantity that can be detected is the intensity $I$ of the radiation incident on the sensor, the irradiance; that is, the time average of the radiation energy which

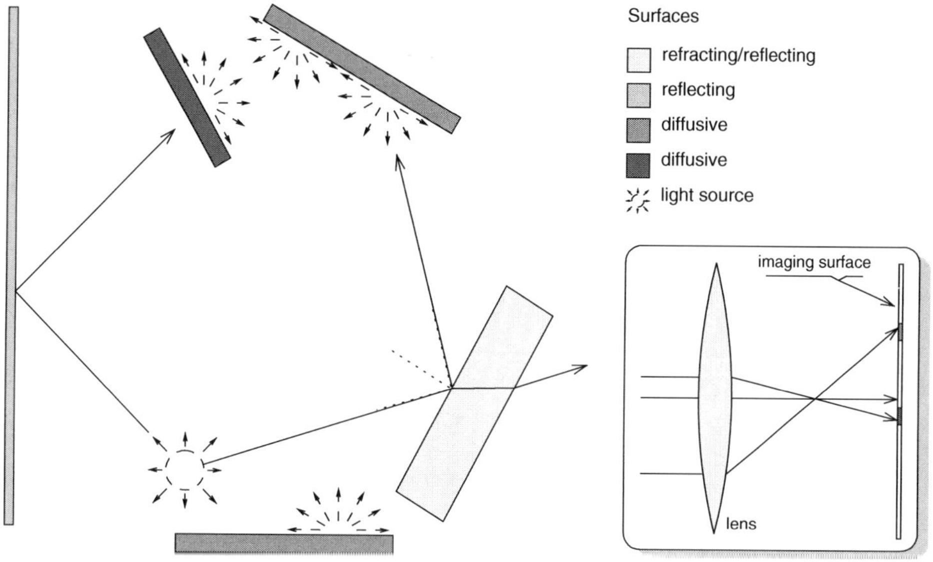


**Figure 2.1.** The process of light propagation. The goal of computer vision is to deduce the path of light from its observation at a sensing surface.

crosses unit area in unit time. For a plane wave

$$I(x_1, x_2) \propto E_0^2 \tag{2.14}$$

where $(x_1, x_2)$ represents a point on the (real or virtual) surface at which we perform the measurement (see Figure 2.1). Deriving information on the world from the two-dimensional intensity map $I(x_1, x_2)$ is the goal of computer vision. Plane waves are not the only solution to Maxwell's equations, spherical waves being another very important one. A spherical wave is characterized by the fact that its components depend only on time and on the distance $r$ from its center, where the light source is located (see Intermezzo 2.3). Energy conservation requires the irradiance of a spherical wave to decay as $r^{-2}$, a fact often appreciated when photographing with a flash unit. A spherical wave can be approximated with a plane wave when $r$ is large; in many cases of interest plane waves can then provide a good approximation of light waves. Two entities (see Figure 2.2) are fundamental for studying light propagation:

**Definition 2.1.1.** *A wavefront is a surface over which an optical disturbance has a constant phase.*

**Definition 2.1.2.** *Rays are lines normal to the wavefronts at every point of intersection.*

The discovery of the photoelectric effect, by which light striking a metal surface ejects electrons whose energy is proportional to the frequency and not the intensity of light, led to quantum field theory: interactions among particles are mediated by other particles, the photon being the mediating particle for the electromagnetic field. Photons have an associated energy

$$E_\lambda = \frac{hc}{\lambda} \tag{2.15}$$

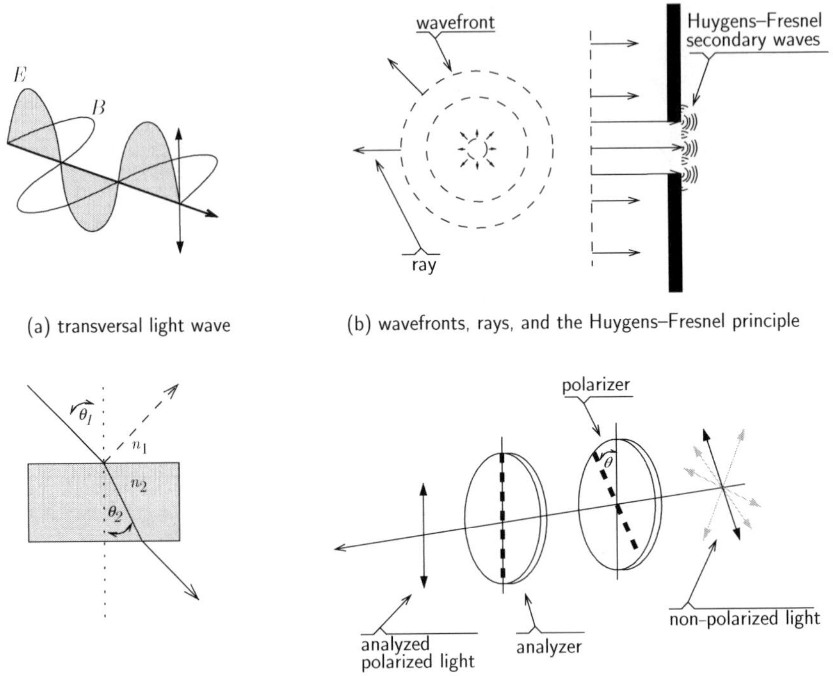


**Figure 2.2.** A graphical illustration of some important optics concepts described in the chapter.

where $h$ is Planck's constant: from a particle point of view, intensity is related to the number of photons. As we will see, both wave and particle aspects of light have important consequences in the development of systems that sense the world using electromagnetic radiation, and they fix some fundamental limits for them. The speed of light depends on the medium of propagation, and in a linear, isotropic, and non-dispersive material is

$$c = \frac{c_0}{n} \tag{2.16}$$

where $n$ is the refractive index of the medium and $c_0$ is the speed of light in vacuum. Usually $n > 1$ and it depends on frequency: it generally decreases with decreasing frequency (increasing wavelength). When light crosses the boundary between media with different refractive indexes it changes its direction and is partially reflected. These effects allow us to control the propagation of light by interposing properly shaped elements of different refractive indexes. The refraction of light crossing the boundary of two different isotropic media results in a change of the propagation direction obeying Snell's law

$$n_1 \sin\,\theta_1 = n_2 \sin\theta_2 \tag{2.17}$$

where $\theta_1$ represents the angle with respect to the normal of the boundaries: when $n_2 > n_1$ the light will be deflected towards the normal (see Figure 2.2). When $n_1 > n_2$, so that light passes

**Intermezzo 2.2.** Polarized light

The vector representing the electric field can be decomposed into two orthogonal components. In the case of a simple harmonic wave the two components have the same frequency but may have different phases. However, if they have the same phase, the direction of the electric vector remains constant and its changes are restricted to a constant plane: the light is linearly polarized. Polarized light may result from the reflection of unpolarized light from dielectric materials (as electric dipoles do not emit in the direction along which they oscillate) or from selective transmission of one of the two components of the electric field (e.g. using a Glan–Thomson prism). A natural cause of partially polarized light is light scattering by small particles. Polarized light is important in computer vision because it can be selectively filtered using polarizers according to Malu's law:

$$I = I_0 \cos^2 \theta_i \tag{2.18}$$

where $I_0$ is the intensity of the polarized light, $I$ the intensity transmitted by the filter, and $\theta_i$ is the angle of the polarized light to that of the polarizer (see Figure 2.2). Reflection preserves polarization while diffusion by scattering surfaces does not, producing instead unpolarized light. Let us consider an inspection problem where the integrity of a non-metallic marker overlying a metallic object must be verified. The visible metallic parts of the object produce a lot of glare that may severely impair the imaging of the marker. However, if we illuminate the specimen with polarized light and we observe the scene with a properly rotated polarizer we may get rid of the reflections from the metal parts (suppressed by Malu's law) while still perceiving a fraction of the unpolarized light from the diffusing surfaces. Another case is the inspection of specimens immersed in water. In this case the water surface reflects partially polarized light. Polarization of reflected light is complete at the Brewster angle $\theta_B = \arctan(n_2/n_1)$, $n_2, n_1$ being the refractive indexes of the materials, that for visible light is approximately 53° to the normal for an air–water interface. These reflections, which would prevent a clear image of objects below the surface, can be reduced using a properly oriented polarizer.

The detection of reflected light polarization is of help in several image understanding applications, including the discrimination of dielectric/metal material, segmentation of specularities, and separation of specular and diffuse reflection components (Wolff 1995).

from a dense to a less dense medium, e.g. from water to air, we may observe total internal reflection: no refracted ray exists.

Different frequencies of oscillation give rise to the different forms of electromagnetic radiation, from radio waves at the lowest frequencies, to visible light at the intermediate frequencies, to gamma rays at the highest frequencies. The whole set of possibilities is known as the electromagnetic spectrum and the nomenclature for the main portions is reported in Table 2.1. The study of the properties and behavior of visible light, with the addition of infrared and ultraviolet light, and of its interaction with matter is the subject of optics, itself an important area of physics. As light is an electromagnetic wave, similar phenomena occur over the complete electromagnetic spectrum and can also be found in the analysis of elementary particles due to wave–particle duality, the fact that matter exhibits both wave-like and particle-like properties.

### 2.1.2. GATHERING LIGHT

Images are created by controlling light propagation with optical devices so that we can effectively detect its intensity without disrupting the information it contains on the world through which it traveled. Refraction of light by means of media with different refractive indexes, the lenses, is the basis of optical systems. Interposition of glass elements of different shapes and refractive indexes, allows us to control the propagation of light so that different rays emitted by a single point in the world can be focused into a corresponding image of

**Table 2.1.** The different portions of the electromagnetic spectrum. The most common unit for wavelength is the angstrom ($10^{-10}$ m).

| Region | Frequency (Hz) | Wavelength (Å) |
|---|---|---|
| Radio | $< 3 \times 10^9$ | $> 10^9$ |
| Microwave | $3 \times 10^9$ to $3 \times 10^{12}$ | $10^9$–$10^6$ |
| Infrared | $3 \times 10^{12}$ to $4.3 \times 10^{14}$ | $10^6$–7000 |
| Visible | $4.3 \times 10^{14}$ to $7.5 \times 10^{14}$ | 7000–4000 |
| Ultraviolet | $7.5 \times 10^{14}$ to $3 \times 10^{17}$ | 4000–10 |
| X-rays | $3 \times 10^{17}$ to $3 \times 10^{19}$ | 10–0.1 |
| Gamma rays | $> 3 \times 10^{19}$ | $< 0.1$ |

**Intermezzo 2.3.** Light sources

There are many sources of light that we encounter daily, such as the sun, incandescent bulbs, and fluorescent tubes. The basic mechanism underlying light emission is atomic excitation and relaxation. When an atom adsorbs energy, its outer electrons move to an excited state from which they relax to the ground state, returning the energy in the form of photons, whose wavelength is related to their energy. The specification of the spectral exitance or spectral flux density, the emitted power per unit area per unit wavelength interval, characterizes a light source. The human eye is sensitive to a limited range of electromagnetic radiation and it perceives different wavelengths as different colors: the longer wavelengths as red, the shorter ones as blue.

Excitation by thermal energy results in the emission of photons of all energies, and the corresponding light spectrum is continuous. The finite time over which all electron transitions happen and atomic thermal motion are responsible for the fact that light emission is never perfectly monochromatic but characterized by a frequency bandwidth $\Delta\nu$ inversely proportional to the temporal extent of the events associated with the emission

$$\Delta\nu \approx \frac{1}{\Delta\tau_c}$$

where $\Delta\tau_c$ is also known as coherence time. For a blackbody, i.e. a perfect absorber, Planck's radiation law holds

$$I(\lambda, t) = \frac{2hc^2}{\lambda^5} \frac{1}{e^{hc(\lambda kt)^{-1}} - 1}$$

and the wavelength of maximum exitance $\lambda_{\text{peak}}$ obeys Wien's law

$$\lambda_{\text{peak}} T = 2.9 \times 10^6 \text{ nm K}$$

so that the hotter the body, the bluer the light. Thermal light sources can then be indexed effectively by their temperature.

When light passes through an absorbing medium, each wavelength may undergo selective absorption: the spectral distribution of light changes. The same thing happens by selective reflection from surfaces; this is the reason why we see them in different colors. In most cases, light-detecting devices integrate light of several wavelengths using a convolution mechanism: they compute a spectrally weighted average of the incoming light. Multiple convolution kernels may operate on different spectral regions, resulting in multichannel, or color, imaging. Different sensors are characterized by different kernels so that their colors may differ: images from different sensors need to be calibrated so that they agree on the color they report. When a single kernel is used, we have monochromatic vision. The light spectrum that a sensor perceives depends then on the light sources and on the media and objects the light interacts with. It is sometimes possible to facilitate the task of computer vision by means of appropriate lighting and environment colors.

the point on a light-detecting device. A typical configuration is composed of several radially symmetric lenses whose symmetry axes are all aligned along the optical axis of the system.

Computations are quite complex in the general case but may be simplified significantly by considering the limit of small angles from the optical axis (the paraxial approximation) and, at the same time, thin lenses with spherical surfaces immersed in air (see Figure 2.3). The main property of a lens is its focal length $f$, the distance on the optical axis from its center to a point onto which collimated light parallel to the axis is focused:

$$\frac{1}{f} \approx (n-1)\left(\frac{1}{R_1} - \frac{1}{R_2}\right) \tag{2.19}$$

where $R_1$ ($R_2$) is the radius of curvature of the lens surface closest to (farthest from) the light source, and $n$ is shorthand for $n_{lm} = n_l/n_m$, the relative refraction index of the lens ($n_l$) and the medium where the lens is immersed ($n_m$). The formula can be derived using the approximation $\sin\theta \approx \theta$ in Snell's law and considering spherical lenses. The focal length and the dimensions $w \times h$ of the imaging surface determine the field of view (FOV) of the optical system:

$$f = {}^{w}_{h} \Big/ \tan\left(\frac{\text{FOV}}{2}\right). \tag{2.20}$$

The reciprocal of the focal length is its optical power. Under the approximations considered the thin lens equation relates the distance $z_o$ of an object from the lens to the distance $z_i$ at which the object is focused

$$\frac{1}{z_o} + \frac{1}{z_i} = \frac{1}{f}. \tag{2.21}$$

This can be most easily seen by considering two rays emanating from points on the object: a ray parallel to the optical axis, which when refracted passes through the focal point, and one passing undeflected through the optical center of the lens (the chief ray). If we put a screen at distance $z_i$ from the lens, a point source at $z_o$ will be imaged as a point on the screen, but if the screen is moved closer (or farther) from the lens, the image of the point is transformed into a disc. The diameter $c$ of the circle corresponding to the out of focus imaging of a point source can be determined using the thin lens equation (see also Figure 2.3) and turns out to be

$$c = A \cdot \frac{|z_o - z_1|}{z_o} \cdot \frac{f}{z_1 - f} = \frac{|z_o - z_1|}{z_o} \cdot \frac{f^2}{N(z_1 - f)} \tag{2.22}$$

where $z_o$ is the distance of the object from the lens, $z_1$ is the distance for which the lens is focused, $A$ is the diameter of the incident light beam as it reaches the lens, the so-called lens aperture, and $N = f/A$ is the so-called $f$-number commonly used in photography. The effective $f$-number $N_e$ is defined as the ratio of the height to the diameter of the light cone that emerges from the lens. The $f$-number is a measure of the light-gathering capability of a lens. A way to reduce the diameter of the circle of confusion is then to reduce the aperture of the optical system. As we will see, due to diffraction effects, there is a limiting useful aperture size below which image quality will deteriorate (see Section 2.1.3). A quantity related to the circle of confusion is the depth of field, a measure of the range over which the circle of confusion is below some critical value. Another drawback of reducing the aperture of the optical system is that the amount of light gathered by the system will decrease, resulting in longer exposure time and increased noise (see Section 2.1.4 and Section 2.3). The pinhole

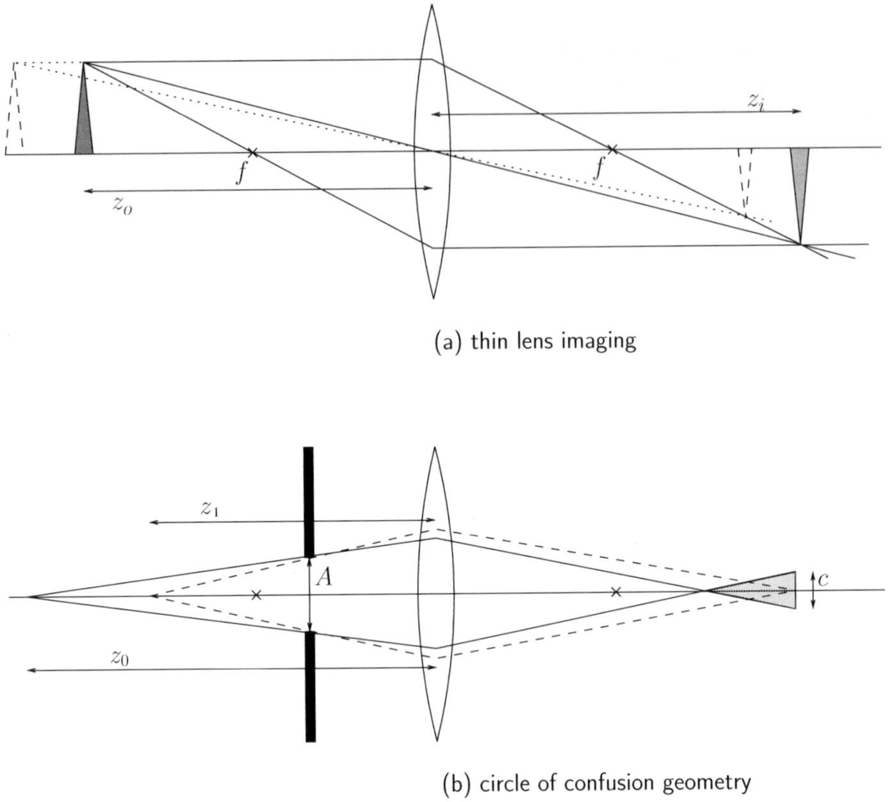


**Figure 2.3.** A basic optical system: the working of a lens under paraxial approximation (a) and the geometry of the circle of confusion for out of focus imaging (b).

camera, a lens-less optical device described in Figure 2.8 and Intermezzo 2.4, is the simplest optical system and it is often used to model the imaging process in computer graphics and computer vision.

The image produced by a point source is called the point spread function (PSF), and we can represent it as a function of the coordinates on the image plane $\mathrm{PSF}(x, y)$. If the only effect of translating a point in the object plane is a proportional translation of the PSF in the image plane, the system is said to be isoplanatic. When image light is incoherent, without a fixed phase relationship, optical systems are linear in intensity. An important consequence is that the image obtained from a linear isoplanatic system is the convolution (see Intermezzo 2.1) of its PSF with the object plane image:

$$I_{\mathrm{image}}(x, y) = \iint I_{\mathrm{object}}(x', y')\mathrm{PSF}(x - x', y - y')\, dx'\, dy' = (I_{\mathrm{object}} * \mathrm{PSF})(x, y). \tag{2.23}$$

If we consider an off-axis point source $P$ with $\theta$ the angle between the optical axis and the principal ray from the point through the center of the aperture, the rays emanating from $P$ will see a foreshortened aperture due to the angular displacement. As a consequence, the light-gathering area is reduced for them, resulting in a falling off of the irradiance $I$ on the sensor with respect to the radiance $L$ on the surface in the direction of the lens:

$$I = L \frac{\pi}{4} \frac{\cos^4 \theta}{N_e^2}. \tag{2.24}$$

This effect is responsible for the radial intensity falloff (vignetting) exhibited by wide-angle lenses.

The simple optical systems considered so far produce perspective images. The image dimensions of equally sized objects are inversely proportional to the corresponding object distances (see Figure 2.3). This variability adversely affects the capability of a computer vision system to recognize objects as it must account for their varying size: the appearance of an object depends on its position within the imaged field and on its distance from the lens. Coupling of simple optical systems and judicious insertion of apertures allow us to build a telecentric system that can produce orthographic images: the appearance of the object does not depend on its distance from the lens or on its position within the field of view (see Figure 2.4). The telecentric design exploits the fact that rays passing through the focal point must emerge (enter) parallel to the optical axis. Let us consider the optical system obtained by removing lens $L_2$ substituting it with the image screen. The aperture stop limits the bundle of rays and the central ray, the one passing through the focal point, is parallel to the optical axis (on the object side). If we move the object towards the camera, the central ray will remain the same and it will reach the screen at the same position: the size of the object does not change. However, if we move the object it goes out of focus as can be easily seen by considering a point on the optical axis. In a telecentric system, in fact, the aperture, besides controlling the ray bundle to obtain an orthographic projection, continues to control the circle of confusion. If we want to focus the new position we must move the screen, but this would change the size of the object. The introduction of the second lens makes the size invariant also to the changes in screen position necessary to focus objects at different distances as the central ray exits from $L_2$ parallel to the optical axis. There is an additional, important advantage in using an image side telecentric design: the irradiance of the sensor plane does not change when we change the focusing distance. The reason is that the angular size of the aperture seen by a point on the image plane is constant due to the fact that rays passing through the same point on the focal plane (where the stop is placed) emerge parallel to each other on the image side of the lens. As a consequence, the effective $f$-number does not change and the intensity on the image plane according to Equation 2.24 does not change. A major limitation of telecentric lenses is that they must be as large as the largest object that needs to be imaged. The effects of focal length and projection type are illustrated in Figure 2.5.

The treatment so far was based on the paraxial approximation that in many real-life situations is not valid. In these cases, an optical system designed using the paraxial approximation exhibits several distortions whose correction requires an approximation up to third order for the $\sin \theta$ function appearing in Snell's law

$$\sin \theta \approx \theta - \frac{\theta^3}{3!} + \frac{\theta^5}{5!}. \tag{2.25}$$

Distortion of first-order optics can be divided into two main groups: monochromatic (Seidel) aberrations and chromatic aberrations, the latter related to the fact that lenses bring different colors of light to a focus at different points as the refractive index depends on wavelength.

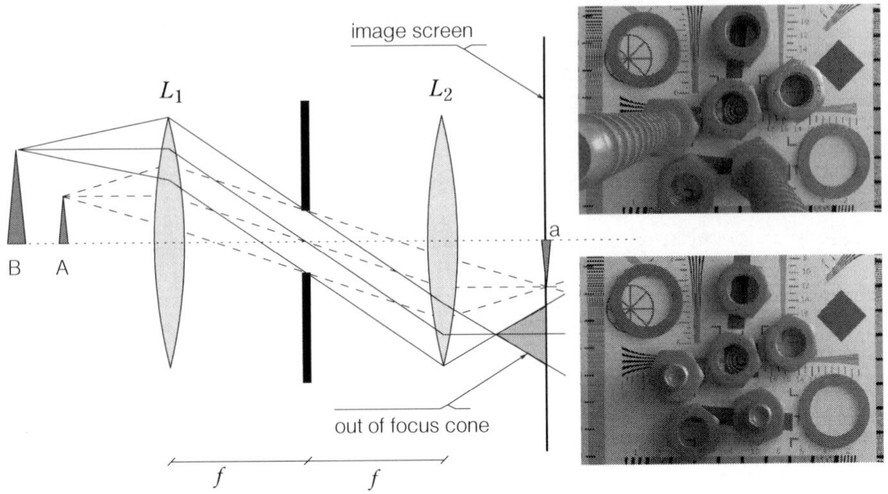


**Figure 2.4.** A doubly telecentric system obtained by using two lenses with one focal point in common and an aperture positioned at it. The two (synthetic) images on the right show the difference between perspective imaging (top) and telecentric imaging (bottom).

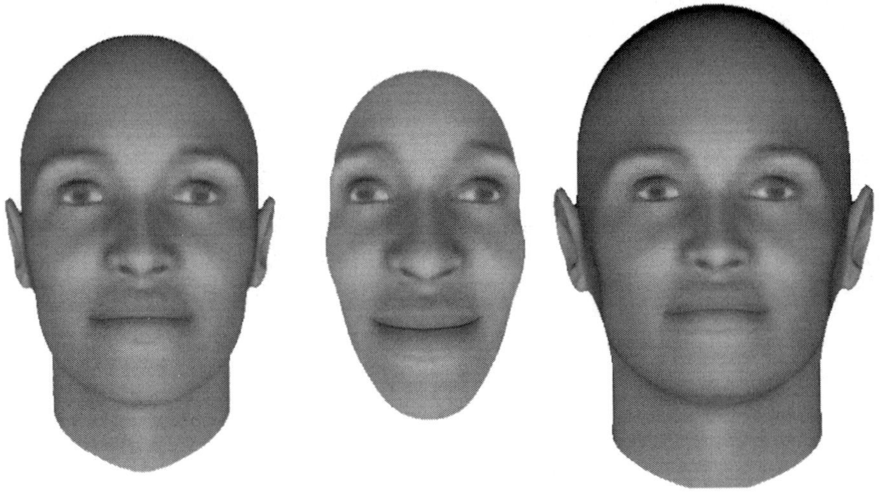

**Figure 2.5.** Camera focal length and projection type have a significant impact on the appearance of objects. From left to right: perspective projections with a field of view of 20° and 60° respectively, and orthographic projection.

Among the monochromatic aberrations we consider only those related to geometrical distortions, the most important in the field of computer vision. The most common cause for distortion is the introduction of a stop, often needed to correct other aberrations (see Figure 2.6). The position of the stop influences the path of the chief ray, the ray passing through the center of the stop. If the stop is positioned at the lens, it will not be refracted and it will leave the lens without changing its angle: the system is orthoscopic and exhibits no

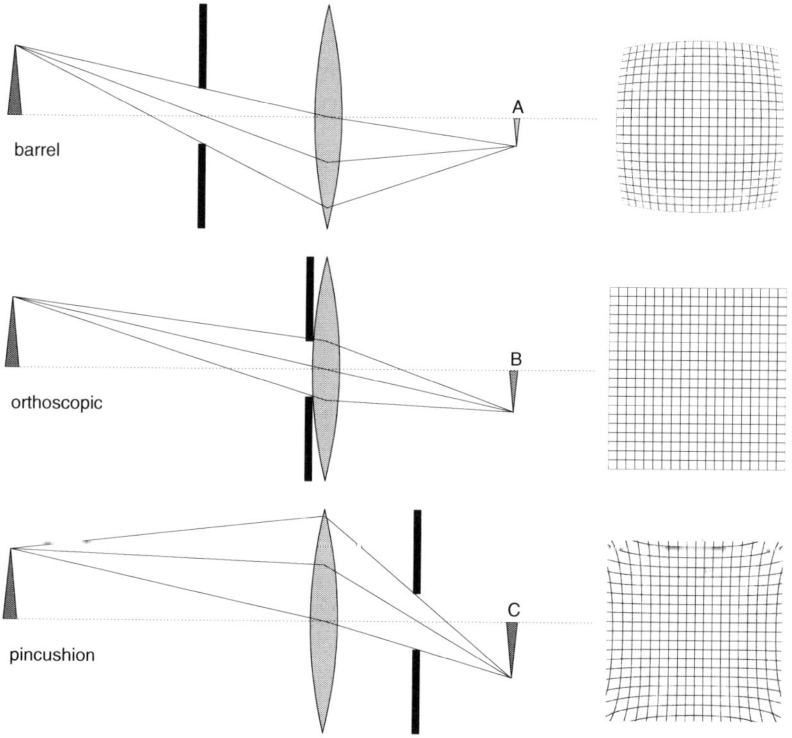


**Figure 2.6.** Geometrical lens distortions are often due to the insertion of stops in the optical system and result in characteristics deformation patterns. The sizes of A, B, and C differ: A < B < C.

distortion. If the stop is before the lens, the chief ray will be refracted closer to the optical axis: the magnification will be smaller and it will depend on off-axis distance. The resulting images will exhibit a typical barrel deformation. The opposite happens when the stop is behind the lens, resulting in pincushion distortion. Distortion is usually quantified as the relative dimensional change of the image plane distance from the optical axis $r'$ with respect to the value $r$ from the paraxial approximation

$$D(r) = \frac{(r' - r)}{r}. \tag{2.26}$$

The type of distortion is related to the slope of $D(r)$, with barrel distortion corresponding to a decreasing $D$. Complex optical systems may exhibit both types of distortions at the same time. In computer vision applications, image distortion can be removed if we know $D(r)$ for the given lens. In fact, if we represent a point on the distorted image as a complex number in polar form $\boldsymbol{x}' = r'e^{i\theta}$, we can rectify the image with the following mapping:

$$\boldsymbol{x}' \to \boldsymbol{x} = r(r')e^{i\theta} \tag{2.27}$$

where $r(r')$ is obtained by inverting Equation 2.26.

### 2.1.3. DIFFRACTION-LIMITED SYSTEMS

As we have already seen, light corresponds (classically) to self-propagating electromagnetic waves that travel with a speed depending upon the traversed medium. While an accurate description can be achieved by means of Maxwell's equations, many optical phenomena can be investigated with the help of simplifying concepts and principles such as the Huygens–Fresnel principle.

**Proposition 2.1.3 (The Huygens–Fresnel Principle).** *Every unobstructed point of a wavefront, at a given instant of time, serves as a source of spherical secondary waves with the same frequency as that of the primary wave. The amplitude of the optical field at any point beyond is the superposition of all these secondary waves considering their amplitudes and relative phases.*

Let us consider what happens when light passes through a small, circular aperture of radius $a$ (see Figure 2.7). According to the Huygens–Fresnel principle, each small area element $dS$ within the aperture $S$ can be considered as covered with secondary, coherent, point sources. As the area element is infinitesimal, hence much smaller than the wavelength $\lambda$ of the emitted light, the contributions of the point sources interact constructively at $\boldsymbol{P}$ independently of the angular position of $\boldsymbol{P}$ with respect to the surface normal. The optical effect at $\boldsymbol{P}$ due to the spherical wave emitted by $dS$ is the real (or imaginary) part of

$$dE = \left(\frac{E_A}{r}\right) e^{i(wt-kr)}\, dS. \tag{2.28}$$

In order to find the effect at $\boldsymbol{P}$ due to the complete aperture we need to integrate the contribution of all small elements. Fraunhofer diffraction assumes that the distance $R$ of $\boldsymbol{P}$ from the aperture is large in comparison to the size of the aperture. This far-field condition allows us to simplify the expression of $r$ as a function of of $\boldsymbol{P} = (X, Y, Z)$ and of the position of $dS$, i.e. $(0, y, z)$, so that we get an approximation of $r$ for the exponent of Equation 2.28 that is accurate enough to model correctly phase effects:

$$r = [X^2 + (Y - y)^2 + (Z - z)^2]^{1/2} \tag{2.29}$$

$$\approx R[1 - (Yy + Zz)/R^2] \tag{2.30}$$

leading to

$$E = \frac{E_A e^{i(wt-kR)}}{R} \int_S e^{ik(Yy+Zz)/R}\, dS. \tag{2.31}$$

The integral can be solved using spherical polar coordinates and the irradiance $I = EE^*/2$, i.e. the squared module of the real part of $E$, at $\boldsymbol{P}$ is found to be

$$I(\theta) = I(0)\left[2\frac{J_1(ka \sin\theta)}{ka}\right]^2 \tag{2.32}$$

where $J_1(\cdot)$ is the zeroth-order Bessel function of the first kind, and is known as the Airy function, after the British astronomer who first derived the result. The resulting irradiance is characterized by a bright disc surrounded by rings of decreasing intensity separated by dark rings reaching zero intensity. The radius of the first dark ring is located at

$$r_1 = 1.22\frac{R\lambda}{2a} \approx 1.22\frac{f\lambda}{A} \quad (2.33)$$

where the approximation is valid for a lens focused on a screen with focal length $f \approx R$ and aperture diameter $A = 2a$. An important consequence of Equation 2.33 is that when $A$ becomes small, the Airy disc becomes large. As the Airy pattern represents the PSF of the optical system considered, the image produced by an object is obtained by convolving the image resulting from geometric optics with the Airy function (see Equation 2.23). When the inner disc gets larger, the image becomes more blurred. An important consequence is that two objects cannot be clearly separated when their Airy discs start to overlap. The commonly used Rayleigh criterion states that two objects are no longer separable when the maximum of one Airy pattern is at the first minimum of the other Airy pattern: $r_1$ is then considered the linear diffraction resolution limit and $r_1/f \approx 1.22\lambda/A$ the angular diffraction limit.

**Intermezzo 2.4.** Real pinhole imaging

The optical system presented in Figure 2.3 is not the simplest one. In fact, we can produce images without any lens, using a pinhole camera, i.e. a box with a small circular hole in one of its walls (see Figure 2.8). In spite of its simplicity, an understanding of a realistic implementation of a pinhole camera requires an understanding of diffraction effects. This should not be surprising after the remarks on Fraunhofer diffraction. Let us consider a plane wave incident on a circular hole and regard the latter as a sequence of circular zones whose radii $R_m$ are

$$R_m^2 = (r_0 + m\lambda/2)^2 - r_0^2 \approx m r_0 \lambda \quad (2.34)$$

where $r_0 \gg \lambda$ is the distance of the detecting screen from the hole and $m$ is not too large. As the distance between the center of two adjacent zones is equal to $\lambda/2$, the light from the two zones interacts destructively. If we remove all even (or odd) zones by masking them, we will obtain a brighter illumination than the one obtained from letting all the light through. If we retain only the first zone, we have a real pinhole camera. The condition reported in Equation 2.34 expresses the radius of the hole as a function of the wavelength and the distance of the detecting screen in order to get a coherent contribution of the light passing through the hole. This is the condition for getting the pinhole to work and $r_0$ corresponds to the focal length of the pinhole camera.

### 2.1.4. QUANTUM NOISE

We now turn to consider another intrinsic limitation of imaging systems related this time to the corpuscular nature of light. Let us consider a stream of uncorrelated photons from a constant light source (such as a chaotic, or thermal, light emitter, where many excited atoms emit photons independently). While we expect an average rate of $r$ photons per unit time, we do not expect them to arrive at regular intervals but randomly. Given a gathering time $\Delta t$, we expect, on average, $\langle n \rangle = r\Delta t$ photons and the probability of detecting $n$ photons is

$$p(n) = e^{-(r\Delta t)}\frac{(r\Delta t)^n}{n!} \quad (2.35)$$

which represents the Poisson distribution. Two remarkable features of this distribution are that its standard deviation $\sigma_p$ equals the square root of the mean and that it approaches the

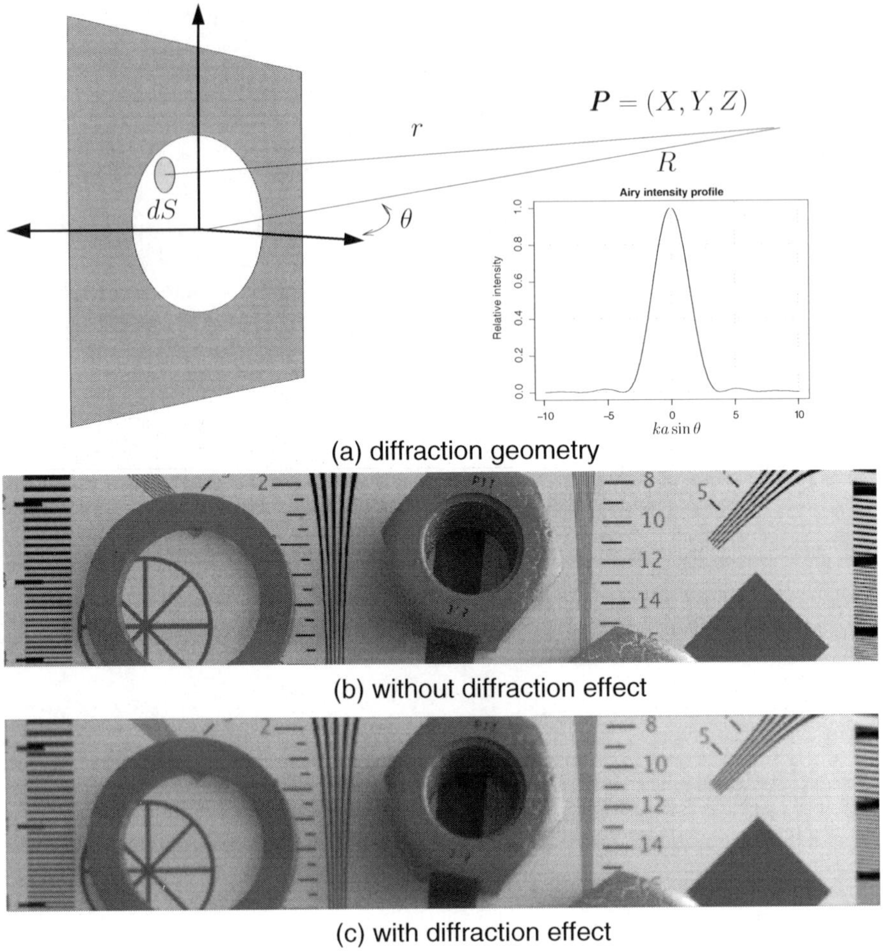


**Figure 2.7.** Diffraction from a circular aperture in the far-field approximation (Fraunhofer diffraction) with the resulting Airy pattern.

Gaussian (normal) distribution for large $n$:

$$\sigma_p = \sqrt{\langle n \rangle} \tag{2.36}$$

$$p(n) \xrightarrow{n \to \infty} N(\langle n \rangle, \sqrt{\langle n \rangle}). \tag{2.37}$$

The major consequence of this type of noise is that the signal to noise ratio (SNR) related to an average measurement of $\langle n \rangle$ photons cannot be greater than $\sqrt{\langle n \rangle}$:

$$\text{SNR} = \frac{I}{\sigma_I} = \frac{\langle n \rangle}{\sqrt{\langle n \rangle}} = \sqrt{\langle n \rangle} \tag{2.38}$$

and will (usually) be lower due to the fact that in order to get the overall standard deviation, we must add to $\sigma_p^2$ the variances from all noise sources (see Figure 2.9 for a comparative presentation of the effects of several noise distributions).

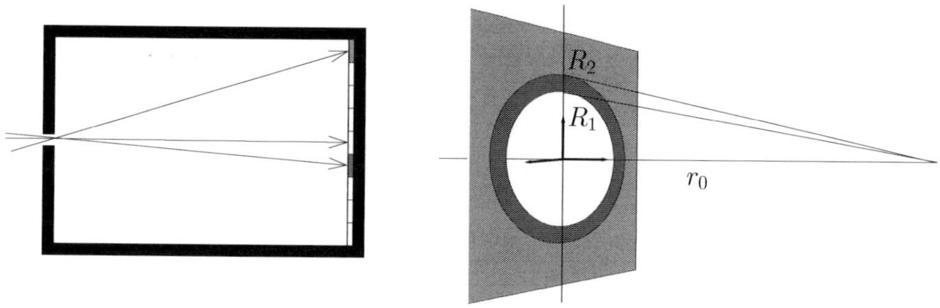


(a) pinhole camera (b) pinhole as a zone plate

**Figure 2.8.** A pinhole camera, the simplest optical system.

Semiclassical and fully quantum calculations provide the same result for the case of coherent light (single mode laser) and for chaotic (thermal) light when the counting time $\Delta t \gg t_c$, the coherence time, which is of the order of $2.3 \times 10^{-15}$ s for tungsten-halogen lamps. There are cases for which the qualitative reasoning presented does not hold. A simple example is that of single molecule fluorescence: the molecule needs some time to re-emit and this prevents close arrival of photons resulting in a different distribution (photon antibunching). If the quantum efficiency of our photon detector is not 1, the final result will still be a Poisson distribution, whose average value is now given by the product of the average number of photons and the efficiency of the detector. Natural and artificial eyes are both affected by photon noise. Poisson noise is not the only kind of noise that can be found in images. An interesting case is that of X-rays images, that are actually obtained by scanning an exposed film. The statistics of the original light source are altered by a logarithmic mapping due to a film exposure mechanism (see Intermezzo 2.5). Yet another different noise distribution is found in magnetic resonance imaging. In this case the complex (as opposed to real) images acquired correspond to the Fourier transform (see Intermezzo 2.7) of the data to be processed. These images are corrupted by Gaussian noise and so are the real and imaginary parts of the inverse Fourier transform due to the orthogonality of the process. However, computing the magnitude of these images is a nonlinear process and the statistics of the noise are changed: Gaussian noise is transformed into Rice distributed noise.

## 2.2. Biological Eyes

In the previous sections we have considered the path of light from its sources onto what we simplistically called the image plane. The result of this analysis is the construction of an irradiance map on the image plane. The next step is to consider how the irradiance map can be transduced into some correlated signal that can be processed by animals or computers. The basic ingredient in the transducing process is the interaction of photons with matter, finally resulting in chemical transformation, as in the case of photographic emulsion, or in electrical signals, driving brains or computer processing. In the following sections we provide some information on the human eye and on the eyes of some other animals that have devised interesting sensing solutions.

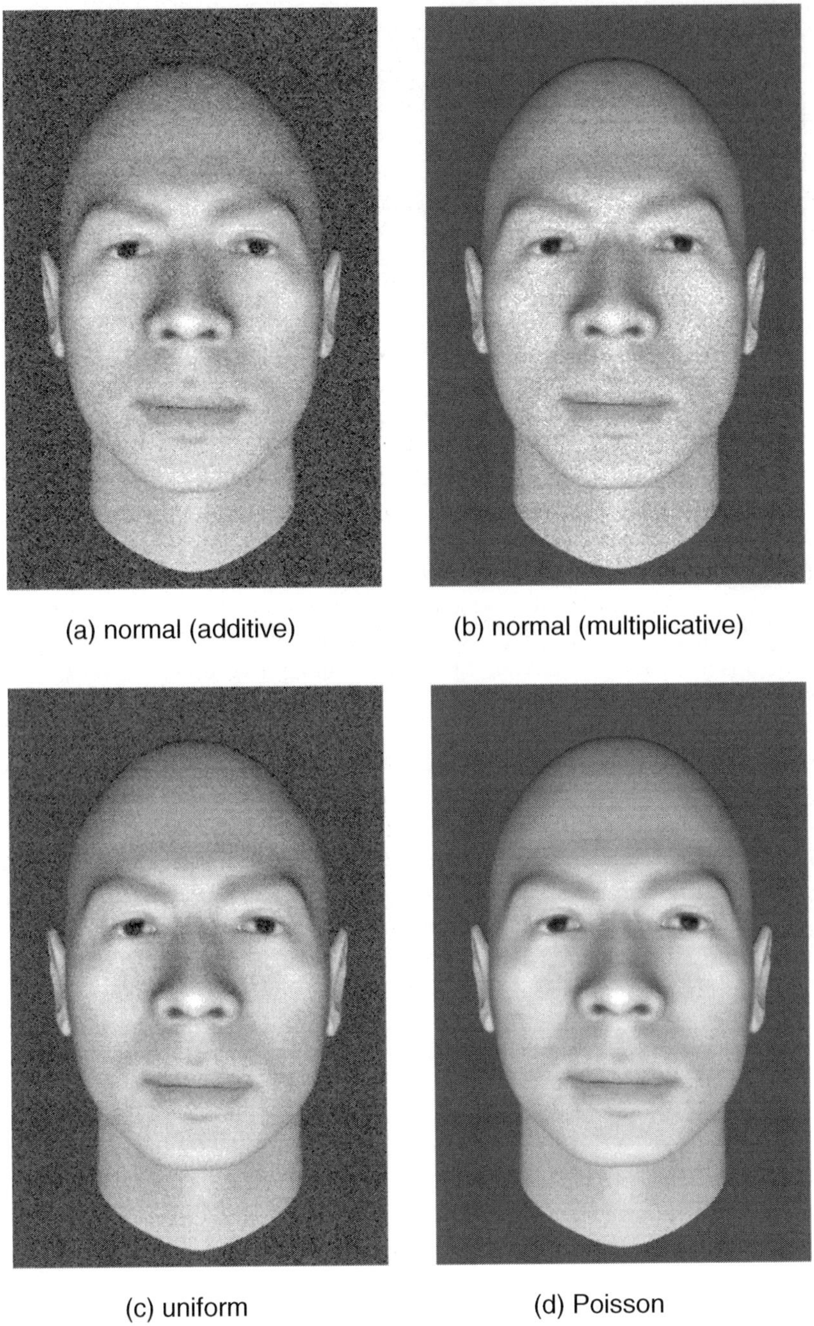


**Figure 2.9.** The effect of different noise distributions on an image.

**Intermezzo 2.5.** Logarithmic intensity scaling of Poisson noise

When we consider images scanned from film, quantum noise no longer exhibits a Poisson distribution. Increase in light exposure causes film darkening that is quantified by the optical density $D$. The relation between the incoming intensity $X$ and $D$ is of the following form:

$$D = D_m - \ln_{10}\left(\frac{X}{X_M}\right) \tag{2.39}$$

where $X_M$ is the maximum intensity passing through the film, $D_m$ is the minimum density corresponding to film support, and $X/X_M$ represents the fraction of light reaching the detecting scanner. If we allow for image value rescaling, the intensity $I$ will be given by

$$I = \gamma D + \delta. \tag{2.40}$$

Given a function $y = f(x)$ the following approximation holds:

$$\sigma_y^2 \approx \left(\frac{\partial f}{\partial x}\right)^2 \sigma_x^2 \tag{2.41}$$

from which, using the fact that for photon noise $\sigma_X^2 = X$, we have

$$\ln(\sigma_I^2) = \left[\ln\left(\frac{\alpha^2\gamma^2}{X_M}\right) - \frac{I_m}{\alpha\gamma}\right] + \frac{I}{\alpha\gamma} \tag{2.42}$$

where $\alpha = \ln_{10}(e)$ and $I_m$ is the intensity corresponding to the minimum density. Image intensity variance due to photon noise varies exponentially with intensity $I$ (Gravel *et al.* 2004).

### 2.2.1. THE HUMAN EYE

The human eye is an example of a single aperture eye, similar to the simple optical systems considered in the previous sections (see Figure 2.10). Its main elements are a flexible lens that is used for focusing, a variable pupil that acts as a stop providing fast control over the amount of incoming light, and the retina, where light is transformed into electrical signals. The field of view is approximately 150° high by 210° wide.

The crystalline lens of the human eye has a variable refractive index that improves its optical power and the curved shape of the retina helps reduce optical aberrations. The transformation of photons into electrical signals, the visual phototransduction process, happens in the retina where we find two distinct classes photoreceptors:

**Rods** are sensitive enough to respond to single photons. They are not selectively sensitive to light wavelength, so they do not provide spectral information, i.e. color. There are approximately $10^8$ rods and they provide low spatial acuity information due to the fact that multiple rods are connected to a single mediating cell that transmits the pooled information.

**Cones** are approximately 100 times less sensitive than rods and are therefore active at higher light levels. They are arranged in a high-density hexagonal mosaic with the highest density at the small region named fovea. Their density outside the fovea decreases and there are approximately $6 \times 10^6$ of them. There are three different classes of cones with selective wavelength response: this is the basis of human trichromatic vision. The three types of cones are named $L$, $M$, and $S$ from having the peak of their sensitivity at long (570 nm), medium (540 nm), and short (430 nm) wavelengths. The response of each class is obtained by summing the product of the sensitivity and the spectral radiance

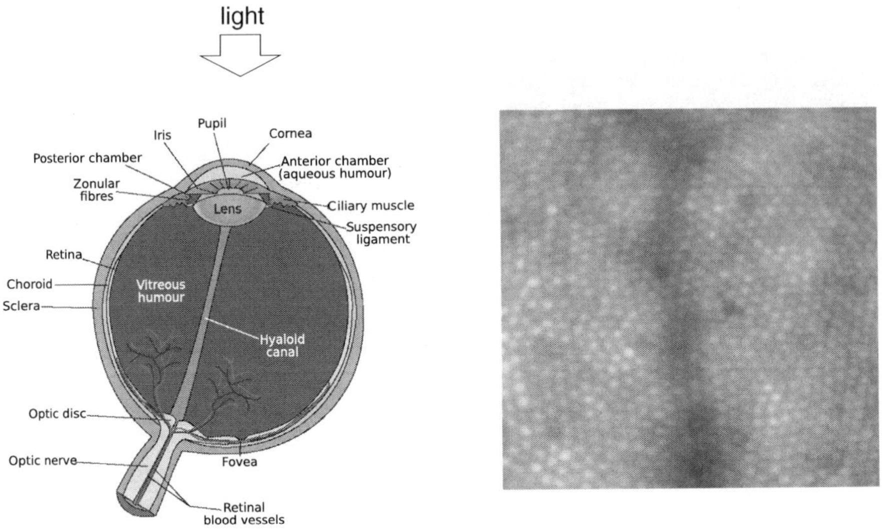


(a) a schematic of the human eye (b) foveal cones mosaic

**Figure 2.10.** The human eye is a complex optical system which employs sophisticated solutions such as a variable refractive index lens. The layout of the photoreceptors in the fovea follows an hexagonal lattice. Source: Roorda and Williams (1999).

of incoming light over all wavelengths. The differential response of the three types of cones to incoming light allows the brain to perceive a multiplicity of colors.

Each photoreceptor type is characterized by different proteins that break down when hit by a photon, initiating a chemical reaction that ends up in sending electrical signals to the brain. While the optical systems analyzed in the previous sections were linear, the response of the human eye is not. The retina has a static contrast ratio of approximately $100:1$ that can be extended dynamically, both chemically and by adjusting the iris, up to $10^6:1$. High-definition vision is obtained by means of rapid eye movements, saccades, by which the high-resolution foveal area is directed towards interesting parts of the scene, whose details are then integrated at the brain level.

The information from the photoreceptors is transmitted to the brain through the optic nerve, which consists of about $1.1 \times 10^6$ nerve cells, suggesting that significant preprocessing of the signals from the $110 \times 10^6$ photoreceptors might take place. In particular, operations akin to smoothing and subsampling are performed by the retinal ganglion cells.

The human eye is affected by Fraunhofer diffraction. As the smallest aperture for the human eye is approximately $f/2.1$, the maximum achievable resolution is $1\,\mu$m: not surprisingly this is approximately the distance between the photoreceptors.

### 2.2.2. ALTERNATIVE DESIGNS

The human eye is an example of a single aperture eye providing high-resolution imaging. There exist many design variations for this type of eye, exploiting pinhole imaging, multiple lenses, mirrors, homogeneous lenses, and more. A particularly interesting example is provided by jumping spiders. They have eight single lens eyes: two high-resolution eyes, two wide-angle ones, and four side eyes. The lenses of their eyes are fixed in the carapace and have a small field of view but they have a movable retina. Muscular contractions allow the retina to move vertically, laterally, and rotationally so that the spider can track its targets without moving the carapace. Some members of this family have tetrachromatic color vision, with sensitivity extending into the ultraviolet range.

A completely different design underlies the compound eye that is found in arthropods (see Figure 2.11). It is a multi-aperture optical sensor of which there are two main variants: apposition and superposition compound eyes. The apposition eye is composed of an array of ommatidia, each a complex of a rhabdom, where photoreceptor cells are located, and an optical section with a corneal lens and a cone that gather light from a small solid angle driving it to the photoreceptor. One of the most sophisticated compound eyes is that of the mantis shrimp: it includes more than 10 different spectral receptor classes spanning the visible spectrum and a portion of the ultraviolet one, including sensitivity to polarization. Polarization sensitivity is important in marine biology because, due to the partial polarization of light in seawater, and the fact that fishes usually reflect unpolarized light, selective attenuation of polarized light can improve the detectability of fishes from the background. The case of mantis shrimps is even more interesting as parts of their body are covered by scales that include micropolarizers: changing the orientation of scales changes the polarization patterns that can be detected by other mantis shrimps. Furthermore, each compound eye is composed of two parts with partially overlapping fields of view, providing monocular stereo capability.

Superposition eyes based on refracting optical elements have a gap between the lenses and the rhabdoms, and no side walls. Each lens takes light at an angle to its axis and reflects it to the same angle on the other side. The result is an image at half the radius of the eye, which is where the tips of the rhabdoms are. In this type of eye the sensory cells of an ommatidium gather light from many optical elements; the resulting superposition image thus gains in brightness (this kind is used mostly by nocturnal insects) but loses in sharpness compared to the apposition image.

The compound eye system does not require and cannot incorporate a mechanism for focusing. The clarity or fuzziness of the image is determined by the fixed number of elements in the eye, and by the insect's distance from the object. The main limitation on resolution in compound eye imaging comes from diffraction effects. The angular sampling $\Delta\phi$ of a single ommatidium can be expressed as

$$\Delta\phi = \frac{A}{R} \tag{2.43}$$

where $A$ is the diameter of the lens and $R$ is the radius of the compound eye. Taking into account the fact that Fraunhofer diffraction limits the maximum angular resolution to

$$\Delta_F\phi \approx 1.22\frac{\lambda}{A} \tag{2.44}$$

we can find the optimal size of a compound eye by imposing that

$$\Delta\phi = \Delta_F\phi \tag{2.45}$$

from which we find

$$A \approx \sqrt{1.22\lambda R} \tag{2.46}$$

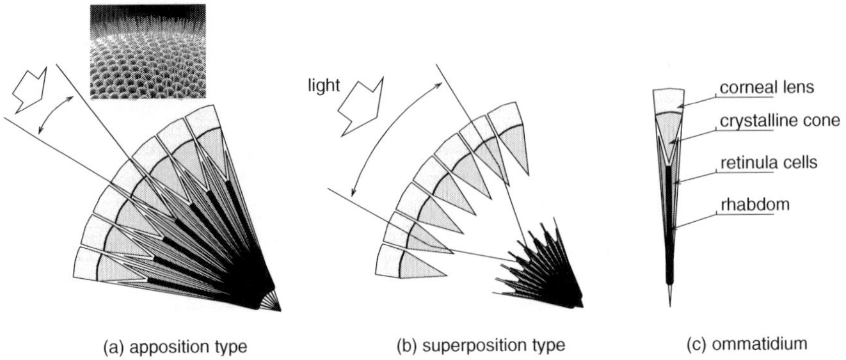


**Figure 2.11.** The two main variants of the compound eye: the apposition type (a) and the superposition type (b). The photographic detail of a scanning electron microscope image in (a) corresponds to a Drosophilidae compound eye. The ommatidium of the apposition compound eye is shown in (c).

and

$$R = 1.22\frac{\lambda}{\Delta\phi^2}. \tag{2.47}$$

A typical interommatidial angle is 0.0175 radians (1°) and $R \approx 0.7$ mm, but if we assume $\Delta\phi = 0.000\,15$ rad, which corresponds to the separation of cones in the human fovea, the radius of the eye would be close to 10 m.

## 2.3. Digital Eyes

It is possible to fabricate devices that transform photon counts over a given area into charge/voltage signals whose amplitude is then numerically coded and fed to a processing unit. The irradiance on the image plane, produced by the imaging optics, can then be transformed into numbers that represent a digital image. A digital imaging sensor is the artificial equivalent of the retina and its photoreceptors: it is a regular lattice, most often an array, of sensitive elements, namely the pixels (see Figure 2.12). Each pixel contains a photodetector that when hit by photons produces a photocurrent. The imaging sensor contains a readout circuit that converts the photocurrent into an electric charge or voltage, finally transforming it into numbers that are read out of the sensor. After the information is read out, the system is reset and the process may start again. The architecture is then very similar to that of biological eyes.

There are two main classes of digital image sensors, characterized by different photodetectors and readout architectures: CCD sensors (from charge-coupled devices) and CMOS sensors (from complementary metal oxide semiconductors). A major advantage of CMOS technology is the possibility of integrating processing capabilities at the level of each single pixel; this means that signals can be processed in real time during integration. An important application is given by high dynamic range imaging, where a single sensor can successfully cope with imaging situations where light intensity exhibits variations over many orders of magnitude. A drawback of CMOS pixels is that the photosensitive portion occupies only a fraction of the pixel area, requiring the use of microlenses to redirect the light hitting the pixel surface onto the photosensitive part.

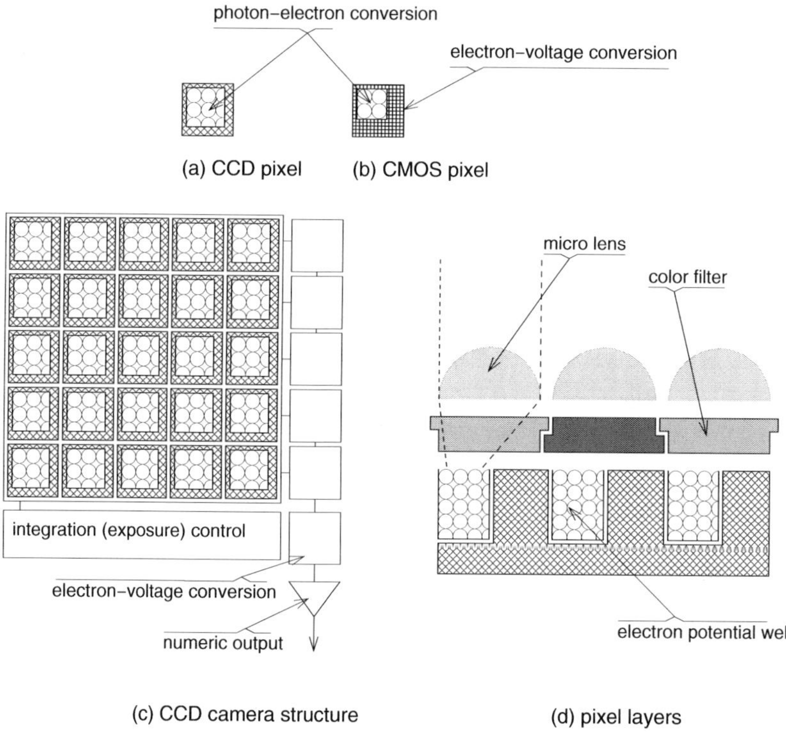


**Figure 2.12.** The structure of digital imaging sensors.

The field of digital imaging is vast but there are a few issues that surface in even the most simple computer vision applications and we think that briefly addressing them will provide valuable information. We know that light is characterized, among other things, by its wavelength. As wavelength is related to the energy of the corresponding photons, it influences their interaction with matter in general, and with photosensitive materials in particular. The photosensitive material used in digital image sensors is sensitive to visible and infrared light even if not evenly over the range. Usually, infrared sensitivity is not required and the light reaching the sensor is carefully filtered to block infrared wavelengths and to selectively absorb the other wavelengths (more where the sensor is more sensitive) in order to obtain a constant response to all wavelengths of visible light, approximating a linear response to irradiance (see Equation 2.14). In particular circumstances, the filter can be removed and infrared sensitivity exploited for selected tasks (see Intermezzo 2.6 and Figure 2.13).

Selective response to wavelength is necessary for color imaging and it is obtained in two different ways: by interposition of absorbing filters in front of each pixel and by prism refraction. In the first case, primary colors are obtained by covering pixels with small filters that let red (or green or blue) light pass through: the response of the pixel will then be proportional to the amount of light at the given wavelength. The alternative is to diffract light by means of a prism, exploiting Snell's law and the wavelength dependency of the refractive index to project different wavelengths onto three different sensors, so that all the pixels of a single chip will be exposed to light of the same wavelength. The second solution is more

**Intermezzo 2.6.** Multispectral face imaging

The intensity reported by a sensor at each pixel is the integrated response over the whole wavelength range to which the sensor is sensitive. It is possible to get information from a narrower range of wavelengths by interposing a filter that transmits only the required portion of the spectrum. Liquid crystal tunable filters provide rapid selection of any wavelength in the visible to infrared (IR) range and can be used to create a spectral imaging camera. The penetration depth exhibited by skin for near-infrared radiation (700–1000 nm) is larger than that for visible light (3.57 mm at 850 nm versus 0.48 mm at 550 nm) enabling the imaging of discriminating subsurface characteristics. The resulting spectral signatures, the reflectance skin spectra under controlled illumination, vary significantly from subject to subject, while exhibiting small within-subject variation at least when acquired on the same day, thereby providing valuable face recognition information (Pan *et al.* 2003). Another possibility is to use a thermal infrared camera, capturing radiation emitted by a face in the thermal infrared range 8–12 μm. The resulting images show a reduced sensitivity to ambient illumination of faces as they capture emitted and not reflected radiation (Kong *et al.* 2007).

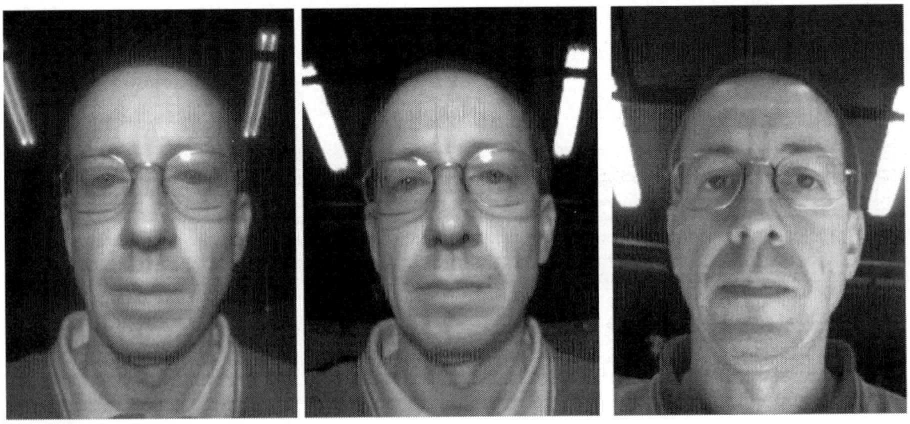

(a) IR illuminator and IR filter (b) IR illuminator, no IR filter (c) standard lighting, no filter

**Figure 2.13.** The use of non-standard light sources and filters can help reduce the distracting effect of the environment. The three pictures present three different combinations of illumination and filtering. The soft focus of the left image is due to the different focal length of the lens when using infrared light.

expensive but provides spatially aligned spectral information, while the first solution usually resorts to filtering pixels of a single sensor by distributing the filters for the different colors in a regular mosaic (see Figure 2.14). The major drawback of this solution is that spectral information is not aligned and in order to recover full color information we must resort to interpolation. This is usually performed automatically by the digital camera, but we must be aware that it can introduce detectable artifacts. Some recent sensors exploit the fact that photons of different energy are absorbed at different depths in the photosensitive material: color information is then derived by the response of three stacked layers of photosensitive material. The images considered in this book, however, will be monochromatic ones and no specific techniques for color image processing will be considered.

The transformation of light into electric charge is integrative: the pixel keeps accumulating the charge produced by the impacting photons for a prescribed amount of time. The integration time, or exposure time, is fixed for all pixels and data are extracted from the sensor after it has expired. The longer the time, the higher the resulting value, and the lower

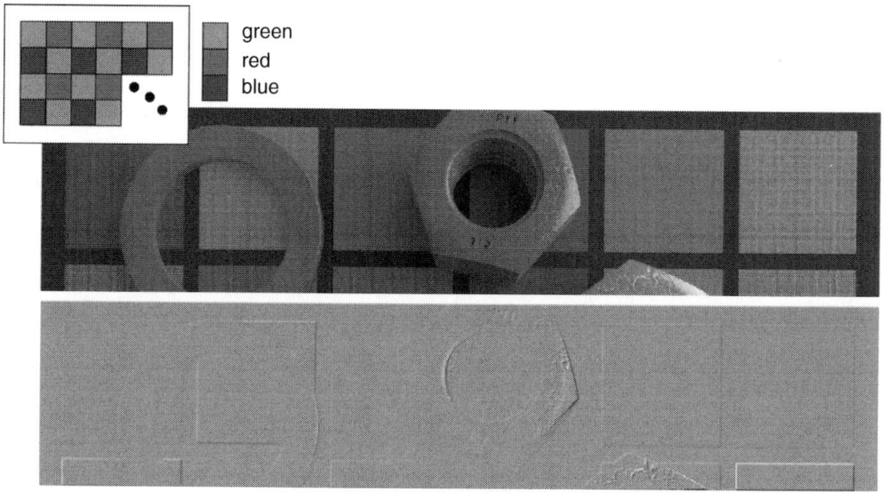


**Figure 2.14.** Color images can be obtained by using small filters overlain on the sensing elements. When such a mosaic is used, no pixel has full color information and a full resolution color image must be reconstructed using interpolation. The resulting image may present significant artifacts that persist even in a derived, monochromatic image. The two synthetic images shows a Bayerized image (upper row) and the difference of the interpolated full color image from the original true color one. The Bayer mosaic is shown in the upper left corner.

the impact of Poisson noise. However, there are several reasons for not keeping the integration time too long. Electric charge must be accumulated at the sensor into potential wells: when the wells are full, the intensity value read out from that pixel will no longer increase and the excess charge may flow from the well into those of the neighboring pixels causing the phenomenon known as blooming. Changing the integration time by means of mechanical or electronic shutters is one way to control the amount of light seen by the sensor, the other being to vary the aperture of the optical system associated with the sensor. A long integration time means that we are blind to intensity variations spanning a shorter time. Another important reason for limiting exposure time is thermal noise, the so-called dark current, a process by which electrons use thermal energy to change state, being collected as those resulting from photon interaction. The result is a (dark) signal corrupted by noise with a Poisson distribution (like photon noise). The signal itself can be estimated and subtracted but the noise variance remains: it can be reduced by cooling the sensor and reducing the integration time.

The output of a digital imaging sensor is an array of integer values, whose range is often limited to the interval [0, 255]. Some digital cameras may offer the option of scaling the electrical signal from the sensor so that it covers the full possible range when its original extension is too small and/or to map it nonlinearly so that it can be better visualized, i.e.

$$I' = [g(I + \eta_p + \eta_r - I_0)]^\gamma + \eta_q \tag{2.48}$$

where $g$ is the electronic gain factor, $\eta_p$ is photon noise, $\eta_q$ is quantization noise due to the discretization of analog to digital conversion, $\eta_r$ is the Gaussian readout noise due to the amplifiers converting photoelectrons to a voltage, and $I_0$ is 0 or the level at which the darkest part of the image is set by optional black level adjustments. In computer vision applications we should keep $\gamma = 1$ to preserve a linear relation with incoming photon counts. The effect of electronic gain $g$ is to amplify both signal and noise, and, whenever it is automatically set by the camera circuitry, it becomes a source of uncontrolled variation. Some cameras provide automatic iris control: the camera changes its aperture (iris) to maximize image intensity without reaching saturation. While useful, it shares one drawback with automatic gain control: it is an uncontrolled source of variation whose action must be inferred from the acquired images.

Finally, we must observe that the effect of Fraunhofer diffraction is particularly relevant in the case of imaging devices with very small pixels. When the radius of the Airy disc at a typical aperture such as $f/8$ becomes comparable to the size of the pixel, it is not possible to increase image resolution by making smaller pixels. Many compact, high-resolution digital cameras are affected by the problem.

An important issue for computer vision applications is the strategy governing the relative timing of pixel exposure and data extraction. Two approaches are commonly used resulting respectively in interlaced and progressive scanning (see Figure 2.15). In the first case, even and odd lines are extracted alternatively from the sensor providing at each time what is usually called a field: at time $t_i$ we get even lines, at $t_{i+1}$ odd lines, at $t_{i+2}$ again even lines. The data from each field are obtained by integrating photon counts over different time intervals. If we are looking at a static scene we may reconstruct a correct full resolution image by combining two consecutive fields. If the scene is not static, combining two consecutive fields does not yield a correct full resolution image as they are representing the image at two different times. A typical effect when imaging moving objects is a comb-like pattern: half of the object is seen as slightly displaced from the other, interleaved half. The second approach, progressive scanning, exposes all pixels at the same time, and, upon completing the exposure, provides all of them. The resulting images are full resolution and do not exhibit comb-like patterns. In spite of the apparent advantage of this approach, the first one is not without merit. Specifically, field interleaving allows pixels to be exposed for a longer interval (interlaced frame integration) or to combine the measures from two vertically adjacent pixels (assuming the rectangular pixel layout of Figure 2.15, $I(i, 2j) + I(i, 2j + 1)$ in one field and $I(i, 2j + 1) + I(i, 2j + 2)$ in the other, interlaced field integration) resulting in increased sensitivity and lower noise.

**Intermezzo 2.7.** The Fourier transform

The $d$-dimensional Fourier transform of a function $f(\boldsymbol{x})$ is defined as

$$F[f](\boldsymbol{\omega}) = (2\pi)^{-d/2} \int f(\boldsymbol{x}) e^{-i\boldsymbol{x}\cdot\boldsymbol{\omega}}\, d\boldsymbol{\omega} \tag{2.49}$$

while the inverse transform is given by

$$F^{-1}[f](\boldsymbol{\omega}) = (2\pi)^{-d/2} \int f(\boldsymbol{x}) e^{i\boldsymbol{x}\cdot\boldsymbol{\omega}}\, d\boldsymbol{\omega}. \tag{2.50}$$

The normalization factors chosen ensure that the Fourier transform is unitary, conserving the norm of the function

$$\int f(\boldsymbol{x}) f(\boldsymbol{x})^*\, d\boldsymbol{x} = \int F[f](\boldsymbol{\omega})^* F[f](\boldsymbol{\omega})\, d\boldsymbol{\omega} \tag{2.51}$$

where $(\cdot)^*$ denotes complex conjugation. The notation for the transformed functions may denote explicitly the transform operation as above or may be implicit, using the same symbol for the function but indicating the $\boldsymbol{\omega}$ argument explicitly. In this chapter the latter option is chosen.

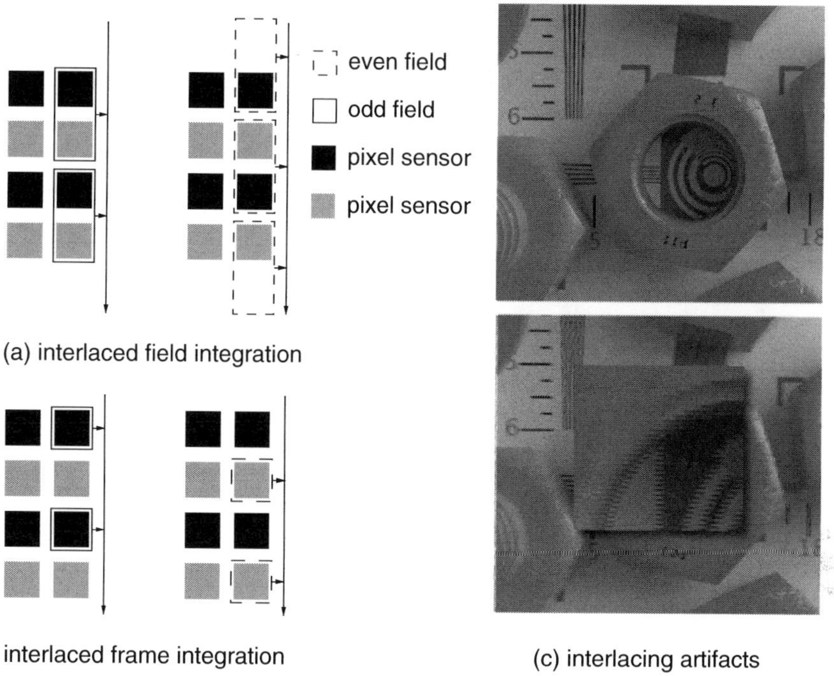


**Figure 2.15.** Interlaced scanning sensors adopting different strategies (a, b). Interlaced images show a characteristic comb artifact in the presence of a camera or object motion (c, bottom). The top (c) image reports the equivalent image from a progressive sensor.

## 2.4. Digital Image Representations

Biological and artificial imaging systems transform the continuous incoming light flow into a discrete representation by means of their photoreceptor systems. An important question is to determine the conditions under which the discrete representation provides all the information contained in the original signal. Another important problem is finding an arrangement of the photoreceptors that can be considered optimal according to some useful criterion. The two issues are related by the sampling theorem, which provides a sufficient (but not necessary) condition for the equivalence of the sampled signal to the original one. After discussing the sampling theorem, and its implications for pixel layout, we will discuss the problem of image resampling, a common operation in many image processing tasks including template matching. Non-regular (i.e. non-translationally invariant) sensing structures, of which the human retina and associated brain structure is an example, offer computational advantages that are briefly discussed.

### 2.4.1. THE SAMPLING THEOREM

The digital acquisition process can be modeled as a sequence of two convolutions followed by a comb sampling. The first convolution is related to the optical system (see Equation 2.23) while the second one refers to the integrative process carried out at each single pixel. The latter can be modeled with a box filter that computes the average of the light falling on each single pixel. We want to derive the conditions under which the discretely sampled signal contains all the information contained in the original signal. We will first consider the one-dimensional case, traditionally associated to time-varying signals, and we will then generalize it to two-dimensional, and more generally to $d$-dimensional signals.

Let $x(t)$ be a continuous signal and $\delta_T$ be the Dirac comb

$$\delta_T = \sum_n \delta(t - nT). \tag{2.52}$$

The sampling operation with a sampling interval $T$ can be described as a convolution

$$x_s(t) = x(t) * \delta_T(t) \tag{2.53}$$

that corresponds to the product of the corresponding Fourier transforms (see Intermezzo 2.7) in the frequency domain

$$x_s(\omega) = \sqrt{2\pi}\, x(\omega)\delta_T(\omega) \tag{2.54}$$

where $f(\omega)$ denotes the Fourier transform of $f(x)$ and the symmetric normalization of the Fourier pair is used. The Fourier transform of the Dirac comb is

$$\delta_T(\omega) = \frac{\sqrt{2\pi}}{T} \sum_k \delta\left(\omega - 2\pi\frac{k}{T}\right) \tag{2.55}$$

so that

$$x_s(\omega) = \frac{2\pi}{T} \sum_n x\left(\omega - 2\pi\frac{n}{T}\right) \tag{2.56}$$

showing that $x_s(\omega)$ is periodic with period $\omega_s = 2\pi/T$ and can then be represented by a Fourier series. If the Fourier transform of the original signal $x(\omega)$ is band limited satisfying

$$x(\omega) = 0 \quad |\omega| \geq \omega_s/2 \tag{2.57}$$

we may recover $x(\omega)$ by filtering $x_s(\omega)$ and then reconstruct the original signal via the inverse Fourier transform (see Figure 2.16). The Fourier series for the sample signal is

$$x_s(\omega) = \sum_n c_n e^{i\omega nT} \tag{2.58}$$

where

$$c_n = \frac{1}{\omega_s} \int_{-\omega_s/2}^{\omega_s/2} x_s(\omega) e^{-in\omega T}\, d\omega \tag{2.59}$$

$$= \frac{1}{\omega_s} \int_{-\infty}^{+\infty} x(\omega) e^{-in\omega T}\, d\omega \tag{2.60}$$

$$= \frac{\sqrt{2\pi}}{\omega_s} x(-nT) \tag{2.61}$$

utilizing Equation 2.57. Equation 2.58 can the be rewritten, introducing $f_s = w_s/2\pi$, as

$$x_s(\omega) = \frac{1}{f_s\sqrt{2\pi}} \sum_n x(-nT) e^{i\omega nT} = \frac{1}{f_s\sqrt{2\pi}} \sum_n x(nT) e^{-i\omega nT}. \tag{2.62}$$

The next step is to filter out all but one of the replicas of the spectrum of $x(\omega)$ contained in the transform of the sampled signal $x_s(\omega)$. This is possible only by virtue of Equation 2.57 which ensures that the copies of the spectra do not overlap. Under these conditions we may simply multiply the transform by the rectangle $\sqcap_{\omega_s}$ function that is zero for $|\omega| > \omega_s/2$ and 1 otherwise:

$$x(\omega) = x_s(\omega)\sqcap_{\omega_s} = \frac{1}{f_s\sqrt{2\pi}} \sum_n x(nT)[e^{-i\omega nT}\sqcap_{\omega_s}]. \tag{2.63}$$

Using the fact that

$$F[\mathrm{sinc}(at)](\omega) = F\left[\frac{\sin at}{at}\right](\omega) = \frac{1}{\sqrt{2\pi a^2}} \sqcap\left(\frac{\omega}{2\pi a}\right) \tag{2.64}$$

and letting $a = \omega_s/2\pi = f_s$, we have

$$x(\omega) = \frac{1}{f_s\sqrt{2\pi}} \sum_n x(nT)\big[\sqrt{2\pi} f_s e^{-in\omega T} F[\mathrm{sinc}(f_s t)]\big](\omega) \tag{2.65}$$

$$= \sum_n x(nT) F[\mathrm{sinc}(f_s(t - nT))](\omega) \tag{2.66}$$

from which we recover the original signal

$$x(t) = \sum_n x(nT)\mathrm{sinc}(f_s(t - nT)) \tag{2.67}$$

concluding the proof of the sampling theorem.

Let us now consider a $d$-dimensional signal. We transform time $t$ into a spatial vector $\boldsymbol{x}$ and $n$ into a (column) vector with integer components $\boldsymbol{n}$. The sampling comb of Equation 2.52 is transformed into

$$\delta_V(\boldsymbol{x}) = \sum_{\boldsymbol{n}} \delta(\boldsymbol{x} - V\boldsymbol{n}) \tag{2.68}$$

where $V$ is a square matrix whose columns represent $d$ linearly independent vectors whose integer linear combinations represent a $d$-dimensional sampling lattice. A bidimensional rectangular sampling lattice with horizontal spacing $l_1$ and vertical spacing $l_2$ corresponds to

$$V = \begin{pmatrix} l_1 & 0 \\ 0 & l_2 \end{pmatrix}. \tag{2.69}$$

The transform of the comb is given by

$$\delta_V(\boldsymbol{\omega}) = \frac{(2\pi)^{d/2}}{|\det V|} \sum_{\boldsymbol{k}} \delta(\boldsymbol{\omega} - U\boldsymbol{k}) \tag{2.70}$$

and $VU^T = 2\pi I$. The definition of $V$ implies that $|\det V|$ represents the spatial area associated to one sample value and its reciprocal is then the sampling density. The sufficient condition is the non-overlapping of the copies of the original spectrum when they are replicated at the lattice points. Let us assume that our signal is bidimensional and the band limit is circular at $\omega_s/2$. If we use a square lattice in the frequency domain, the frequency space (polar) sampling lattice must be defined as

$$U_r = \begin{pmatrix} \omega_s & 0 \\ 0 & \omega_s \end{pmatrix} \tag{2.71}$$

from which we have

$$V_r = \begin{pmatrix} \frac{2\pi}{\omega_s} & 0 \\ 0 & \frac{2\pi}{\omega_s} \end{pmatrix} \tag{2.72}$$

and a sampling density $\rho_r = \omega_s^2/\pi^2$. Under the same band limit, if we use a hexagonal lattice

$$U_h = \begin{pmatrix} \omega_s & \omega_s \\ \sqrt{3}\omega_s & -\sqrt{3}\omega_s \end{pmatrix} \tag{2.73}$$

we have

$$V_h = \begin{pmatrix} \frac{\pi}{\omega_s} & \frac{\pi}{\omega_s} \\ \frac{\pi}{\sqrt{3}\omega_s} & -\frac{\pi}{\sqrt{3}\omega_s} \end{pmatrix} \tag{2.74}$$

and a sampling density $\rho_h = \sqrt{3}\omega_s^2/(2\pi^2) < \rho_r$. Hexagonal sampling is more efficient than rectangular sampling (Figure 2.16): the representation of a circularly band-limited signal requires 13.4% fewer samples with the hexagonal lattice. The storage advantage implies a computational advantage as fewer samples must be processed. Furthermore, each pixel in a hexagonal lattice has six nearest neighbors at the same distance while a pixel in a rectangular lattice has four nearest neighbors and four diagonal neighbors located farther away. The more regular structure can be advantageous in many processing tasks and hexagonal sampling is exploited in several biological and artificial systems. The mosaic of the cones in the human eye and the compound insect eye are hexagonal and digital sensors with a hexagonal pixel layout are available as consumer electronics. Let us note that bandwidth limitation cannot be achieved with digital filtering after the signal has been sampled: frequency aliasing is an irreversible operation.

### 2.4.2. IMAGE RESAMPLING

A digital image processing task where the implications of the sampling theorem are particularly relevant is the correct implementation of image transforms based on a linear transformation of the coordinates of the image domain

$$\boldsymbol{x} \to \boldsymbol{x}' = A\boldsymbol{x} \tag{2.75}$$

where $\boldsymbol{x} = (x_1, x_2)^T$ represents the coordinates of an image point and $A \in \mathbb{R}^{2\times 2}$ is the transformation matrix representing rotation, decimation, and zooming of images:

$$A = \begin{pmatrix} s\cos\theta & -s\sin\theta \\ s\sin\theta & s\cos\theta \end{pmatrix}. \tag{2.76}$$

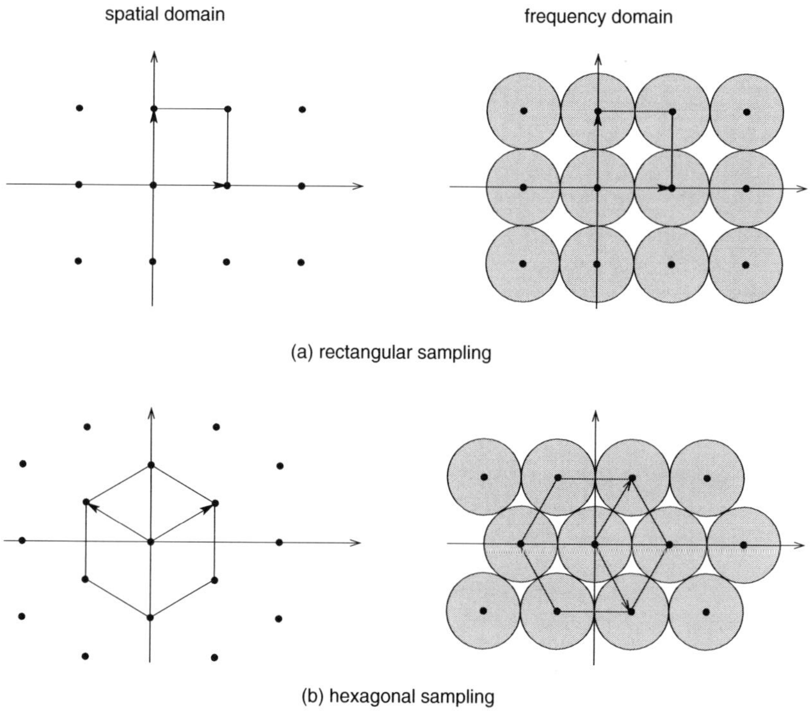


**Figure 2.16.** Periodic replication of the spectrum of the original signal: reconstruction from the sampled version requires that the replicas be non-overlapping (shaded circles in the frequency domain).

The correct way to compute the transformed image $I'(\boldsymbol{x}')$ is by means of the relation

$$I'(\boldsymbol{x}') = I(A^{-1}\boldsymbol{x}') = I(\boldsymbol{x}) \tag{2.77}$$

so that we fill all values of image $I'$ by back projecting its pixels onto the original image $I$: the mapping based on the back projection guarantees that all pixels of $I'$ get the corresponding value. In some cases, such as rotation, the most important aspect is that of proper reconstruction of the original signal that must then be resampled at the same resolution. Other cases, such as image decimation, require that frequency content be properly conditioned in order to satisfy the requirements of the sampling theorem: if the number of samples used to represent the image is reduced, so should the frequency content. The kernel used in the proof of the sampling theorem to reconstruct the original signal via convolution with its sampled version has infinite terms (see Equation 2.67). In practice, a truncated version must be used, limiting the summation to a finite number of terms. Unfortunately, truncation of the support of the sinc function results in significant reconstruction errors due to the slowly decreasing amplitude of the sinc oscillations. This effect is mainly related to the requirements of using a perfect box filter, the rectangle $\sqcap_{\omega_s}$, in the frequency domain. If we relax the requirements and accept a trapezoidal shape, where the discontinuity at $\omega_s/2$ is replaced with a linear response, it is possible to obtain a filter that approximates well the original sinc function but has finite support. The resulting filter takes the form of a windowed sinc function and is known as the Lanczos filter, whose one-dimensional version is

$$L_1(x) = \begin{cases} r_a \sin(2\pi f_s x) \sin\left(2\pi f_s \frac{x}{r_a}\right)(2\pi f_s x)^{-2} & |x| < r_a \\ 0 & |x| \geq r_a \end{cases} \tag{2.78}$$

where $r_a$ controls the size of the kernel and $f_s$ corresponds to the required frequency limit. Good results can be obtained with $r_a = 2$ (higher values tend to produce ringing, i.e. intensity oscillations near step edges). Considering the multidimensional sampling theorem, we may simply treat the two-dimensional case as operating independently in the two dimensions

$$L_2(\boldsymbol{x}) = L_1(x)L_1(y). \tag{2.79}$$

In the interpolation case, $f_s = 1/2$ and we see that if we resample the image under the identity transformation no changes will occur due to the placement of the zero of $L(r)$. If we need to undersample (decimate) the image we must decrease $f_s$. Let us consider reducing the resolution by a factor of 2. From Equation 2.67, we see that we may use it to go from the higher resolution to the lower one enforcing bandwidth limitation in a single step: we simply perform the computation with $f_s = 1/4$ and consider the samples at $\boldsymbol{x} = (2n_1, 2n_2)$. This means that we sample the Lanczos kernel more densely than in the interpolation case, gaining bandwidth limitation.

### 2.4.3. LOG-POLAR MAPPING

So far we have considered image sampling using regular lattices and the sampling theorem has provided useful insights on how dense the sampling should be. However, regular sampling lattices are not the solution chosen by higher vertebrates where space-variant sampling is ubiquitous. Let us consider in particular the human visual system. We have already noted that the density of the photoreceptors decreases from the fovea towards the periphery of the retina, providing an example of spatio-variant sampling. Interestingly, the information provided by the retina is mapped in a specific area of the brain, the visual cortex, where we may find multiple continuous topographic maps of each visual hemi-field: points that are close in the hemi-field are projected to points close in the visual cortex. The size of the visual cortex patches associated to a given angular extent of the hemi-field is not constant over the whole hemi-field. The cortical magnification factor, defined as the distance in the cortex assigned to the representation of a 1° step in visual space, is approximately isotropic in the so-called V1 brain area: it does not depend on the direction of the step. Under the assumption that the function mapping the visual field onto the cortex is complex analytic (points are considered as complex numbers), the isotropicity of its derivative, i.e. the cortical magnification factor, implies that the function represents a conformal mapping, i.e. an angle-preserving transformation. As the reciprocal of the V1 magnification factor is approximately linear, a natural guess for the mapping function of a single hemi-field is

$$w = \log(z + a) \quad \mathrm{Re}(z) > 0, a > 0 \tag{2.80}$$

where $z$ is the polar representation of a retinal point, and $a \in \mathbb{R}$ is introduced to remove the singularity of the logarithm at $z = 0$, moving it to $z = -a$. The other hemi-field is treated similarly. The resulting mapping is found experimentally to be approximately correct.

Summing up, we can see that in our visual system we are employing a spatio-variant sampling scheme, with decreasing resolution from the center to the periphery of our field of view: the representation we use for the incoming visual information is of the log-polar type reported in Equation 2.80. Let us look at some characteristics of this mapping (Figure 2.17). If we represent a point $(x, y)$ in the plane as a complex number $z = \rho e^{i\theta}$ the transformed point is

$$w = \log(z) = \ln(\rho) + i\theta \tag{2.81}$$

where we have assumed for simplicity that $a = 0$ and to avoid the singularity we restrict ourselves to $\rho \geq 1$. If we apply to $z$ a rotation by $\Delta\theta$ or scaling by $s$ the transformed value $w$ undergoes a translation: $w' = w + (\ln s + i\Delta\theta)$. This is useful in image processing tasks: if we keep the origin of our coordinate system fixed on a salient object point, the effect of changing the distance from it or rotating around the axis joining the object point and the center of our sensor is simply a translation of the transformed pattern. This comes at the expense of a marked sensitivity to translations of the origin of the reference system: if we fix on another point of the object, the image pattern will be completely different. While generally a nuisance, this sensitivity may be useful for keeping the sensor perfectly aligned to a target in spite of distance variations and rotations.

We obtain our new image representation (see Figure 2.17) by regularly sampling the transformed coordinates $w = (\ln \rho, \theta) = (\eta, \xi)$:

$$(i_\rho \Delta\eta, i_\theta \Delta\, \xi) = (e^{i_\rho \Delta\eta}, i_\theta \Delta\theta). \tag{2.82}$$

The most effective way to get a log-polar image is to use a sensor that samples the incoming light accordingly. While sensors of this type have been built, in many cases we need to generate the log-polar image by resampling a rectangular lattice image. As we are resampling an image, a major problem is related to aliasing: the farther we are from the origin, the larger the distance of a sample from its nearest neighbors. If we consider the sampling theorem, we see that we should progressively reduce the frequency content with the distance from the origin in order to get $I(i_\rho \Delta\eta, i_\theta \Delta\, \xi)$. A quick, very approximate solution to the problem can be obtained by averaging the intensity over the set $C(i_\rho, i_\theta)$ of all the points in the original image that are projected onto $(i_\rho \Delta\eta, i_\theta \Delta\, \xi)$ in the log-polar representation:

$$I(i_\rho, i_\theta) = \frac{1}{n_C} \sum_{\boldsymbol{i} \in C(i_\rho, i_\theta)} I(\boldsymbol{i}). \tag{2.83}$$

As the computation of $C$ depends solely on the chosen mapping and not on the specific image, the computation can be performed effectively by scanning the original image and accumulating the values in the transformed image. Besides the useful property of transforming scale variations and rotations into image plane translations, the log-polar image representation is efficient in pixel usage with respect to a regular rectangular sampling if we require their peak resolutions to match. Assuming a constant angular resolution of $n_\theta$ pixels and a maximum image field radius of $R$, the area of the image disc is covered by $\pi R^2$ pixels of unit area. For the log-polar image we require that pixels be approximately squared: the side of a pixel at the boundary of the image disc is then $l = (2\pi R)/n_\theta$. The next inner ring must be located at $R - l$ giving a reduction factor $\alpha = (R - l)/R$ for the radii of the pixel rings:

$$R_i = R\alpha^i \tag{2.84}$$

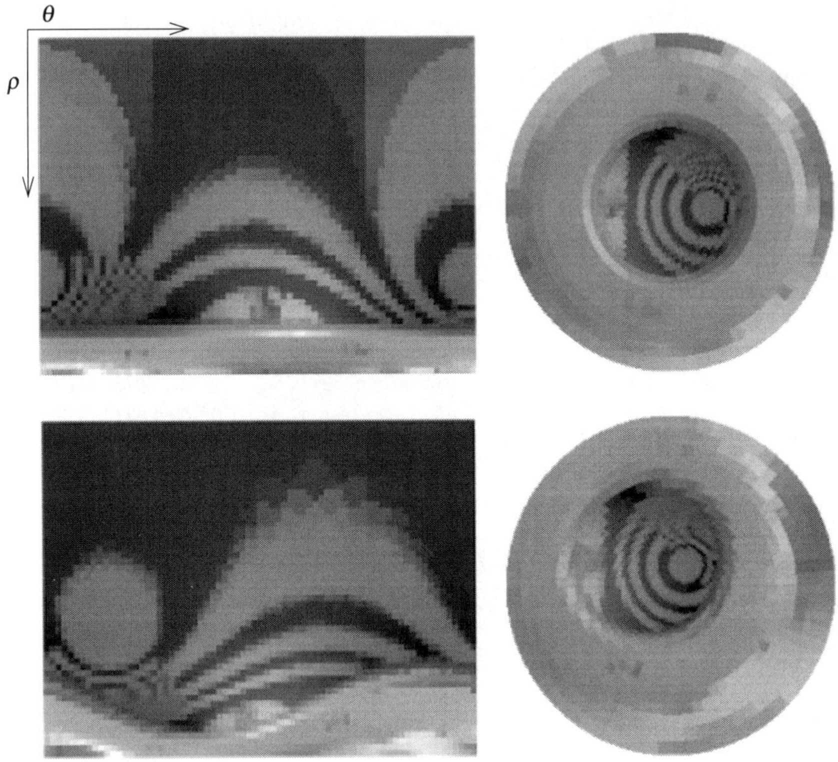


**Figure 2.17.** Log-polar mapping (left) provides an efficient way of representing images employing space-variant resolution (right). While changing the fixation point (top and bottom rows) results in complex image deformation, scaling and rotation are transformed into rotations when the origin is kept on the same object point.

and we stop when $R_m = R\alpha^m = 1$, the central pixel where we have the maximum resolution matching that of regular rectangular sampling. As $m = -\log(R)/\log(\alpha)$ we have that for $R = 128$ and $n_\theta = 64$ the log-polar sensor needs only 3008 pixels while the one based on a square lattice would require 51 472 pixels.

## 2.5. Bibliographical Remarks

A clear and comprehensive reference textbook on optics is the one by Hecht (1987). Coherence time of thermal light sources is discussed in Fercher *et al.* (2000) and photon statistics for different light sources are reviewed in Baltz (2005). Telecentric design is considered in the paper by Watanabe and Nayar (1997) where a simple method is presented to turn commercially available lenses into image side telecentric systems.

Evolution of eyes is considered by Fernald (2000) while the limitations imposed by physics on sensation and perception are considered by Bialek (1987). A detailed, computational analysis of the human eye is presented by Deering (2005). An interesting optical analysis of the fovea can be found in the paper by Pumphrey (1948). The characteristics of compound insect eyes are analyzed by Land (1997) and the possibility of reusing their design in miniaturized imaging systems is explored by Volkel *et al.* (2003). The eye of the mantis shrimp is described by Cronin and Marshall (2001). The idea of improving sensor resolution by moving it is investigated by Ben-Ezra *et al.* (2005).

The optical systems considered in the chapter are of classical refractive design. Large field of views are sometimes necessary and wide-angle lenses can be used to obtain them. However, extreme fields of view can be obtained also using reflective optical systems coupled to sensors with space-variant resolution optimized for each optical design (Baker and Nayar 2001; Gaspar *et al.* 2002; Sandini *et al.* 2002).

The sampling theorem is a basic signal processing result and its application to hexagonal sampling lattices is discussed by Mersereau (1979) who also presents several processing algorithms for hexagonally sampled data. Lanczos filtering is discussed by Duchon (1979) and a survey of resampling techniques is presented in van Ouwerkerk (2006).

A comprehensive review on CMOS sensors, which also compares them to CCD ones, is the paper by El Gamal and Eltoukhy (2005). Noise estimation is important in many computer applications and the paper by Liu *et al.* (2006) presents a technique to estimate an upper bound of the noise level using a single digital image. The problem of the characterization of noise for modern digital cameras is considered also in Hytti (2005) and denoising techniques for digital images from CCD sensors are discussed in Faraji and MacLean (2006a). An extensive introduction to digital color imaging can be found in Shama and Trussell (1997).

Space-variant, retinotopic imaging is a fascinating subject and it is addressed from several perspectives by many papers. Design considerations for sensors with complex logarithmic geometry are presented by Rojer and Schwartz (1990) while a very detailed discussion can be found in Wallace *et al.* (1994) where a graph-based description of log-polar images is introduced. A generalization of the Fourier transform for log-polar images is discussed in Bonmassar and Schwartz (1997).

## References

Baker S and Nayar S 2001 Single viewpoint catadioptric cameras. *Panoramic Vision: Sensors, Theory and Applications*, Monographs in Computer Science. Springer, pp. 39–71.

Baltz R 2005 Photons and photon statistics: from incandescent light to lasers. In *Frontiers of Optical Spectroscopy: Investigating Extreme Physical Conditions with Advanced Optical Techniques* (ed. di Bartolo B and Forte O). Kluwer.

Ben-Ezra M, Zomet A and Nayar S 2005 Video super-resolution using controlled subpixel detector shifts. *IEEE Transactions on Pattern Analysis and Machine Intelligence* **27**, 977–987.

Bialek W 1987 Physical limits to sensation and perception. *Annual Review of Biophysics and Biophysical Chemistry* **16**, 455–478.

Bonmassar G and Schwartz E 1997 Space-variant Fourier analysis: the exponential chirp transform. *IEEE Transactions on Pattern Analysis and Machine Intelligence* **19**, 1080–1089.

Cronin T and Marshall J 2001 Parallel processing and image analysis in the eyes of mantis shrimps. *Biological Bulletin* **200**, 177–183.

Deering M 2005 A photon accurate model of the human eye. *ACM Transactions on Graphics* **24**, 649–658.

Duchon C 1979 Lanczos filtering in one and two dimensions. *Journal of Applied Meteorology* **18**, 1016–1022.

El Gamal A and Eltoukhy H 2005 CMOS image sensors. *IEEE Circuits and Devices Magazine* **21**, 6–20.

Faraji H and MacLean W 2006a CCD noise removal in digital images. *IEEE Transactions on Image Processing* **15**, 2676–2685.

Fercher A, Hitzenberger C, Sticker M, Leitgeb EMR, Drexler W and Sattmann K 2000 A thermal light source technique for optical coherence tomography. *Optics Communications* **185**, 57–64.

Fernald R 2000 Evolution of eyes. *Current Opinion in Neurobiology* **10**, 444–450.

Gaspar J, Decco C, Okamoto J and Santos-Victor J 2002 Constant resolution omnidirectional cameras. *Proceedings of the 3rd Workshop on Omnidirectional Vision*, pp. 27–34.

Gravel P, Beaudoin G and De Guise J 2004 A method for modeling noise in medical images. *IEEE Transactions on Medical Imaging* **23**, 1221–1232.

Hecht E 1987 *Optics* 2nd edn. Addison-Wesley.

Hytti H 2005 Characterization of digital image noise properties based on RAW data. *Image Quality and System Performance III*, vol. 6059 of *Proceedings of SPIE*.

Kong S, Heo J, Boughorbel F, Zheng Y, Abidi B, Koschan A, Yi M and Abidi M 2007 Multiscale fusion of visible and thermal IR images for illumination-invariant face recognition. *International Journal of Computer Vision* **71**, 215–233.

Land M 1997 Visual acuity in insects. *Annual Review of Entomology* **42**, 147–177.

Liu C, Freeman W, Szeliski R and Kang S 2006 Noise estimation from a single image. *Proceedings of the IEEE Conference on Computer Vision and Pattern Recognition (CVPR'06)*, vol. 1, pp. 901–908.

Mersereau R 1979 The processing of hexagonally sampled two-dimensional signals. *Proceedings of the IEEE* **67**, 930–949.

Pan Z, Healey G, Prasad M and Tromberg B 2003 Face recognition in hyperspectral images. *IEEE Transactions on Pattern Analysis and Machine Intelligence* **25**, 1552–1560.

Pumphrey R 1948 The theory of the fovea. *Journal of Experimental Biology* **25**, 299–312.

Rojer A and Schwartz E 1990 Design considerations for a space-variant visual sensor with complex-logarithmic geometry. *Proceedings of the 15th IAPR International Conference on Pattern Recognition (ICPR'00)*, vol. 2, pp. 278–285.

Roorda A and Williams D 1999 The arrangement of the three cone classes in the living human retina. *Nature* **397**, 520–522.

Sandini G, Santos-Victor J, Pajdla T and Berton F 2002 Omniviews: direct omnidirectional imaging based on a retina-like sensor. *Proceedings of IEEE Sensors*, vol. 1, pp. 27–30.

Shama G and Trussell H 1997 Digital color imaging. *IEEE Transactions on Image Processing* **6**, 901–932.

van Ouwerkerk J 2006 Image super-resolution survey. *Image and Vision Computing* **24**, 1039–1052.

Volkel R, Eisner M and Weible K 2003 Miniaturized imaging systems. *Microelectronic Engineering* **67–68**, 461–472.

Wallace R, Ong P, Bederson B and Schwartz E 1994 Space variant image processing. *International Journal of Computer Vision* **13**, 71–90.

Watanabe M and Nayar S 1997 Telecentric optics for focus analysis. *IEEE Transactions on Pattern Analysis and Machine Intelligence* **19**, 1360–1365.

Wolff L 1995 Polarization vision: a new sensory approach to image understanding. *Image and Vision Computing* **15**, 81–93.

# 3 TEMPLATE MATCHING AS TESTING

But tell me true, will't be a match?

*Two Gentlemen of Verona*
WILLIAM SHAKESPEARE

This chapter formally introduces template matching as a hypothesis testing problem. The Bayesian and frequentist approaches are considered with particular emphasis on the Neyman–Pearson paradigm. Matched filters are introduced from a signal processing perspective and simple pattern variability is addressed with the normalized Pearson correlation coefficient. Hypothesis testing often requires the statistical estimation of the parameters characterizing the associated decision function: some subtleties in the estimation of covariance matrices are discussed.

## 3.1. Detection and Estimation

While a single template can be aptly represented by means of a single image, a fruitful way to represent the variability of templates belonging to a given class is by means of a probability distribution. Let us assume that all templates can be represented by a set of $n_d$ pixels at given positions, not necessarily arranged in a rectangular region, within a monochrome image. We consider each of them as an $n_d$-dimensional random vector: each pixel represents a random variable whose value corresponds to the image intensity level. Each template class is then represented by a probability density

$$p_T(I_{i_1}, \ldots, I_{i_{n_d}}) \tag{3.1}$$

where $\boldsymbol{i}_j$ represents a pixel in the image plane, $I_{\boldsymbol{i}_j}$ the corresponding random variable, and $I(\boldsymbol{i}_j)$ a specific value of the random variable.

The problem of template detection fits within the statistical fields of detection and estimation theory which in turn may be considered as particular cases of game theory. Broadly speaking, decision and estimation theory can be considered as a two-person statistical game between nature and a computational agent. The game is characterized by a triple $(\Theta, \Delta, C)$ supplemented by a random observable $X$ defined over a sample space $\mathbb{X}$. The distribution of $X$ depends on the state $\theta \in \Theta$ chosen by nature. A decision rule $\phi : \mathbb{X} \to \Delta$ maps the observations that are probabilistically conditioned by $\theta$ into the decision space $\Delta$, associating to each observation $\boldsymbol{x} \in \mathbb{X}$ a decision (or action) $\delta \in \Delta$. A randomized decision rule $\phi : \mathbb{X} \to \phi_\Delta(\cdot|\boldsymbol{x})$ maps observations into probability distributions over the decision space, quantifying

*Template Matching Techniques in Computer Vision: Theory and Practice* Roberto Brunelli

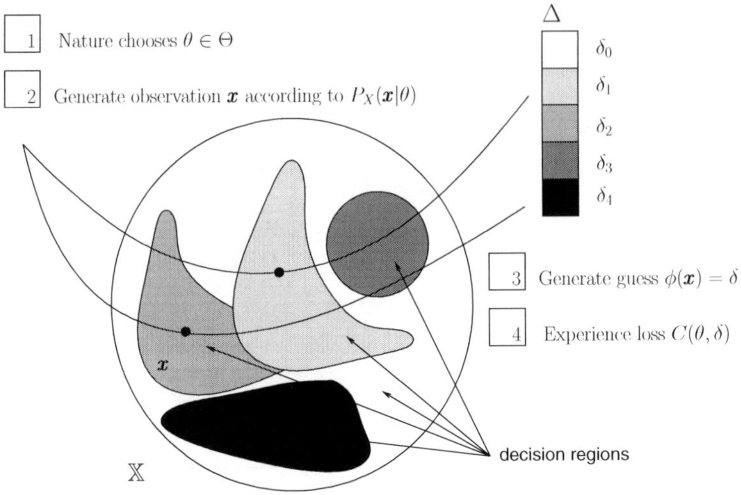


**Figure 3.1.** Hypothesis testing as a game against nature.

the probability of selecting each possible action based on the observable $\boldsymbol{x}$. If the mass of the probability distribution is concentrated on a single element of $\Delta$ the rule is said to be non-randomized. The game proceeds along the following steps illustrated in Figure 3.1:

1. Nature chooses a state $\theta \in \Theta$.

2. A hint $x$ is generated according to the conditional distribution $P_X(x|\theta)$.

3. The computational agent makes its guess $\phi(x) = \delta$.

4. The agent experiences a loss $C(\theta, \delta)$.

Two very important issues are the choice of the decision function $\phi(\cdot)$ and the assessment of its quality. Two cases of the above general game are relevant to the problem of template matching:

1. $\Delta = \{\delta_0, \delta_1, \ldots, \delta_{K-1}\}$, which corresponds to hypothesis testing, and in particular the case $K = 2$, corresponding to binary hypothesis testing. Many problems of pattern recognition fall within this category.

2. $\Delta = \mathbb{R}^n$, corresponding to the problem of point estimation of a real parameter vector, a typical problem being that of model parameter estimation.

The reason why both categories are required for template matching is that the decision function in the game against nature is usually of a parametric kind, and, in order to be of practical use, it requires the estimation of appropriate values for all of the parameters in its definition.

## 3.2. Hypothesis Testing

In the many cases we will consider in this book, each observation can be represented by a set of random vectors $\{\boldsymbol{x}_i\}$, $i = 1, \ldots, N$, $\boldsymbol{x}_i \in \mathbb{R}^{n_d}$, and the vectors themselves are considered to be independent and identically distributed (iid).

Let us consider the case of binary hypothesis testing $\Delta = \{\delta_0, \delta_1\}$: we postulate that the data we observe are distributed according to one of two possible models. The first model, let it be $H_0$, assumes that data are distributed according to the probability density function $p_0(\boldsymbol{x})$, while the other model, let it be $H_1$, assumes that data are distributed differently, according to $p_1(\boldsymbol{x})$. The two different models, usually termed hypotheses, correspond respectively to $\delta_0$ and $\delta_1$ and are denoted

$$H_0 : \boldsymbol{x}_i \sim p_0(\boldsymbol{x}), \quad i = 1, \ldots, N \tag{3.2}$$

$$H_1 : \boldsymbol{x}_i \sim p_1(\boldsymbol{x}), \quad i = 1, \ldots, N. \tag{3.3}$$

A hypothesis test is a rule that associates to each measurement $\{\boldsymbol{x}_i\}$ the model that best explains it. The particular case considered, when there are only two hypotheses, is named the binary hypothesis test, a particular case of the more general $M$-ary hypothesis test. Binary hypothesis testing is singled out due to its simplicity, prevalence in applications, and the fact that many theoretical results for the binary case do not apply to the more general $M$-ary case. Hypotheses are distinguished in two categories: simple, when the corresponding probability distribution is fully known and no parameters must be estimated from data; and composite in the other cases. This can be formalized as

$$H_0 : \boldsymbol{x} \sim p_\theta(\boldsymbol{x}), \theta \in \Theta_0 \tag{3.4}$$

$$H_1 : \boldsymbol{x} \sim p_\theta(\boldsymbol{x}), \theta \in \Theta_1 \tag{3.5}$$

and a hypothesis is simple whenever $\Theta$ contains a single element. In binary hypothesis tests, one of the two competing hypotheses often describes the situation where a feature is absent: it is commonly denoted by $H_0$ and called the null hypothesis. The other hypothesis, commonly termed the alternative hypothesis, is then denoted by $H_1$. In the general $M$-ary hypothesis testing problem for real-valued data, a hypothesis test is a mapping $\phi$

$$\phi : (\mathbb{R}^{n_d})^N \to \{0, \ldots, M-1\}. \tag{3.6}$$

The test $\phi$ returns a hypothesis for every possible input, partitioning the input space into a disjoint collection $R_0, \ldots, R_{M-1}$:

$$R_k = \{(\boldsymbol{x}_1, \ldots, \boldsymbol{x}_N) | \phi(\boldsymbol{x}_1, \ldots, \boldsymbol{x}_N) = k\}. \tag{3.7}$$

The sets $R_k$ are termed decision regions and the boundary separating two of them is called a decision boundary. In pattern classification tasks the test $\phi$ is also termed a classifier and each hypothesis corresponds to the assignment of the observation to a different class.

Template detection can be formalized as a binary hypothesis test. A typical situation is searching for a specific pattern within a (larger) image possibly containing it (see Figure 1.2). We consider in turn each position $\boldsymbol{i}$ within the image: we imagine our reference template at $\boldsymbol{i}$ and we hypothesize whether the aligned image content $\boldsymbol{x}(\boldsymbol{i})$ is an instance of our template or not. We must then perform many hypotheses tests, one for each image position. Let our template be represented in vector form as $\boldsymbol{f} = (f_1, f_2, \ldots, f_{n_d})^T \in \mathbb{R}^{n_d}$. If the template is a portion of an image $T$ of width $w$ and height $h$, the vector form can be obtained by considering the pixels in scan line order:

$$f_{(i_2 w + i_1)+1} = T(i_1, i_2), \ 0 \leq i_1 < w, \ 0 \leq i_2 < h. \tag{3.8}$$

We want to detect the template when it is corrupted by additive white Gaussian noise $\boldsymbol{\eta} \sim N(\mathbf{0}, \sigma^2 I)$ where $\sigma$ is known and $I$ represents the identity matrix (see Figure 2.9). If $\boldsymbol{x}$ represents the data, the testing problem can be summarized as

$$H_0 : \boldsymbol{x} = \boldsymbol{\eta} \tag{3.9}$$

$$H_1 : \boldsymbol{x} = \boldsymbol{f} + \boldsymbol{\eta} \tag{3.10}$$

where the null hypothesis $H_0$ corresponds to the presence of noise only. In this case both hypotheses are simple as all parameters are known. In many template matching tasks the signal appearing in the image may be a transformed version of $\boldsymbol{f}$ such as $\boldsymbol{f}' = \alpha \boldsymbol{f} + \boldsymbol{o}$ where the scaling $\alpha$ and the offset $\boldsymbol{o}$ are unknown. In this case $H_1$ would be a composite hypothesis

$$H_1 : \boldsymbol{x} = \alpha \boldsymbol{f} + \boldsymbol{o} + \boldsymbol{\eta}, \tag{3.11}$$

and is considered in Section 3.5. Let us assume that, in a binary hypotheses test, one of the two hypotheses represents the real state of the world. If $H_0$ is true but the test returns $H_1$, an error of type I (or false alarm) occurs. If $H_1$ is true but $H_0$ is chosen, an error of type II (or miss) occurs. If $H_0$ is a simple hypothesis the false alarm probability $P_F$ (also called the size of the test and denoted $\alpha$) is given by

$$\alpha = P_F = P(\phi = 1 | H_0) \tag{3.12}$$

while in the case of $H_0$ composite

$$\alpha = P_F = \sup_{\theta} P(\phi = 1 | \theta \in \Theta_0). \tag{3.13}$$

The detection probability $P_D$ (also called the power of the test and denoted $\beta$) is given by

$$\beta(\theta) = P_D = P(\phi = 1 | \theta \in \Theta_1), \tag{3.14}$$

and is a function of $\theta$ when $H_1$ is composite. The probability of a type II error, or miss probability $P_M$, is

$$P_M = 1 - P_D. \tag{3.15}$$

The design of the hypothesis test (or detector) $\phi$ is often cast as an optimization problem. There are two main classes of optimality criteria distinguished by the underlying modeling of probability: Bayesian and frequentist. The former approach is represented by the Bayes risk criterion, the latter by the Neyman–Pearson criterion. They are both important and they will be considered separately in the following sections.

### 3.2.1. THE BAYES RISK CRITERION

The Bayes approach is characterized by the assumption that the occurrence probability $\pi_i$ of each hypothesis is known a priori. An $M$-ary hypothesis test on pattern $\boldsymbol{x}$ may result in $M^2$ different couples

$$(\phi(\boldsymbol{x}) = i,\ H_j(\boldsymbol{x}))$$

depending on the chosen hypothesis $\phi(\boldsymbol{x})$ and the true one, $H_j(\boldsymbol{x})$, under which $\boldsymbol{x}$ is generated: to each outcome we associate a cost $C_{ij}$. From a Bayesian perspective, the optimal test (or classifier) is the one that minimizes the Bayes risk $C_B$ defined as

$$C_B = \sum_{i,j} C_{ij} P(\phi(\boldsymbol{X}) = i | H_j)\pi_j. \tag{3.16}$$

The optimal decision rule can be constructed explicitly for a binary hypothesis test with simple hypotheses. The optimal test is fully specified once the correct decision region $R_0$ is chosen, region $R_1$ being identified as the complement of $R_0$. The Bayes risk $C_B$ can be rewritten as

$$C_B = \sum_{i,j} C_{ij}\left(\int_{R_i} p_j(\boldsymbol{x})\, d\boldsymbol{x}\right)\pi_j \tag{3.17}$$

$$= \int_{R_0} (C_{00}\pi_0 p_0(\boldsymbol{x}) + C_{01}\pi_1 p_1(\boldsymbol{x}))\, d\boldsymbol{x}$$

$$+ \int_{R_1} (C_{10}\pi_0 p_0(\boldsymbol{x}) + C_{11}\pi_1 p_1(\boldsymbol{x}))\, d\boldsymbol{x}. \tag{3.18}$$

The input space is partitioned by $R_0$ and $R_1$ and any pattern $\boldsymbol{x}$ must belong to exactly one of the two decision regions. We can then minimize the Bayes risk by assigning each possible $\boldsymbol{x}$ to the region whose integrand at $\boldsymbol{x}$ is smaller. Specifically, we assign $\boldsymbol{x}$ to $R_0$ when

$$C_{00}\pi_0 p_0(\boldsymbol{x}) + C_{01}\pi_1 p_1(\boldsymbol{x}) < C_{10}\pi_0 p_0(\boldsymbol{x}) + C_{11}\pi_1 p_1(\boldsymbol{x}) \tag{3.19}$$

and to $R_1$ otherwise. The rule can then be formalized as

$$L(\boldsymbol{x}) \equiv \frac{p_1(\boldsymbol{x})}{p_0(\boldsymbol{x})} \mathop{\gtrless}_{H_0}^{H_1} \frac{\pi_0(C_{10} - C_{00})}{\pi_1(C_{01} - C_{11})} \equiv \nu \tag{3.20}$$

where $L(\boldsymbol{x})$ is called the likelihood ratio and the test itself the likelihood ratio test. In many cases there is no cost associated with correct decisions: $C_{00} = C_{11} = 0$. An important, and common, special case is when $C_{00} = C_{11} = 0$ and $C_{10} = C_{01} = 1$. The rule of Equation 3.20 is then termed the minimum probability of error decision rule and it has the following simplified expression:

$$L(\boldsymbol{x}) \equiv \frac{p_1(\boldsymbol{x})}{p_0(\boldsymbol{x})} \mathop{\gtrless}_{H_0}^{H_1} \frac{\pi_0}{\pi_1} \equiv \nu. \tag{3.21}$$

Figure 3.2 illustrates the false alarm and detection probability associated to $\nu = 1$ when $\pi_0 = \pi_1$. The rule implied by Equation 3.21 can be reformulated as

$$\phi(\boldsymbol{x}) = \operatorname*{argmax}_{i \in \{0,1\}} \pi_i p_i(\boldsymbol{x}) \tag{3.22}$$

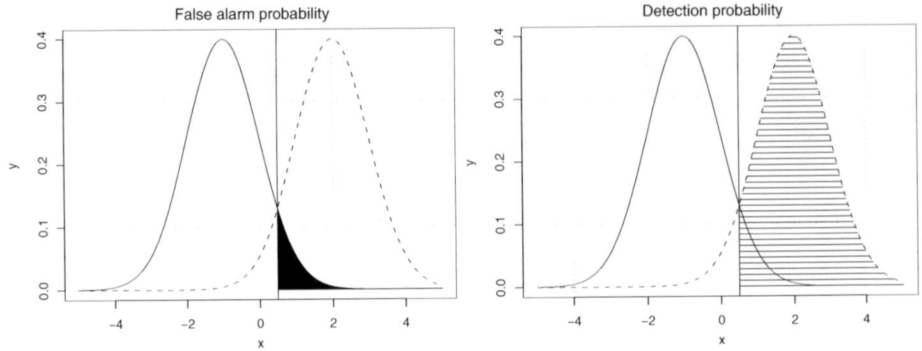


**Figure 3.2.** A graphical illustration of the false alarm and detection probability for a likelihood ratio classifier.

and, in the new formulation, it is known as the maximum a posteriori (MAP) decision rule. The name derives from the Bayes rule:

$$P(H_i|\boldsymbol{x}) = \frac{P(H_i)p(\boldsymbol{x}|H_i)}{p(\boldsymbol{x})}; \tag{3.23}$$

in our case $P(H_i) = \pi_i$ and $p(\boldsymbol{x}|H_i) = p_i(\boldsymbol{x})$ and, as the maximization is performed over $i$, the unconditional density $p(\boldsymbol{x})$ can be considered as a constant. An important advantage of the MAP interpretation is that it can be extended in a straightforward way to $M$-ary hypothesis testing:

$$\phi(\boldsymbol{x}) = \operatorname*{argmax}_{i\in\{0,1,\ldots,M-1\}} \pi_i p_i(\boldsymbol{x}). \tag{3.24}$$

### 3.2.2. THE NEYMAN–PEARSON CRITERION

The Bayes decision rules illustrated in the previous section have proven to be very useful in practice and the resulting tests are optimal by definition if the criterion is that of minimum error probability. A potential problem with them is that they depend on knowledge of the a priori probability of the different hypotheses. The alternative to Bayesian hypothesis testing is based on the Neyman–Pearson criterion and follows a classic, frequentist approach. We can rewrite the false alarm probability and the detection probability emphasizing the role of decision regions as

$$P_F = \int_{R_1} p_0(\boldsymbol{x})\, d\boldsymbol{x} \tag{3.25}$$

$$P_D = \int_{R_1} p_1(\boldsymbol{x})\, d\boldsymbol{x}. \tag{3.26}$$

As the densities are positive, shrinking $R_1$ reduces both probabilities, and enlarging $R_1$ increases them both. If we want to increase $P_D$ we cannot avoid increasing $P_F$ unless the two distributions do not overlap: balancing $P_F$ and $P_D$ represents the key trade-off of hypothesis testing. The relation between the two probabilities, specifically $P_D$ as a function of $P_F$, is usually represented with a receiver operating characteristic (ROC) curve that is extensively discussed in Appendix C.

The Neyman–Pearson criterion differs fundamentally from the Bayes rules in that it is based exclusively on $P_F$ and $P_D$, stating that we should design the decision rule in order to maximize $P_D$ without exceeding a predefined bound on $P_F$:

$$\hat{R}_1 = \underset{R_1 : P_F \leq \alpha}{\operatorname{argmax}} \; P_D(R_1). \tag{3.27}$$

Stated differently, we search for the most powerful test of size not exceeding $\alpha$. The problem can be solved with the method of Lagrange multipliers maximizing the following Lagrangian:

$$\mathcal{L} = P_D + \lambda(P_F - \alpha') \tag{3.28}$$

$$= \int_{R_1} p_1(\boldsymbol{x})\, d\boldsymbol{x} + \lambda\Big(\int_{R_1} p_0(\boldsymbol{x})\, d\boldsymbol{x} - \alpha'\Big) \tag{3.29}$$

$$= -\lambda\alpha' + \int_{R_1} (p_1(\boldsymbol{x}) + \lambda p_0(\boldsymbol{x}))\, d\boldsymbol{x} \tag{3.30}$$

where $\alpha' \leq \alpha$. In order to maximize $\mathcal{L}$, the integrand should be positive leading to the following condition:

$$\frac{p_1(\boldsymbol{x})}{p_0(\boldsymbol{x})} \overset{H_1}{>} -\lambda \tag{3.31}$$

as we are considering region $R_1$. Interestingly, even if the Neyman–Pearson criterion is completely different from the Bayes one, the decision function is based on the same likelihood ratio.

The Neyman–Pearson approach is the most appropriate for those real-world scenarios where the false alarm probability must not exceed some predefined value. An important scenario of this type is that of fraud detection, including biometric systems, i.e. identification systems such as those based on fingerprint or face recognition. In many cases a classification system that cannot meet a predefined limit on the false alarm rate is of no practical use. The most important result for Neyman–Pearson testing is the following lemma:

**Theorem 3.2.1 (Neyman–Pearson Lemma).** *Consider the binary hypothesis testing problem*

$$H_0 : \boldsymbol{x} \sim p_0(\boldsymbol{x}) \tag{3.32}$$

$$H_1 : \boldsymbol{x} \sim p_1(\boldsymbol{x}) \tag{3.33}$$

*where $p_0$ and $p_1$ are both probability density functions or probability mass functions. If $\alpha_0 \in [0, 1)$ is the size (or false alarm probability) constraint, the most powerful test of size $\alpha \leq \alpha_0$ is given by the decision rule*

$$\phi(\boldsymbol{x}) = \begin{cases} 1 & \text{if } L(\boldsymbol{x}) > \nu \\ \gamma & \text{if } L(\boldsymbol{x}) = \nu \\ 0 & \text{if } L(\boldsymbol{x}) < \nu \end{cases} \tag{3.34}$$

*where $\nu$ is the largest constant for which*

$$P(L(\boldsymbol{x}) \geq \nu | H_0) \geq \alpha_0 \tag{3.35}$$

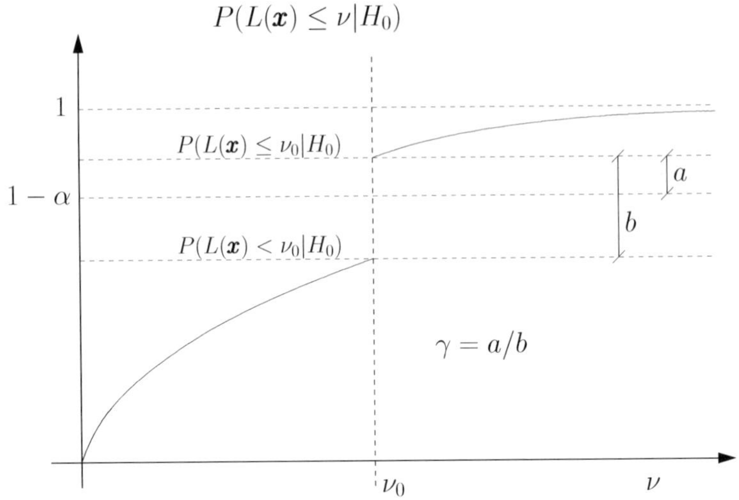


**Figure 3.3.** Discontinuities in $P(L(x) \leq \nu|H_0)$ require the adoption of randomized decision rules.

*and*

$$P(L(\boldsymbol{x}) \leq \nu|H_0) \geq 1 - \alpha_0 \tag{3.36}$$

*and* $\gamma$ *is any number chosen so that*

$$E_0\{\phi(\boldsymbol{x})\} = P(L(\boldsymbol{x}) > \nu|H_0) + \gamma P(L(\boldsymbol{x}) = \nu|H_0) = \alpha_0 \tag{3.37}$$

*and is unique if*

$$P(L(\boldsymbol{x}) = \nu|H_0) > 0. \tag{3.38}$$

*The test is unique up to sets of probability zero under* $H_0$ *and* $H_1$.

The randomized decision rule is necessary when there are discontinuities in $P(L(x) \leq \nu|H_0)$ in order to attain the prescribed size $\alpha$ (see Figure 3.3). The most natural way to interpret $\gamma$ is in fact that of considering it as the probability with which we choose hypothesis $H_1$ when $L(\boldsymbol{x}) = \nu$.

## 3.3. An Important Example

Let us consider in detail the problem of discriminating between two deterministic multi-dimensional signals corrupted by zero average Gaussian noise. The example is important because it allows us to illustrate the two detection paradigms and because it is the foundation for the developments of other template matching techniques that will be presented later on. The binary Gaussian detection problem can be formalized as

$$H_0 : \boldsymbol{x} \sim N(\boldsymbol{\mu}_0, \Sigma), \tag{3.39}$$
$$H_1 : \boldsymbol{x} \sim N(\boldsymbol{\mu}_1, \Sigma), \tag{3.40}$$

so that we have

$$p_0(\boldsymbol{x}) = \frac{1}{(2\pi)^{n/2}|\Sigma|^{1/2}} \exp\left[-\frac{1}{2}(\boldsymbol{x} - \boldsymbol{\mu}_0)^T \Sigma^{-1} (\boldsymbol{x} - \boldsymbol{\mu}_0)\right] \tag{3.41}$$

$$p_1(\boldsymbol{x}) = \frac{1}{(2\pi)^{n/2}|\Sigma|^{1/2}} \exp\left[-\frac{1}{2}(\boldsymbol{x} - \boldsymbol{\mu}_1)^T \Sigma^{-1} (\boldsymbol{x} - \boldsymbol{\mu}_1)\right]. \tag{3.42}$$

When $\boldsymbol{\mu}_0 = \mathbf{0}$, hypothesis $H_0$ corresponds to the presence of noise, while hypothesis $H_1$ corresponds to a signal corrupted by additive, not necessarily white, Gaussian noise. The exponents can be expressed in terms of the Mahalanobis distance

$$d^2_\Sigma(\boldsymbol{x}, \boldsymbol{y}) = (\boldsymbol{x} - \boldsymbol{y})^T \Sigma^{-1} (\boldsymbol{x} - \boldsymbol{y}) \tag{3.43}$$

leading, for instance, to the following expression for $p_1(\boldsymbol{x})$:

$$p_1(\boldsymbol{x}) = \frac{1}{(2\pi)^{n/2}|\Sigma|^{1/2}} \exp\left[-\frac{1}{2} d^2_\Sigma(\boldsymbol{x}, \boldsymbol{\mu}_1)\right] \tag{3.44}$$

so that minimizing the Mahalanobis distance corresponds to maximizing $p_1(\boldsymbol{x})$. When $\Sigma = \sigma^2 I$, the case of white Gaussian noise, the Mahalanobis distance is proportional to the usual $L_2$ norm

$$\sigma^2 d^2_\Sigma(\boldsymbol{x}, \boldsymbol{y}) = \|\boldsymbol{x} - \boldsymbol{y}\|^2. \tag{3.45}$$

The likelihood ratio, upon which both Bayes and Neyman–Pearson classification rules are based, is then given by

$$L(\boldsymbol{x}) = \frac{p_1(\boldsymbol{x})}{p_0(\boldsymbol{x})} = \exp\left[(\boldsymbol{\mu}_1 - \boldsymbol{\mu}_0)^T \Sigma^{-1} \boldsymbol{x} + \frac{1}{2}(\boldsymbol{\mu}_1 + \boldsymbol{\mu}_0)^T \Sigma^{-1} (\boldsymbol{\mu}_0 - \boldsymbol{\mu}_1)\right] \tag{3.46}$$

and the log-likelihood ratio by

$$\log(L(\boldsymbol{x})) = \Lambda(\boldsymbol{x}) = (\boldsymbol{\mu}_1 - \boldsymbol{\mu}_0)^T \Sigma^{-1} (\boldsymbol{x} - \boldsymbol{x}_0) \tag{3.47}$$

where

$$\boldsymbol{x}_0 = \frac{1}{2}(\boldsymbol{\mu}_1 + \boldsymbol{\mu}_0). \tag{3.48}$$

If we let $\boldsymbol{w} = \Sigma^{-1}(\boldsymbol{\mu}_1 - \boldsymbol{\mu}_0)$ the log-likelihood ratio can be written in a very compact form as

$$\log(L(\boldsymbol{x})) = \Lambda(\boldsymbol{x}) = \boldsymbol{w}^T (\boldsymbol{x} - \boldsymbol{x}_0) \tag{3.49}$$

so that the set of points for which $\log(L(\boldsymbol{x})) = 0$, i.e. the points at which the probability densities for the two hypotheses are equal, is represented by a plane orthogonal to $\boldsymbol{w}$, a vector parallel to the segment joining the means of the two distributions, passing through $\boldsymbol{x}_0$, which represents the midpoint of the segment. The decision based on the log-likelihood ratio is

$$\phi(\boldsymbol{x}) = \begin{cases} 1 & \boldsymbol{w}^T (\boldsymbol{x} - \boldsymbol{x}_0) \geq \nu_\Lambda \\ 0 & \boldsymbol{w}^T (\boldsymbol{x} - \boldsymbol{x}_0) < \nu_\Lambda. \end{cases} \tag{3.50}$$

In order to apply the Neyman–Pearson hypothesis test we need to determine the associated performance. Being a linear combination of Gaussian random variables, $\Lambda(\boldsymbol{x})$ is itself a scalar

Gaussian random variable whose expected average and variance are given by

$$\mu_0(\Lambda(\boldsymbol{x})) = -\frac{1}{2}(\boldsymbol{\mu}_1 - \boldsymbol{\mu}_0)^T \Sigma^{-1} (\boldsymbol{\mu}_1 - \boldsymbol{\mu}_0) = -\frac{1}{2}\boldsymbol{w}^T \Sigma \boldsymbol{w}, \tag{3.51}$$

$$\mu_1(\Lambda(\boldsymbol{x})) = -\mu_0(\Lambda(\boldsymbol{x})), \tag{3.52}$$

$$\sigma_0^2(\Lambda(\boldsymbol{x})) = \sigma_1^2(\Lambda(\boldsymbol{x})) = \boldsymbol{w}^T \Sigma \boldsymbol{w}. \tag{3.53}$$

As the distribution of the log-likelihood under the two competing hypotheses is known, we can compute the false alarm rate $P_F$ and the detection rate $P_D$:

$$P_F = P(\Lambda(\boldsymbol{x}) \geq \nu | H_0) = Q\left(\frac{\nu + \sigma_0^2/2}{\sigma_0}\right) = Q(z) \tag{3.54}$$

$$P_D = P_1(\Lambda(\boldsymbol{x}) \geq \nu) = Q\left(\frac{\nu - \sigma_0^2/2}{\sigma_0}\right) = Q(z - \sigma_0) \tag{3.55}$$

where $z = \nu/\sigma_0 + \sigma_0/2$ and the Q-function is described in Intermezzo 3.1. We then see that having fixed $P_F$, we also fixed $z$: the best detection probability depends only on $\sigma_0$, the distance of the means of the two classes normalized by the amount of noise, which is a measure of the SNR of the classification problem (see Figure 3.4). As $Q(\cdot)$ is monotonically decreasing, the best achievable detection rate increases with increasing $\sigma_0$ as shown by Equation 3.55. An important case, often encountered in practice, is when $\Sigma = \sigma^2 I$. In this case all signal samples are independent, as the covariance matrix is diagonal, and the expression for the log-likelihood reduces to

$$\Lambda(\boldsymbol{x}) = \frac{1}{\sigma^2}(\boldsymbol{\mu}_1 - \boldsymbol{\mu}_0)^T (\boldsymbol{x} - \boldsymbol{x}_0) \tag{3.56}$$

**Intermezzo 3.1.** The Q-function

A recurrent need when working with Gaussian (or normal) distributions in probability theory is that of computing the corresponding right tail probabilities. The Q-function, $Q(x) : \mathbb{R} \to [0, 1]\}$, is defined as the probability that a zero-mean, unit-variance normal random variable exceeds value $x \in \mathbb{R}$

$$Q(x) = \frac{1}{\sqrt{2\pi}} \int_x^{\infty} e^{-t^2/2}\, dt. \tag{3.57}$$

$Q(x)$, also known as the complementary cumulative distribution function, is monotonically decreasing and satisfies $Q(-\infty) = 1$ and $Q(\infty) = 0$. The error function erf, defined as

$$\operatorname{erf}(x) = \frac{2}{\sqrt{\pi}} \int_0^x e^{-t^2}\, dt, \tag{3.58}$$

is related to the Q-function by

$$Q(x) = \frac{1}{2}\left(1 - \operatorname{erf}\left(\frac{x}{\sqrt{2}}\right)\right). \tag{3.59}$$

whose variance is given by

$$\sigma_0^2 = \frac{1}{\sigma^2}\|\boldsymbol{\mu}_1 - \boldsymbol{\mu}_0\|^2 = \frac{d^2}{\sigma^2} \tag{3.60}$$

where $d = \|\boldsymbol{\mu}_1 - \boldsymbol{\mu}_0\|$ is the usual $L_2$ (Euclidean) norm. This case corresponds to the discrimination of two signals corrupted by additive white Gaussian noise. When $\boldsymbol{\mu}_0 = \mathbf{0}$ this

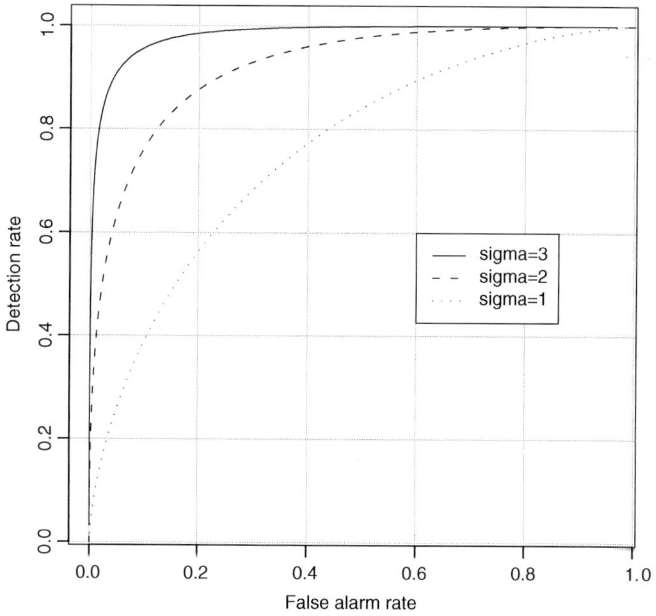


**Figure 3.4.** The dependency of ROC curves on class separation as quantified by $\sigma_0$ in Equations 3.54–3.55.

amounts to the discrimination between noise and the corrupted signal

$$\Lambda(\boldsymbol{x}) = \boldsymbol{\mu}_1^T \left( \boldsymbol{x} - \frac{\boldsymbol{\mu}_1}{2} \right) = \boldsymbol{\mu}_1^T \boldsymbol{x} - \frac{\|\boldsymbol{\mu}_1\|^2}{2}. \tag{3.61}$$

Pattern $\boldsymbol{x}$ contributes only by means of its projection onto the signal, which corresponds to the unnormalized correlation $r_u$ of the signal with the template

$$r_u = \boldsymbol{\mu}_1^T \boldsymbol{x} = \sum_{i=1}^{n_d} \mu_{1,i} x_i. \tag{3.62}$$

This results in the following decision rule:

$$r_u = \boldsymbol{\mu}_1^T \boldsymbol{x} \underset{H_0}{\overset{H_1}{\gtrless}} \nu'_\Lambda \tag{3.63}$$

where the last term of Equation 3.61 has been incorporated into the threshold $\nu'_\Lambda$. We obtain another simplified form when the norms of the two signals are equal: $\|\boldsymbol{\mu}_0\| = \|\boldsymbol{\mu}_1\|$. The log-likelihood further simplifies to

$$\Lambda(\boldsymbol{x}) = \frac{1}{\sigma^2} (\boldsymbol{\mu}_1 - \boldsymbol{\mu}_0)^T \cdot \boldsymbol{x}. \tag{3.64}$$

We may then incorporate $\sigma^2$ into the decision threshold

$$\phi(\boldsymbol{x}) = \begin{cases} 1 & (\boldsymbol{\mu}_1 - \boldsymbol{\mu}_0)^T \cdot \boldsymbol{x} \geq \nu'_\Lambda \\ 0 & (\boldsymbol{\mu}_1 - \boldsymbol{\mu}_0)^T \cdot \boldsymbol{x} < \nu'_\Lambda \end{cases} \tag{3.65}$$

and classification is performed by projecting the pattern onto the segment joining the means of the two classes.

The problem can be also considered from a Bayesian perspective. In order to do this, we must know the (a priori) probability of the two cases:

$$P(H_1) = P(\boldsymbol{\mu} = \boldsymbol{\mu}_1) = p \tag{3.66}$$

$$P(H_0) = P(\boldsymbol{\mu} = \boldsymbol{\mu}_0) = 1 - p. \tag{3.67}$$

If we consider the minimum probability of error criterion (see Equation 3.21) we obtain the same classification rule reported in Equation 3.50 but with a different threshold $\nu_\Lambda^B$

$$\nu_\Lambda^B = \log \frac{1-p}{p} \tag{3.68}$$

which depends on the a priori information. Having fixed the decision threshold $\nu_\Lambda^B$ we may use Equations 3.54–3.55 to compute the corresponding probabilities and the probability of error $C_B$, rewriting Equation 3.16 using $P_D$ and $P_F$:

$$C_B = p(1 - P_D) + (1 - p)P_F \tag{3.69}$$

$$= p(1 - Q(z - \sigma_0)) + (1 - p)Q(z) \tag{3.70}$$

where now $z = \eta_B/\sigma_0 + \sigma_0/2$.

## 3.4. A Signal Processing Perspective: Matched Filters

The probabilistic approach to template matching presented in the previous section is not the only one possible and it is instructive to consider the problem from a signal processing perspective. One of the reasons for which template matching (or signal detection) by means of correlation is widespread is that correlation can be shown to be the optimal linear operation for detecting a deterministic signal corrupted by white Gaussian noise.

Let us restrict ourselves to the case of a one-dimensional signal

$$g(t) = f(t) + \eta(t) \tag{3.71}$$

where $f(t)$ is the original uncorrupted signal and $\eta(t)$ is noise with power spectrum $s(\omega)$. We further assume noise to be wide-sense stationary with zero average:

$$E\{\eta(t)\} = 0, \quad E\{\eta(t+\tau)\eta(t)\} = R(\tau) \tag{3.72}$$

where $E\{\cdot\}$ represents the expected (average) value, and $R(\cdot)$ the autocorrelation function. We want to detect the signal by applying to $g(t)$ a linear filter with impulse response $h(t)$ and system function $h(\omega)$:

$$z(t) = g(t) * h(t) = \int_{-\infty}^{\infty} g(t - \tau)h(\tau)\, d\tau \tag{3.73}$$

$$= z_f(t) + z_\eta(t). \tag{3.74}$$

Using the convolution theorem for the Fourier transform (unitary version, see Intermezzo 3.2) we can rewrite the convolution of the clean signal with the filter as

$$z_f(t) = \int_{-\infty}^{\infty} f(t-\tau)h(\tau)\,d\tau \tag{3.75}$$

$$= \int_{-\infty}^{\infty} f(\omega)h(\omega)e^{i\omega t}\,d\omega \tag{3.76}$$

where Fourier-transformed functions are implicitly identified by their argument $\omega$. We want to determine $h(\omega)$ to maximize the SNR at the location of the signal $t_0$ defined as the ratio of the filter response at the uncorrupted signal and at the noise

$$r^2_{\text{SNR}} = \frac{|z_f(t_0)|^2}{E\{z^2_\eta(t_0)\}}. \tag{3.77}$$

An important characteristic of $r^2_{\text{SNR}}$ is that it is defined at the true location of the signal and does not take into account the response of the filter for $t \neq t_0$: maximization of $r^2_{\text{SNR}}(t_0)$ does not constrain $r^2_{\text{SNR}}(t)$ for $t \neq t_0$.

**Intermezzo 3.2.** The convolution and correlation theorems

Under suitable conditions, the Fourier transform of the convolution of two functions $f_1$ and $f_2$ is the point-wise product of Fourier transforms so that

$$f_1 * f_2 = \alpha_F \mathcal{F}^{-1}\{\mathcal{F}\{f_1\}\mathcal{F}\{f_2\}\} \tag{3.78}$$

and the Fourier transform of a correlation is the point-wise product of the complex conjugate of one Fourier transform by the other transform

$$f_1 \otimes f_2 = \alpha_F \mathcal{F}^{-1}\{\mathcal{F}\{f_1\}\mathcal{F}\{f_2\}^*\} \tag{3.79}$$

where $\alpha_F = \sqrt{2\pi}$ if the unitary version of the Fourier transform is used and $\alpha = 1$ for the non-unitary convention or for ordinary frequency. The continuous convolution and correlation theorems have a discrete counterpart where the Fourier transform is replaced by the discrete Fourier transform (DFT)

$$(\text{DFT}^{-1}\{s_{\omega,1}s_{\omega,2}\})(j) = z(j) = (s_1 * s_2)(j) = \sum_{i=0}^{N-1} s_1(i)s_2((j-i) \mod N) \tag{3.80}$$

$$(\text{DFT}^{-1}\{s_{\omega,1}s^*_{\omega,2}\})(j) = z(j) = (s_1 \otimes s_2)(j) = \sum_{i=0}^{N-1} s_1(i)s_2((j+i) \mod N). \tag{3.81}$$

Even the discrete case extends in a straightforward way to a multidimensional setup: the multidimensional DFT is a sequence of mono-dimensional transforms, and fast Fourier transform algorithms can be extended accordingly.

Two cases of particular interest are those of white and colored noise. White noise is characterized by a flat energy spectrum

$$s(\omega) = |f(\omega)|^2 = s_0. \tag{3.82}$$

The Schwartz inequality states that

$$\left| \int_a^b f(t)g(t)\,dt \right|^2 \leq \int_a^b |f(t)|^2\,dt \int_a^b |g(t)|^2\,dt \tag{3.83}$$

and the equality holds if and only if $f(t) = kg^*(t)$, where $(\cdot)^*$ denotes complex conjugation, so that we can bound the SNR

$$r_{\text{SNR}}^2 \leq \frac{\int |f(\omega)e^{i\omega t_0}|^2 \, d\omega \int |h(\omega)|^2 \, d\omega}{s_0 \int |h(\omega)|^2 \, d\omega} \tag{3.84}$$

from which

$$r_{\text{SNR}}^2 \leq \frac{E_f}{s_0} \tag{3.85}$$

where

$$E_f = \int |f(\omega)|^2 \, d\omega \tag{3.86}$$

represents the energy of the signal. From the Schwartz inequality, equality holds only if

$$h(\omega) = kf^*(\omega)e^{-i\omega t_0}. \tag{3.87}$$

The spatial domain version of the filter is then the mirror image of the signal

$$h(t) = kf(t_0 - t) \tag{3.88}$$

which implies that convolution of the signal with the filter can be expressed as the cross-correlation with the signal: it is a matched spatial filter (MSF).

If the spectrum of noise is not flat, noise is said to be colored. In this case the following holds:

$$\begin{aligned} z_f(t) &= \int f(\omega)h(\omega)e^{i\omega t} \, d\omega, \\ |z_f(t)|^2 &= \left| \int \frac{f(\omega)}{\sqrt{s(\omega)}} \sqrt{s(\omega)} h(\omega) e^{i\omega t} \, d\omega \right|^2 \\ &\leq \int \frac{|f(\omega)e^{i\omega t}|^2}{s(\omega)} \times \int s(\omega)|h(\omega)|^2 \, d\omega, \end{aligned}$$

hence

$$r_{\text{SNR}}^2 \leq \int \frac{|f(\omega)e^{i\omega t}|^2}{s(\omega)} \, d\omega,$$

with equality holding only when

$$\sqrt{s(\omega)}h(\omega) = k \frac{f^*(\omega)e^{-i\omega t}}{\sqrt{s(\omega)}}.$$

The main consequence of the color of noise is that the optimal filter corresponds to a modified version of the signal

$$h(\omega) = k \frac{f^*(\omega)e^{-i\omega t_0}}{s(\omega)} \tag{3.89}$$

which emphasizes the frequencies where the energy of the noise is smaller. The optimal filter can also be considered as a cascade of a whitening filter $s^{-1/2}(\omega)$ and the usual filter based on the transformed signal.

## 3.5. Pattern Variability and the Normalized Correlation Coefficient

The only source of variability for the patterns considered so far in the framework of hypothesis testing has been the addition of Gaussian noise. However, the details of the imaging process considered in Chapter 2 suggest that a common source of signal variability is its scaling by an unknown gain factor $\alpha$ possibly coupled to a signal offset $\beta$

$$\boldsymbol{x}' = \alpha \boldsymbol{x} + \beta \mathbf{1} \tag{3.90}$$

where $\mathbf{1} = (1, 1, \ldots, 1)$ and all vectors belong to $\mathbb{R}^{n_d}$. Although simple, the signal transform of Equation 3.90 substantially modifies the classification task: hypothesis $H_1$ considered in the tests turns from simple to composite as some parameters of the probability densities are now unknown. Instead of entering the complex field of composite hypothesis testing, we address the problem from the perspective of invariant tests: we compute a signal statistic that is invariant to the transformation described by Equation 3.90, and we develop a test based on it. A practical strategy is to normalize both the reference signal and the pattern to be classified to zero average and unit variance:

$$\boldsymbol{x}' = \frac{(\boldsymbol{x} - \bar{x})}{\sigma_x} \tag{3.91}$$

$$\bar{x} = \frac{1}{n_d} \sum_{i=1}^{n_d} x_i \tag{3.92}$$

$$\sigma_x = \frac{1}{n_d} \sum_{i=1}^{n_d} (x_i - \bar{x})^2 = \frac{1}{n_d} \sum_{i=1}^{n_d} x_i^2 - \bar{x}^2. \tag{3.93}$$

The $r_u$ of Equation 3.62 for two normalized vectors can be expressed using the coordinates of the original (unnormalized) vectors and is known as the normalized correlation coefficient, or Pearson's $r$, hereafter $r_P$:

$$r_P(\boldsymbol{x}, \boldsymbol{y}) = \frac{\sum_i (x_i - \mu_x)(y_i - \mu_y)}{\sqrt{\sum_i (x_i - \mu_x)^2}\sqrt{\sum_i (y_i - \mu_y)^2}}. \tag{3.94}$$

The Pearson coefficient can also be used as an estimator of the correlation parameter of a bivariate normal distribution (see Intermezzo 3.3). The coefficient $r_P$ and its variants will be discussed extensively in our journey through template matching techniques. Among the properties of $r_P$ we find

$$r_P(\boldsymbol{x}, \boldsymbol{y}) \in [-1, 1]\ \forall \boldsymbol{x}, \boldsymbol{y} \tag{3.95}$$

$$r_P(\boldsymbol{x}, \boldsymbol{y}) = 1 \iff \boldsymbol{x}' = \boldsymbol{y}' \tag{3.96}$$

$$r_P(\boldsymbol{x}, \boldsymbol{y}) = -1 \iff \boldsymbol{x}' = -\boldsymbol{y}' \tag{3.97}$$

$$r_P(\boldsymbol{x}, \boldsymbol{y}) = r_P(\boldsymbol{y}, \boldsymbol{x}) \tag{3.98}$$

$$r_P(\boldsymbol{x}, \boldsymbol{y}) = r_P(\boldsymbol{x}, a\boldsymbol{y} + b), \quad \forall a, b \in \mathbb{R}, a > 0. \tag{3.99}$$

The normalized correlation coefficient can be interpreted geometrically as the cosine of the angle between the two normalized vectors. A related similarity measure that does not require zero average normalization is given by the cosine of the angle between the vectors representing the two patterns

$$r_{\cos} = \frac{\boldsymbol{x} \cdot \boldsymbol{y}}{\|\boldsymbol{x}\| \, \|\boldsymbol{y}\|} \tag{3.100}$$

and is named cosine similarity. From the point of view of maximum likelihood estimation, we can consider the scale and offset as two additional parameters to be estimated. A solution to this variability is the design of invariant tests, i.e. tests whose distribution does not depend on unavoidable and unknown transformations of data. By its very definition the normalized correlation coefficient is not modified by arbitrary data offset and scaling and can be used to develop an invariant test or an invariant similarity measure (see Figure 3.5).

**Intermezzo 3.3.** What does the Pearson coefficient estimate?

If we consider the coordinates of two vectors $\boldsymbol{x}$ and $\boldsymbol{y}$ as samples of two random variables $X$ and $Y$, the numerator of the Pearson coefficient corresponds to the sample estimate of

$$E\{XY\} - E\{X\}E\{Y\} \tag{3.101}$$

which, in the case of two independent variables, is null. While independence implies $r_{\mathrm{P}} = 0$ the converse is not true: uncorrelated variables are not necessarily independent. The Pearson correlation coefficient $r_{\mathrm{P}}$ is an estimator of the correlation parameter $\rho$ of the following bivariate normal distribution with zero averages and unit variances:

$$b(x, y; \rho) = N(x, y; 0, 1, \rho) = \frac{1}{2\pi\sqrt{1-\rho^2}} \exp\left[\frac{(x^2 - 2\rho xy + y^2)}{2(1-\rho^2)}\right]. \tag{3.102}$$

It is an asymptotically unbiased estimator and a nearly unbiased version for samples of size $n$ is given by

$$r'_{\mathrm{P}} = r_{\mathrm{P}}\left[1 + \frac{(1 - r_P^2)}{2(n-3)}\right]. \tag{3.103}$$

The amount of bias is then dependent on estimator value and sample size while the values at which it is maximum ($\rho = \pm 1/\sqrt{3}$) do not depend on sample size. The distribution of $r_{\mathrm{P}}$ for a given value of $\rho$ is not normal but it is possible to transform $r_{\mathrm{P}}$ in such a way that the distribution of the transformed estimator is approximately normal:

$$z = \frac{1}{2} \ln\left(\frac{1 + r_{\mathrm{P}}}{1 - r_{\mathrm{P}}}\right) \tag{3.104}$$

when the size of the sample is moderately large ($> 10$). In the case of images, the values of the pixels represent the samples of the two random variables $X$ and $Y$ and, in general, it is not possible to assume that they are normally distributed. It is sometimes possible (and useful) to transform the intensity values of an image so that their distribution matches a target one, the normal distribution being just one possibility (see Intermezzo 5.1).

The square of $r_{\mathrm{P}}$ is also known as the coefficient of determination and is related to the fraction of the variance in $y$ accounted for by a linear fit of $x$ to $y$

$$r_P^2 = 1 - \frac{s_{y|x}^2}{s_y^2} \tag{3.105}$$

where $s_{y|x}^2$ is the average square error of the least squares fit $\hat{\boldsymbol{y}} = \hat{a}\boldsymbol{x} + \hat{b}$

$$s_{y|x}^2 = \sum_{i=1}^{n_d} (y_i - \hat{y}_i)^2 = \sum_{i=1}^{n_d} (y_i - \hat{a}x_i - \hat{b})^2 \tag{3.106}$$

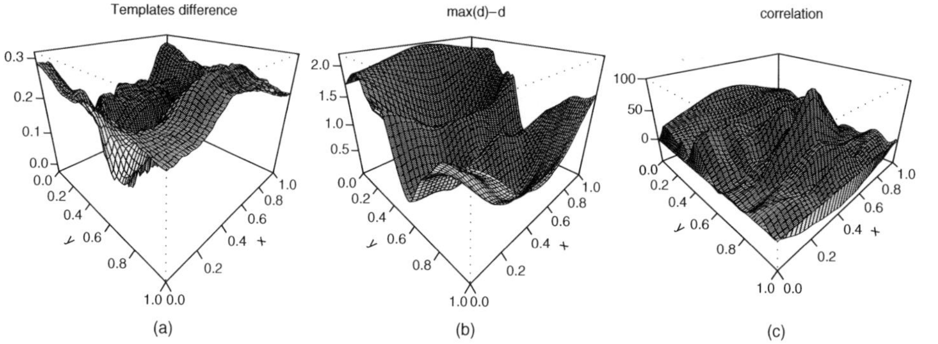


**Figure 3.5.** Template differences due to scale and average variations have a major, negative, impact on template comparison by means of the sum of squared differences. The intensity of the mouth template from Figure 1.2 was divided by 2 and offset by 10 and its difference from the original is reported in (a). The effect on matching using the sum of squared differences is reported in (b): the maximum value no longer corresponds to the correct position. Template changes have no impact on correlation matching as shown in (c): the maximum value corresponds to the position of the template.

and

$$s_y^2 = \sum_{i=1}^{n_d} (y_i - \bar{y})^2 \tag{3.107}$$

represents the norm of the centered signal. Let us consider the case of a signal $\boldsymbol{x}$ with unknown scale and offset (and no noise) with respect to a reference version $\boldsymbol{y}$. We further assume that the reference template is characterized by a white Gaussian distribution, which we do not consider as noise but as true template specific variability. Least squares determination of the unknown parameters corresponds to a maximum likelihood estimate of them: as the distribution of the reference template is Gaussian, the least squares fit corresponds to maximization of the log-likelihood. We can then express the probability of a given pattern under hypothesis $H_1$ as

$$p_1(\boldsymbol{x}) = \frac{1}{\sqrt{2\pi}\sigma} e^{-(s_y^2/2\sigma^2)(1-r_P^2)} \tag{3.108}$$

while its probability under hypothesis $H_0$ is

$$p_0(\boldsymbol{x}) = \frac{1}{\sqrt{2\pi}\sigma} e^{-\|\boldsymbol{x}\|^2/2\sigma^2} \tag{3.109}$$

leading to the following likelihood ratio test:

$$\frac{e^{-(s_y^2/2\sigma^2)(1-r_P^2)}}{e^{-\|\boldsymbol{x}\|^2/2\sigma^2}} \underset{H_0}{\overset{H_1}{\gtrless}} \nu \tag{3.110}$$

which no longer depends on signal scale and offset which have been substituted by their maximum likelihood estimates. If we consider the coordinates of the two vectors $\boldsymbol{x}$ and $\boldsymbol{y}$ as samples of two random variables $X$ and $Y$ and we assume $X$ and $Y$ to be independent, if $Y$ is normally distributed it is possible to prove that the distribution of $r_P$ does not depend on that of $X$ and is given by

$$p(\hat{r}_P) = \begin{cases} \dfrac{\Gamma[(n_d-1)/2]}{\Gamma(1/2)\Gamma[(n_d-2)/2]}(1-\hat{r}_P^2)^{(n_d-4)/2}, & -1 < \hat{r}_P < 1 \\ 0, & \text{elsewhere} \end{cases} \tag{3.111}$$

which corresponds to the fact that $T = \sqrt{n_d - 2}\,\hat{r}_P / \sqrt{1 - \hat{r}_P^2}$ has a $t$ distribution with $n_d - 2$ degrees of freedom. This is useful when we consider the problem of detecting a signal immersed in white Gaussian noise: noise is normally distributed and we may leverage on knowledge of the distribution of $r_P$ to reject the null hypothesis with a given degree of confidence.

## 3.6. Estimation

In all the examples considered so far, the probability distributions were known: both the functional form of the densities and the actual values of their parameters were assumed to be known. Unfortunately, this is a rather uncommon situation: in the vast majority of cases, the parameters are unknown, and in many cases also the functional forms of the probability density functions are unknown. The two cases are quite different and require different approaches. When the functional form of the probability densities is unknown, from a pattern classification perspective, the way to proceed is to estimate directly the classifier function $\phi(\cdot)$ rather than to estimate the underlying probability density functions (pdfs) and build the classifier upon them. This case will be treated in some detail in Chapter 12. In the next sections we address the first case: the functional form is known but some of its parameters are not and they must be estimated from available data. Even in this case we may distinguish two different approaches to estimation: classical and Bayesian. The first one leads to maximum likelihood estimators (MLEs), a class of estimators possessing attractive (asymptotic) properties. As far as Bayesian estimation is concerned, we will only present the philosophy behind it. However, we will discuss in some detail a class of estimators, known as James–Stein or shrinkage estimators, that outperform MLEs when the amount of data upon which estimation must be based is small and the favorable asymptotic properties of MLEs are not yet valid.

Let us introduce some fundamental estimation concepts for scalar quantities. We assume that we have $N$ iid scalar samples $x_i$ from a random variable $X$ and that we want to estimate a parameter $\theta \in \mathbb{R}$ of the distribution according to which they are distributed:

$$x \sim p(x; \theta).$$

To estimate $\theta$ we use an estimator $\hat{\theta}(x_1, \ldots, x_N)$ that is a function of our observations and we follow the usual practice of omitting the dependency on the observations. Being a function of random variables, it is a random variable itself and it is then appropriate to consider its average value $E(\hat{\theta})$ and its variance $\mathrm{var}(\hat{\theta}) = E((\hat{\theta} - E(\hat{\theta}))^2$. One of the most desirable properties of an estimator is that of being correct on average: such an estimator is called unbiased.

**Definition 3.6.1.** *The bias of an estimator* $\hat{\theta}$ *is*

$$\mathrm{bias}(\hat{\theta}) = E(\hat{\theta}) - \theta \tag{3.112}$$

*where* $\theta$ *represents the true value. If* $\mathrm{bias}(\hat{\theta}) = 0$ *the operator is said to be unbiased.*

Another important characteristic of an estimator is its mean squared error.

**Definition 3.6.2.** *The mean squared error (MSE) of an estimator is*

$$\mathrm{MSE}(\hat{\theta}) = E((\hat{\theta} - \theta)^2). \tag{3.113}$$

A simple but very important result is the following decomposition of the mean squared error of an estimator into a bias and variance term:

$$\mathrm{MSE}(\hat{\theta}) = \mathrm{var}(\hat{\theta}) + \mathrm{bias}^2(\hat{\theta}). \tag{3.114}$$

Errors in the estimation of the parameters of distributions impact on the performance of classifiers that use them (see Figure 3.6). An unbiased estimator does not necessarily have a small mean squared error: we may be better off with a biased estimator with a smaller variance. As we will see in detail when discussing the James–Stein effect, the advantage can be substantial. Another important concept associated with the variance of an unbiased estimator is that of efficiency. For an unbiased estimator it is possible to derive a lower bound on its variance, the so-called Rao–Cramer bound,

$$\mathrm{var}_\theta(\hat{\theta}; X_1, \ldots, X_N) \geq E\left[\left(\frac{\partial \log p(X_1, \ldots, X_N; \theta)}{\partial \theta}\right)^2\right]^{-1} \tag{3.115}$$

that depends on the true value of the parameter $\theta$ at which the estimate is performed and on the size of the sample used in the computation of the estimate. An estimator is said to be efficient if it achieves the Rao–Cramer lower bound.

### 3.6.1. MAXIMUM LIKELIHOOD ESTIMATION

The most desirable estimator, for a parameter given a set of data, is the minimum variance unbiased estimator, MVUE for short: an operator whose average value over all datasets of any given cardinality equals the correct value of the parameter to be estimated and whose variance around the correct value is minimal with respect to that of any other unbiased estimator. The computation of the MVUE is often difficult, so an alternative class of estimators is often used in practice. Let $\boldsymbol{\theta} \in \mathbb{R}^q$ be the parameter vector that controls the distribution of data $\boldsymbol{x}$ via the density function $p(\boldsymbol{x}|\boldsymbol{\theta})$. The likelihood function is defined as

$$l(\boldsymbol{\theta}|\boldsymbol{x}^N) = \prod_{i=1}^{N} p(\boldsymbol{x}_i|\boldsymbol{\theta}) \tag{3.116}$$

where $\boldsymbol{x}^N = \{\boldsymbol{x}_i\}_{i=1}^N$ is our (fixed) dataset and $l(\cdot)$ is considered to be a function of $\boldsymbol{\theta}$. The likelihood is not a probability density: it is not necessarily integrable (hence not necessarily renormalizable) and $\boldsymbol{\theta}$ is not considered as a random variable so that it cannot have a probability density unless we switch to a Bayesian approach to be discussed in the next section.

The MLE $\hat{\boldsymbol{\theta}}$ is defined as

$$\hat{\boldsymbol{\theta}} = \operatorname*{argmax}_{\boldsymbol{\theta}} \, l(\boldsymbol{\theta}|\boldsymbol{x}^N) \tag{3.117}$$

resulting in the parameter that maximizes the likelihood of our observations. The definition of the MLE does not provide us with any information on how close our estimate is to the

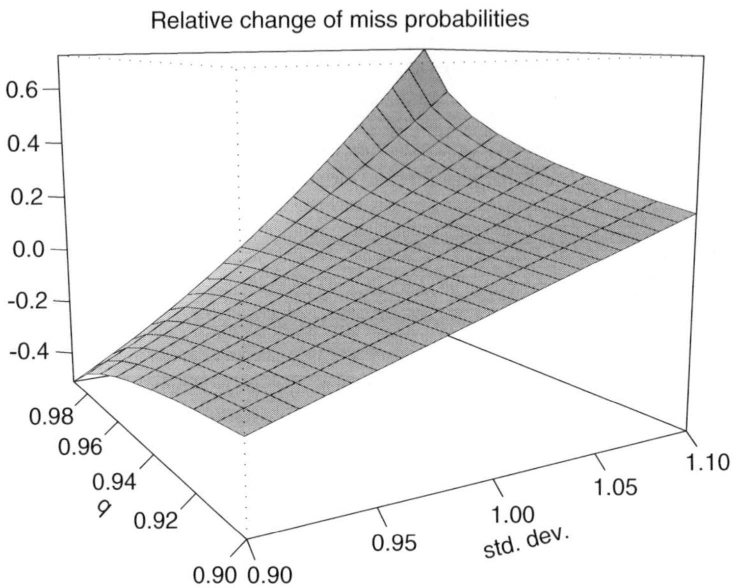


**Figure 3.6.** Errors in the estimation of the standard deviation (covariance matrix) have a significant impact on classification (even) in the case of normally distributed data. The plot shows $(1 - P_1(\theta; \sigma'))/(1 - P_1(\theta; 1)) - 1$ (the relative miss probabilities for normal distributions $N(0, \sigma'^2)$ and $N(0, 1)$) for $\theta$ corresponding to the quantiles $q$ of $P_1(x; 1)$.

real value of the parameter, and how large its variance is. The computation of the MLE for a differentiable likelihood, or log-likelihood, is based on solving

$$\frac{\partial}{\partial \theta_i}(\log l(\boldsymbol{\theta}|\boldsymbol{x}^N)) = 0 \quad i = 1, \ldots, q. \tag{3.118}$$

When the pdf belongs to an exponential family of distributions, examples being normal distributions, $\Lambda$ distributions, binomial, Poisson, and exponential distributions, the solution of Equation 3.117 can be found using calculus and linear algebra. When no direct solution is available, we must resort to numerical techniques, the most important of which is the expectation–maximization algorithm, commonly known as EM. The EM algorithm, like the other available techniques, is of an iterative nature and improves an initial solution that must be provided as a starting point of the algorithm. The solution found is a local optimum, not a global one, and the algorithm should be restarted several times starting from a different initial hypothesis to maximize the probability of finding the global optimum. MLEs are used often because of the many useful asymptotic properties they possess:

1. The MLE is asymptotically unbiased, i.e. its bias tends to zero as the number of samples increases to infinity.

2. The MLE is asymptotically efficient, i.e. it achieves the Cramer–Rao lower bound when the number of samples tends to infinity. This means that, asymptotically, no unbiased estimator has lower mean squared error than the MLE.

3. The MLE is asymptotically normal: as the number of samples increases, the distribution of the MLE tends to the Gaussian distribution

$$p(\hat{\theta}) \to G(\theta, v(\theta)/N) \tag{3.119}$$

where $v(\theta)$ is the asymptotic variance and $N$ the number of samples: the more samples used, the smaller is the variance of the estimate.

The MLEs of the average vector and of the covariance matrix of a Gaussian distribution are discussed in Intermezzo 3.4.

**Intermezzo 3.4.** MLE of the covariance matrix

Let us consider a random vector $X \in \mathbb{R}^d$ with a Gaussian distribution

$$p(\boldsymbol{x}) \propto \det(\Sigma)^{-1/2} \exp\left[\frac{1}{2}(\boldsymbol{x} - \boldsymbol{\mu})^T \Sigma^{-1} (\boldsymbol{x} - \boldsymbol{\mu})\right]. \tag{3.120}$$

We want to estimate $(\boldsymbol{\mu}, \Sigma)$ from a set of $N$ samples. We switch to the log-likelihood of the dataset

$$\ln L(\boldsymbol{\mu}, \Sigma) = \text{const} - \frac{N}{2} \ln \det(\Sigma) - \frac{1}{2}\text{tr}\left[\Sigma^{-1} \sum_{i=1}^{N} (\boldsymbol{x}_i - \boldsymbol{\mu})(\boldsymbol{x}_i - \boldsymbol{\mu})^T\right] \tag{3.121}$$

whose differential is given by

$$\begin{aligned} d \ln L(\boldsymbol{\mu}, \Sigma) = -\frac{N}{2}\text{tr}[\Sigma^{-1} d\Sigma] - \frac{1}{2}\text{tr}\Bigg[ & -\Sigma^{-1}(d\Sigma)\Sigma^{-1} \sum_{i=1}^{N} (\boldsymbol{x}_i - \boldsymbol{\mu})(\boldsymbol{x}_i - \boldsymbol{\mu})^T \\ & - 2\Sigma^{-1} \sum_{i=1}^{N} (\boldsymbol{x}_i - \boldsymbol{\mu}) d\boldsymbol{\mu}^T \Bigg]. \end{aligned} \tag{3.122}$$

Setting to zero the terms multiplying $d\boldsymbol{\mu}$ we obtain

$$\sum_{i=1}^{N} (\boldsymbol{x}_i - \boldsymbol{\mu}) = 0 \tag{3.123}$$

from which we recover the sample mean as the MLE:

$$\hat{\boldsymbol{\mu}} = \frac{1}{N} \sum_{i=1}^{N} \boldsymbol{x}_i. \tag{3.124}$$

The unbiased sample scatter matrix

$$S = \frac{1}{N-1} \sum_{i=1}^{N} (\boldsymbol{x}_i - \hat{\boldsymbol{\mu}})(\boldsymbol{x}_i - \hat{\boldsymbol{\mu}})^T \tag{3.125}$$

is the sum of $N$ rank 1 matrices and we have rank $S \leq N - 1$ as $\hat{\boldsymbol{\mu}}$ is estimated from data. Setting $d\Sigma$ to zero we find

$$-\frac{1}{2}\text{tr}(\Sigma^{-1}(d\Sigma)[NI - \Sigma^{-1} S]) = 0 \tag{3.126}$$

where $I$ is the appropriate identity matrix, so that

$$\hat{\Sigma} = \frac{N-1}{N} S \tag{3.127}$$

which is then a biased (but asymptotically unbiased) estimator of the covariance matrix.

### 3.6.2. BAYES ESTIMATION

When we introduced the likelihood function we remarked that it cannot be considered as a probability because from a classical estimation perspective the estimated parameter is not a random variable. Limiting ourselves to the estimation of a scalar parameter, the Bayesian approach distinguishes itself from the classical approach by considering $\theta$ as a random variable with an associated pdf $p(\theta)$. It is therefore possible to relate its conditional distribution to data by means of the Bayes rule

$$p(\theta|\boldsymbol{x}^N) = \frac{p(\boldsymbol{x}^N|\theta)p(\theta)}{\int p(\boldsymbol{x}^N|\theta)p(\theta)\,d\theta}, \tag{3.128}$$

a fact that allows us to get the causes, i.e. the parameters of our data generation model, from the effects, namely the observations. A Bayesian statistical model is then composed of the classic data generation model $p(\boldsymbol{x}|\theta)$ and a prior distribution on the parameter $p(\theta)$. The latter models our a priori knowledge of the distribution of the parameter and modulates the value used by classical estimation

$$p(\theta|\boldsymbol{x}^N) = \frac{p(\boldsymbol{x}^N|\theta)p(\theta)}{p(\boldsymbol{x}^N)} \propto p(\boldsymbol{x}^N|\theta)p(\theta) \tag{3.129}$$

as $p(\boldsymbol{x}^N)$ is a constant, $\boldsymbol{x}^N$ being known. The choice of the prior is important and two different classes of priors are commonly distinguished:

- informative priors, which express our knowledge of the distribution of the parameters to be estimated;
- non-informative priors, which only express very general constraints on the parameters, usually derived from invariance arguments.

In order to facilitate the computation of the posterior probability in closed form, the prior is chosen in such a way that the posterior distribution belongs to the same family. A very common case is the choice of Gaussian priors when working with normally distributed data. Once the posterior probability of the parameters given the data is known, we may introduce a cost (or loss) function $C(\theta, \breve{\theta})$ and compute the estimate $\hat{\theta}$ of the parameter as

$$\hat{\theta} = \operatorname*{argmin}_{\breve{\theta}} \int_{\theta} C(\theta, \breve{\theta}) p(\theta|\boldsymbol{x}^N)\,d\theta. \tag{3.130}$$

Among the loss functions commonly used we find:

- the minimum squared error (MSE)

$$C_{\text{MSE}}(\theta, \breve{\theta}) = (\theta - \breve{\theta})^2 \tag{3.131}$$

  from which we obtain as estimator the posterior mean

$$\hat{\theta} = \int \theta p(\theta|\boldsymbol{x}^N)\,d\theta = E(\theta|\boldsymbol{x}^N); \tag{3.132}$$

- the absolute error loss

$$C_{\mathrm{AEL}}(\theta, \breve{\theta}) = |\theta - \breve{\theta}| \tag{3.133}$$

from which we obtain as estimator the posterior median

$$\hat{\theta} : P(\theta < \hat{\theta}|\boldsymbol{x}^N) = P(\theta > \hat{\theta}|\boldsymbol{x}^N); \tag{3.134}$$

- the 0–1 loss

$$C_{0/1}(\theta, \breve{\theta}) = \begin{cases} 0 & \text{if } \breve{\theta} = \theta \\ 1 & \text{if } \breve{\theta} \neq \theta. \end{cases} \tag{3.135}$$

As

$$E[L_{0/1}(\theta, \hat{\theta})] = E[I_{\{\hat{\theta} \neq \theta\}}] = P(\hat{\theta} \neq \theta|\boldsymbol{x}^N), \tag{3.136}$$

we obtain the maximum a posteriori estimator by choosing $\hat{\theta}$ as the value with the highest posterior probability.

An example of Bayesian estimation is presented in Intermezzo 3.5.

**Intermezzo 3.5.** MAP estimation of the average of a normal variate

Let $X_1, \ldots, X_n$ denote a random sample from the normal distribution $G(\theta, \sigma^2)$ where $\sigma$ is known. The joint density for the given dataset is

$$p(x_1, \ldots, x_N) = \prod_{i=1}^{N} \frac{1}{\sqrt{2\pi}\sigma} \exp\left[-\frac{(x_i - \theta)^2}{2\sigma^2}\right]. \tag{3.137}$$

If we assume a Gaussian prior $G(0, \sigma_\theta^2)$ for $\theta$

$$p(\theta) = \frac{1}{\sqrt{2\pi}\sigma_\theta} \exp\left[-\frac{\theta^2}{2\sigma_\theta^2}\right] \tag{3.138}$$

we find that

$$\frac{\partial \log\ p(\theta|\boldsymbol{x})}{\partial\theta} = \frac{1}{\sigma^2}\sum_{i=1}^{N}(x_i - \theta) - \frac{\theta}{\sigma_\theta^2} = 0 \tag{3.139}$$

leading to the following estimate $\hat{\theta}$:

$$\hat{\theta}_{\mathrm{MAP}} = \left[1 + \frac{1}{N}\frac{\sigma^2}{\sigma_\theta^2}\right]^{-1} \frac{1}{N}\sum_i x_i. \tag{3.140}$$

This means that when our prior knowledge of $\theta$ is uncertain, i.e. $\sigma_\theta \to \infty$, or when the amount of data becomes significant, i.e. $N \to \infty$, the estimator converges to the maximum likelihood one, based solely on data evidence. When $N$ or $\sigma_\theta$ is small, $\hat{\theta}_{\mathrm{MAP}}$ provides a smaller, shrunk, estimate with respect to $\hat{\theta}_{\mathrm{MLE}}$.

### 3.6.3. JAMES–STEIN ESTIMATION

Asymptotic variance of MLEs is particularly relevant from the point of view of template matching as classification. It is often the case that a pattern in a space of high dimension $n_d$ must be classified as the template searched for or not. In the case of images, the dimensionality of patterns corresponds to the number of pixels and can be very large.

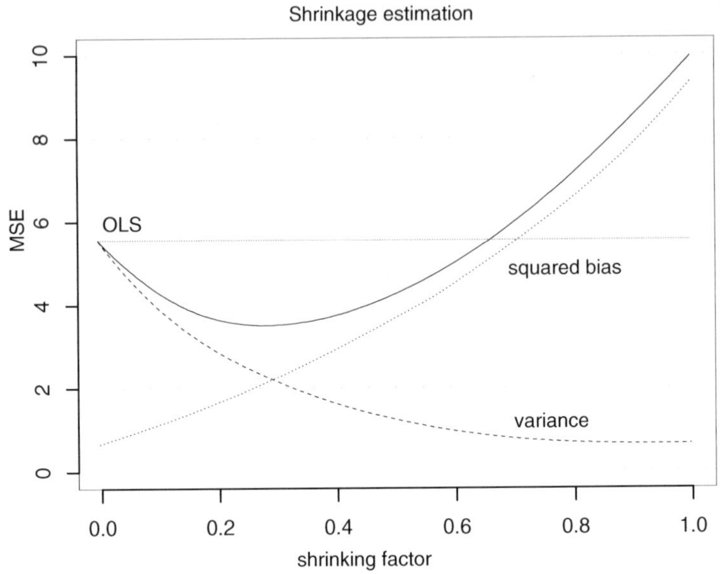


**Figure 3.7.** Trading bias for variance can result in estimators that, while biased, feature a lower MSE.

We have seen that the most basic way to perform classification is by means of log-likelihood ratios that, in the simplest (but common) cases, are based on the assumption of Gaussian distributions. However, in many cases, the number of available samples $N$ is small with respect to the dimensionality of the patterns $n_d$. Apart from the fact that when $N < n_d$ the sample covariance matrix is singular (as its rank is at most equal to the number of independent samples), a reliable estimate of $\Sigma$ requires $N \geq 10n_d$. In many practical cases the asymptotic properties of the MLE are not reached and biased estimators, with smaller variance than the MLE, may perform better in the $N/n_d$ regime characteristic of a specific application. The reason underlying this possibility can be understood by inspecting Equation 3.113. If the operator is unbiased, the only way to improve its MSE is via variance reduction. Methods exist but they scale badly with the dimensionality of pattern space. An alternative is to decrease variance at the cost of introducing bias (see Figure 3.7). As an extreme example, if we want to estimate the average of a set of numbers, we may use a single constant value $a$: its variance is zero, but its bias is very high (and it happens to be zero only for data distributed around $a$).

It is possible to guess the general form that such an operator can assume by a heuristic argument. Let $\boldsymbol{X} \in \mathbb{R}^{n_d}$ be an observation vector with uncorrelated components providing an estimate for a parameter vector $\boldsymbol{\theta}$. It could be a single pattern from a multivariate normal distribution, or it could be a vector of estimated averages. Let us further assume that the components of $\boldsymbol{X}$ have equal variance $\sigma^2$. If $E\{\boldsymbol{X}\} = \boldsymbol{\theta}$, i.e. $\boldsymbol{X}$ is an unbiased estimator for $\boldsymbol{\theta}$, we have that $E\{(\boldsymbol{X} - \boldsymbol{\theta})^T \boldsymbol{\theta}\} = 0$ and we may consider $\boldsymbol{X} - \boldsymbol{\theta}$ to be orthogonal to $\boldsymbol{\theta}$. The expected size of $\boldsymbol{X}$ is given by $E\{\|\boldsymbol{X}\|^2\} = n_d \sigma^2 + \|\boldsymbol{\theta}\|^2$ making $\boldsymbol{X}$ an oversized estimate of $\boldsymbol{\theta}$, suggesting that a better estimator $\delta_\theta$ for $\boldsymbol{\theta}$ can be obtained by shrinking $\boldsymbol{X}$:

$$\delta_\theta(a) = (1 - a)\boldsymbol{X} \tag{3.141}$$

for an appropriately estimated $a$. A simple way to estimate $a$ is by explicit minimization of the loss represented by the expected quadratic error incurred by $\delta_a$ over all possible $\boldsymbol{\theta}$:

$$C(\boldsymbol{\delta}(a, \boldsymbol{\theta})) = \mathrm{MSE}(\boldsymbol{\delta}(a, \boldsymbol{\theta})) = E_{\boldsymbol{\theta}}\{\|(1-a)\boldsymbol{X} - \boldsymbol{\theta}\|^2\}. \tag{3.142}$$

It is instructive to express the expected quadratic loss $C$ as the sum of a variance term $\sum_{i=1}^{n_d}(\delta_i - \bar{\delta}_i)^2$ and a bias term $\sum_{i=1}^{n_d}(\bar{\delta}_i - \theta_i)^2$:

$$C(\boldsymbol{\delta}(a, \boldsymbol{\theta})) = \sum_{i=1}^{n_d}(\delta_i - \bar{\delta}_i)^2 + \sum_{i=1}^{n_d}(\bar{\delta}_i - \theta_i)^2 \tag{3.143}$$

$$= n_d\sigma^2(1-a)^2 + a^2\|\boldsymbol{\theta}\|^2. \tag{3.144}$$

The value of $a$ minimizing the expected squared error is readily found by equating to zero the derivative of the loss with respect to $a$:

$$\frac{d}{da}C = -2n_d\sigma^2(1-a) + 2a\|\boldsymbol{\theta}\|^2 = 0 \tag{3.145}$$

so that

$$a = \frac{n_d}{n_d + \|\boldsymbol{\theta}\|^2} \tag{3.146}$$

and noting that the second derivative at this point is positive. Exploiting the fact that $E\{\|\boldsymbol{X}\|^2\} = n_d\sigma^2 + \|\boldsymbol{\theta}\|^2$, we make the substitution $\|\boldsymbol{X}\|^2 = n_d\sigma^2 + \|\boldsymbol{\theta}\|^2$ obtaining

$$\boldsymbol{\delta} = \left(1 - \frac{n_d\sigma^2}{\|\boldsymbol{X}\|^2}\right)\boldsymbol{X}. \tag{3.147}$$

Let us consider the estimation of the mean vector of a multivariate normal distribution by means of a single data point $\boldsymbol{x}$. The MLE in this case is simply $\hat{\boldsymbol{\theta}} = \boldsymbol{X}$ and Equation 3.147 tells us that a smaller value would be a better estimate. While the derivation of Equation 3.147 is general but heuristic, a similar result can be proved.

**Theorem 3.6.3 (James–Stein Estimator).** *Let $\boldsymbol{X}$ be distributed according to an $n_d$-variate normal distribution $N(\boldsymbol{\theta}, \sigma^2 I)$. Under the squared loss, the usual estimator $\boldsymbol{\delta}(\boldsymbol{X}) = \boldsymbol{X}$ exhibits a higher MSE loss for any $\boldsymbol{\theta}$, being therefore dominated, than*

$$\boldsymbol{\delta}_a(\boldsymbol{X}) = \boldsymbol{\theta}_0 + \left(1 - \frac{a\sigma^2}{\|\boldsymbol{X} - \boldsymbol{\theta}_0\|^2}\right)(\boldsymbol{X} - \boldsymbol{\theta}_0) \tag{3.148}$$

*for $n_d \geq 3$ and $0 < a < 2(n_d - 2)$, and $a = n_d - 2$ gives the uniformly best estimator in the class. The MSE loss of $\delta_{n_d-2}$ at $\boldsymbol{\theta}_0$ is constant and equal to $2\sigma^2$ (instead of $n_d\sigma^2$ of the usual estimator).*

Before applying these results to the estimation of covariance matrices let us note that:

- The use of this estimator is advantageous when all elements of the average vector must be estimated: the dominance relation with respect to the usual estimator is with respect to the total error.

- The MSE advantage of this operator is robust to the underlying distributional hypothesis (normality) and also to the estimation of the expected variances of the single components: the advantage persists even when they are unknown and are therefore estimated from data.
- While using this estimator is always advantageous, the maximal benefit is obtained when the true parameter corresponds to the shrinkage point $\boldsymbol{\theta}_0$. If we correctly guess the shrinking point, the improvement can be dramatic when $n_d \gg 1$.

Let us apply this line of thought to the estimation of covariance matrices. When we estimate the covariance matrix of a set of data distributed according to a multivariate normal, we obtain the (single) estimate $\hat{\Sigma} = S$ that corresponds to the usual estimator $\boldsymbol{X}$ mentioned in Theorem 3.6.3. The distribution of $S$ is not normal (it is a Wishart distribution) but the generality of the heuristic derivation of the shrinkage estimator and the robustness of the James–Stein result to the underlying distributional hypothesis justify a shrinkage approach. A natural generalization of the square loss in the case of matrices is the Frobenius norm that is the $L_2$ norm of the linearized matrix

$$\|A\|_F = \sum_i \sum_j |A_{ij}|^2 \tag{3.149}$$

that corresponds to the usual $L_2$ norm applied to the vectorized matrix. Let $S_s = aT + (1 - a)S$ be the shrunk estimator, $T$ the shrinking target, and $S$ the empirical covariance. We want to find the value of $a$ that minimizes the average Frobenius loss:

$$E\{C(a)\} = E\{\|[aT + (1 - a)S] - \Sigma\|_F^2\} = \sum_i \sum_j [aT_{ij} + (1 - a)S_{ij} - \Sigma_{ij}]^2. \tag{3.150}$$

It is possible to prove that such a value always exists and that it is unique. If $S$ is an unbiased estimator, the optimal value $\hat{a}$ is given by

$$\hat{a} = \frac{\sum_{i=1}^{n_d} \sum_{j=1}^{n_d} [\mathrm{var}(S_{ij}) - \mathrm{cov}(T_{ij}, S_{ij})]}{\sum_{i=1}^{n_d} \sum_{j=1}^{n_d} E\{(T_{ij} - S_{ij})^2\}} \tag{3.151}$$

from which we see that:

1. The smaller the variance of $S$, the smaller the valued $\hat{a}$: increasing sample size reduces the influence of the shrinking target.
2. The more the correlation of $S$ and $T$, the smaller the shrinkage.
3. The larger the discrepancy of $S$ and $T$, the smaller the shrinkage.

The value of the optimal shrinking factor $\hat{a}$ of Equation 3.151 is based on non-observable values: the expected values, including the variances and covariances, are averages over the distribution at the true parameter value. It is therefore necessary to estimate them from the data in order to compute $\hat{a}$. As noted earlier, the choice of the shrinking target is important to maximize the benefit of the estimation procedure. In many cases, an appropriate target is given by the diagonal, unequal variance model: $T_{ij} = \delta_{ij} S_{ii}$. The resulting shrunk covariance estimate will automatically be positive definite being a convex combination of a positive

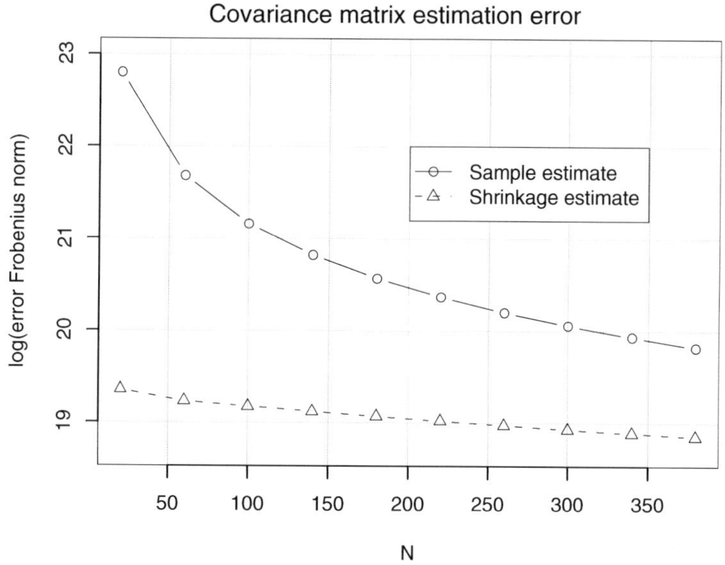


**Figure 3.8.** The plots shows the MSE in the estimation of a random covariance matrix for $n_d = 625$ and a varying number of samples, a typical situation in many low-resolution face processing tasks. The advantage of the shrinkage estimator over the empirical one is marked (see also Figure 8.4).

matrix $T$ and a positive semidefinite matrix $S$. A variant of the James–Stein result says that it is possible to shrink towards a subspace: in our case we may limit shrinking to the off-diagonal elements. We can then characterize the elements of the covariance matrix in terms of variances $S_{ij}$ and correlations $R_{ij}$ obtaining $S_{ij} = R_{ij}\sqrt{S_{ii}S_{jj}}$. As we are not shrinking the diagonal elements, the shrinking will act only on the correlations and the minimization of $C(a)$ from Equation 3.150 leads to

$$a = \frac{\sum_{i \neq j} \widehat{\text{Var}}(R_{ij})}{\sum_{i \neq j} R_{ij}^2}. \tag{3.152}$$

Let $X \in \mathbb{R}^{N \times n_d}$ be the matrix whose rows represent the $N$ samples from which the covariance matrix must be estimated. In order to compute the unbiased empirical covariances we denote by $\bar{X}_i = \frac{1}{N}\sum_{k=1}^{N} X_{ki}$ the empirical mean of the $i$th coordinate. The unbiased empirical covariance is then

$$\widehat{\text{Cov}}(X_i, X_j) = S_{ij} = \frac{N}{N-1}\bar{W}_{ij} \tag{3.153}$$

where $\bar{W}_{ij} = N^{-1}\sum_{k=1}^{N} W_{kij}$ and $W_{kij} = (X_{ki} - \bar{X}_i)(X_{kj} - \bar{X}_j)$. The empirical unbiased variances of the entries of $S$ are computed as

$$\widehat{\text{Var}}(S_{ij}) = \frac{N^2}{(N-1)^2}\widehat{\text{Var}}(\bar{W}_{ij}) = \frac{N}{(N-1)^2}\widehat{\text{Var}}(W_{ij}) = \frac{N}{(N-1)^3}\sum_{k=1}^{N}(W_{kij} - \bar{W}_{ij}). \tag{3.154}$$

The corresponding expressions for the correlation coefficients $R_{ij}$ can be computed using the same formula applied to the standardized data matrix, where the $i$th coordinate is divided

**Intermezzo 3.6.** A more flexible, regularized covariance estimator method

In pattern classification tasks, the estimate of the covariance matrix is instrumental to classification and an approach that more directly targets classification may provide better results. Let us consider the case of $M$-ary classification under the assumption of approximately Gaussian distribution of the patterns within each class. The ideas of shrinkage and regularization underlie the following estimator of the class covariance matrix:

$$\hat{\Sigma}_i(\alpha, \beta) = \alpha S_i + (1 - \alpha) S + \beta I \tag{3.155}$$

where

$$S_i = \frac{1}{n_i - 1} \sum_{j \in C_i} (\boldsymbol{x}_j - \boldsymbol{\mu}_i) \cdot (\boldsymbol{x}_j - \boldsymbol{\mu}_i)^T \tag{3.156}$$

is the sample class scatter matrix computed using the samples belonging to class $C_i$,

$$S = \frac{1}{n - 1} \sum_{j=1}^{n} (\boldsymbol{x}_j - \boldsymbol{\mu}) \cdot (\boldsymbol{x}_j - \boldsymbol{\mu})^T \tag{3.157}$$

is the sample total scatter matrix and in both cases the unbiased version is used. The parameters $\alpha$ and $\beta$ in Equation 3.155 apply to all classes. The addition of a multiple of the identity matrix corresponds to the effect of additive white Gaussian noise. The effects due to the addition of the same amount of noise to all classes can be corrected with the introduction of an additional set of class-specific parameters $\gamma_i$ to be used in the classification stage based on the Bayes MAP decision rule. The standard rule that assigns pattern $\boldsymbol{x}$ to class $\hat{\imath}$

$$\hat{\imath} = \underset{i}{\operatorname{argmin}}\ d_i(\boldsymbol{x}) \tag{3.158}$$

where

$$d_i(\boldsymbol{x}) = (\boldsymbol{x} - \boldsymbol{\mu}_i)^T \Sigma_i (\boldsymbol{x} - \boldsymbol{\mu}_i) + \ln |\Sigma_i| - 2 \ln \pi_i \tag{3.159}$$

is transformed into

$$\hat{\imath} = \underset{i}{\operatorname{argmin}}\ d_i(\boldsymbol{x}) + \gamma_i \tag{3.160}$$

where $\gamma_1 = 0$. The parameters $\alpha$, $\beta$, and $\gamma_i$ for $i > 1$ must be estimated using cross-validation techniques (see Section C.2) from the available data as no closed form expressions for them exists. The results reported in the literature show that classification performance is significantly increased in many face analysis tasks including face/non-face discrimination and smile/neutral expression discrimination resulting in a 5- to 10-fold reduction in classification error. In many cases the scaled identity matrix ($\beta$) contributes significantly: as it does not rotate the covariance matrix but changes the scales of the axes, we conclude that principal component analysis (see Section 8.1), which rotates the coordinate reference system in such a way that the resulting covariance matrix is diagonal, is appropriate for selecting the most significant descriptive directions, but is not so reliable in sphering the data for classification purposes (Robinson 2005).

by the corresponding scale estimator $\sqrt{S_{ii}}$. An example comparing empirical and shrinkage estimators is reported in Figure 3.8. Even when the MSE advantage is more limited, the facts that the covariance matrix from the shrinkage estimator is guaranteed to be positive definite and that its inverse, the matrix needed in the classification task, can be computed much more efficiently than from the sample (empirical) estimates, favor the shrinkage approach. In pattern recognition applications the estimation of the covariance matrix can be driven by the goal of improving the performance of the classifier: this approach is discussed in Intermezzo 3.6.

## 3.7. Bibliographical Remarks

A concise source of mathematical definitions and important results is the *Encyclopedic Dictionary of Mathematics* (Mathematical Society of Japan 1993). The presentation of template detection as hypothesis testing is based heavily on the book by Moon and Stirling

(2000) where many interesting proofs and examples can be found. The book also provides an extensive treatment of the expectation–maximization algorithm.

Basic statistical concepts can be found in Hogg and Craig (1978), including some of the characteristics of the correlation coefficient considered in the chapter. The bias of the Pearson correlation coefficient is discussed by Zimmerman *et al.* (2003), whose paper also considers the case of rank correlation that is described in Chapter 5.

The description of matched filters is based on Brunelli and Poggio (1997), a paper that contains additional material to be covered in Chapter 6 on techniques to match variable patterns, and in Chapter 8 on low-dimensional pattern representation and matching.

The material on James–Stein estimation is taken from several papers. The paper from Efron and Morris (1973) presents this class of estimators from an empirical Bayes approach, according to which the (hyper)parameters controlling the prior distribution must be estimated from data. Several examples are presented in Efron and Morris (1975). A modern presentation with proofs and the variants considered in the text can be found in Brandwein and Strawderman (1990).

The problems associated with the estimation of the covariance matrix in the large-dimension/small-sample case is discussed from a pattern classification perspective by Friedman (1987). An application of James–Stein estimation to quadratic discrimination is presented by Piper *et al.* (1994). The explicit formulas for the computation of the shrunk estimate of the covariance matrix can be found in the papers by Ledoit and Wolf (2002) and Schafer and Strimmer (2005).

## References


Brandwein A and Strawderman W 1990 Stein estimation: the spherically symmetric case. *Statistical Science* **5**, 356–369.

Brunelli R and Poggio T 1997 Template matching: matched spatial filters and beyond. *Pattern Recognition* **30**, 751–768.

Efron B and Morris C 1973 Stein's estimation rule and its competitors – an empirical Bayes approach. *Journal of the American Statistical Association* **68**, 117–130.

Efron B and Morris C 1975 Data analysis using Stein's estimator and its generalizations. *Journal of the American Statistical Association* **70**, 311–319.

Friedman J 1987 Regularized discriminant analysis. *Journal of the American Statistical Association* **84**(405), 165–175.

Hogg R and Craig A 1978 *Introduction to Mathematical Statistics*. Macmillan.

Ledoit O and Wolf M 2002 Improved estimation of the covariance matrix of stock returns with an application to portfolio selection. *Journal of Empirical Finance* **10**, 603–621.

Mathematical Society of Japan 1993 *Encyclopedic Dictionary of Mathematics* 2 edn. MIT Press.

Moon T and Stirling W 2000 *Mathematical Methods and Algorithms for Signal Processing*. Prentice Hall.

Piper J, Poole I and Carothers A 1994 Stein's paradox and improved quadratic discrimination of real and simulated data by covariance weighting. *Proceedings of the 12th IAPR International Conference on Pattern Recognition (ICPR'94)*, vol. 2, pp. 529–532.

Robinson J 2005 Covariance matrix estimation for appearance-based face image processing *Proceedings of the British Machine Vision Conference (BMVC'05)*, vol. 1, pp. 389–398.

Schafer J and Strimmer K 2005 A shrinkage approach to large-scale covariance matrix estimation and implications for functional genomics. *Statistical Applications in Genetics and Molecular Biology* **4**(1), Article 32, Berkeley Electronic Press.

Zimmerman D, Zumbo B and Williams R 2003 Bias in estimation and hypothesis testing of correlation. *Psicologica* **24**, 133–158.

# 4 ROBUST SIMILARITY ESTIMATORS

> If it be possible for you to displace it with your little finger.
>
> *Coriolanus*
> WILLIAM SHAKESPEARE

A major issue in template matching is the stability of similarity scores with respect to noise, including unmodeled phenomena. Many commonly used estimators suffer from a lack of robustness: small perturbations in the data can drive them towards uninformative values. This chapter addresses the concept of estimator robustness in a technical way presenting applications of robust statistics to the problem of pattern matching.

## 4.1. Robustness Measures

We have seen in Chapter 3 that comparing a reference template to an image using the $L_2$ distance is related to computing the probability of the mismatch under the assumption that pixel differences are normally distributed (see Equation 3.45). Let us consider some slightly different but related problems.

The first one is that of image registration. We would like to find the best possible parameters of the aligning transformation. Let us also assume for simplicity that the image contains a perfect copy of the template, apart from Gaussian noise. When the template is perfectly aligned to its image version, the correct probability of match is given by the $L_2$ distance. When the template is close to alignment, we are not (usually) comparing it to Gaussian noise but to an incomplete version of itself plus additional image content, possibly noise. When the template does not overlap its image instance, we are comparing it to other image content, which might be noise whose distribution we know. This means that the relation between distance and likelihood holds only at the aligned position: in the other cases the distribution of pixel differences is not necessarily Gaussian.

Another important application of template matching is stereo vision. When we look at a scene from two different viewpoints (e.g. our left and right eye) we perceive two different images. The ability to estimate depth is based on finding the correspondence of the pixels of one image in the other one: the distance at which the correspondences are found is related to the distance of the corresponding point in the real world from the observer. Due to the different viewpoints, the appearance of the corresponding image patches will differ. Let us compute the pixel differences of a local image patch from the left image to a neighborhood of the corresponding patch in the right image: will the distribution turn out to be Gaussian?

*Template Matching Techniques in Computer Vision: Theory and Practice* Roberto Brunelli

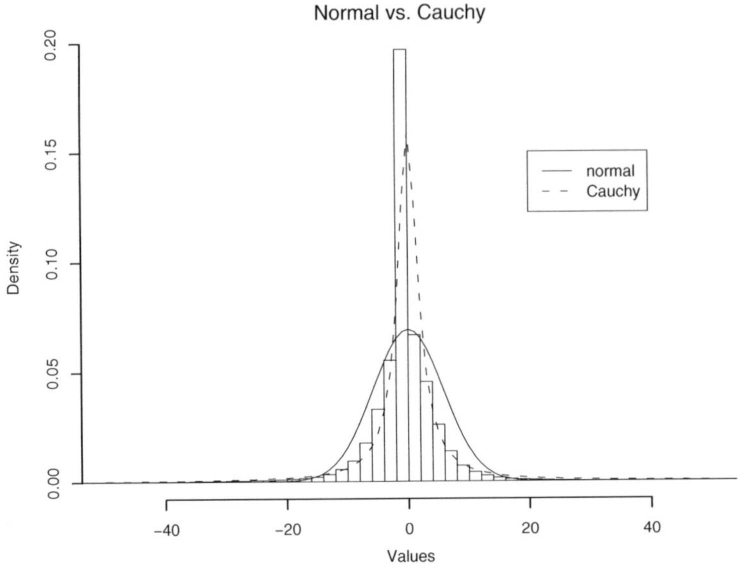


**Figure 4.1.** The distribution of pixel differences between two slightly misaligned images of the same pattern (in this case a face) are better modeled with a Cauchy distribution than with a normal one.

Another similar problem is tracking a deformable pattern in time using a stream of images. An example is tracking the lips and eyes of a talking person. Again, let us compute the distribution of the differences between the pixels of corresponding image patches as a function of time: will the distribution be Gaussian? Will it change based on the temporal distance of the images used for the tracking? The experimental answer (see Figure 4.1 for an example) is that in all three cases the distribution of the errors is definitely not Gaussian, nor exponential, but turns out to be a Cauchy distribution:

$$C(x; \mu_C, \sigma_C) = \frac{1}{\pi}\left[\frac{\sigma_C}{(x-\mu_C)^2 + \sigma_C^2}\right] \tag{4.1}$$

(see Intermezzo 4.1 on the comparison of distributions). An important feature of the Cauchy distribution (also known as the Lorentz distribution) is that its moments, including the mean and variance, are not defined as the corresponding integrals fail to converge (see Intermezzo 4.2). This is due to its tails decaying very slowly as opposed to those of other distribution like the Gaussian (see Figure 4.2 for a visual comparison).

The experimental finding that, in the cases considered, the differences between the reference template and its matching signal are described by a Cauchy distribution is important for two reasons:

- Due to the way the error distribution in the above examples is computed, the Cauchy distribution appears to provide a good model for the process of image registration and matching in the local neighborhood of the true template position.

- It tells us that errors of significant size are not as uncommon as the Gaussian distribution would lead us to think.

**Intermezzo 4.1.** Are two distributions the same?

As hypothesis tests are based on specific distributional hypotheses, it is necessary to verify that they actually hold. The way to proceed is, again, via an appropriate hypothesis test: we hypothesize a specific distribution $F$ and we check whether our hypothesis should be rejected or not. The rejection test of the null hypothesis depends on the specific distributional hypothesis, but it is possible, given two distributions $F$ and $G$, to compute a statistic, the chi-square statistic $\chi^2$, whose distribution, under reasonable assumptions, has an approximate chi-square distribution. There are two different cases:

1. The model distribution $F$ is not binned.
2. Both distributions are binned.

In the first case we have

$$\chi^2 = \sum_{i=1}^{N_b} \frac{(n_i(G) - n_i(F))^2}{n_i(F)} \tag{4.2}$$

where $i$ runs over the histogram bins, and $n_i(F)$ represents the number of events for bin $i$ according to distribution $F$. If the number of bins $N_b$ or the number of events per bin is large, the $\chi^2$ statistic follows a $P(\chi^2; \nu)$ distribution with $\nu = N_b - 1$ degrees of freedom if the $n_i(F)$ are normalized to match the data

$$n_j(F) \to \frac{\sum_i n_i(G)}{\sum_i n_i(F)} n_j(F). \tag{4.3}$$

In the second case we have

$$\chi^2 = \sum_{i=1}^{N_b} \frac{(n_i(G) - n_i(F))^2}{n_i(F) + n_i(G)} \tag{4.4}$$

and this follows a $P(\chi^2; \nu)$ distribution with $\nu = N_b - 1$ degrees of freedom if counts are matched.

**Intermezzo 4.2.** A statistical joke

**Question**: What's Cauchy's least favorable question?
**Answer**: Got a moment?

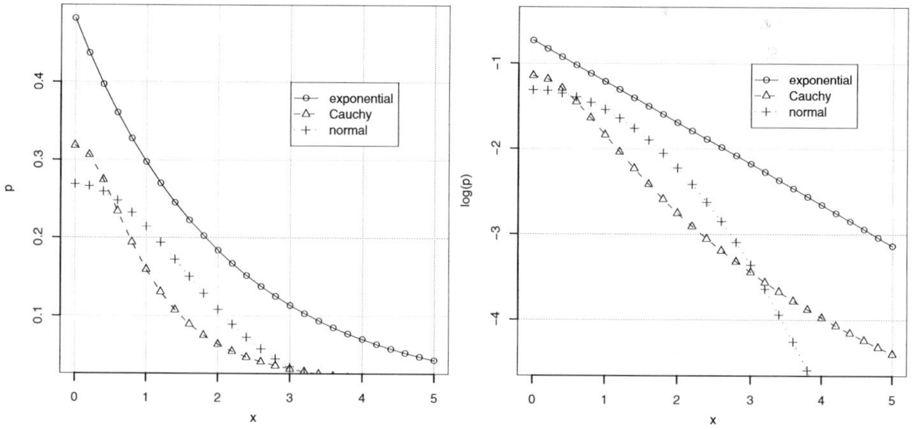


**Figure 4.2.** A comparison of the Cauchy, normal, and exponential distributions with the same scale factor as determined by the mad(·) estimator of Equation 4.14.

In Chapter 3 we focused on the discrimination between signal and noise, but this does not completely solve the problem of optimal template localization as in the proximity of the correct position we are not comparing signal to noise but signal to (translated and noisy) signal. The first finding lets us extend the distance–likelihood link from the matching position to its neighborhood, and suggests that minimizing the $L_2$ distance no longer corresponds to maximizing the likelihood. The second finding suggests that using similarity measures (or likelihoods) tailored to a Gaussian distribution may over-penalize matches that are actually correct: a (moderately) large error at a single pixel can drive the matching estimate towards zero due to the fact that it is much more unexpected under a Gaussian error distribution than under a Cauchy distribution. Limited sensitivity to large matching errors restricted to a limited number of image locations is particularly important in computer vision applications where spatially localized phenomena, such as specularities or sensor defects, and even the occlusion effects of stereo, act as arbitrary perturbations of the distribution otherwise governing the data (see Figure 4.3). The resulting factionalism, multiple phenomena within the same image processing window, can be formally described as

$$F = (1 - \epsilon)F_1 + \epsilon G \tag{4.5}$$

where $F_1$ is a known distribution, termed the inlier (or target) distribution, e.g. the Cauchy distribution, $G$ is an arbitrary distribution, termed the outlier distribution, and $\epsilon$ represents the (small) amount of contamination. As we saw in Chapter 3, the development of a classifier for template detection relies heavily on knowledge of the data distribution. The fact that, in most cases, $F_1$ cannot be known exactly and that, additionally, it can be arbitrarily corrupted by outliers belonging to $G$ leads us to search for methods that are robust to errors in the estimation of $F_1$ (model robustness) and to arbitrary perturbations (outlier robustness). Finding a useful approximation to the target distribution $F_1$ is usually not too difficult and the resulting approximations are often under control and not critical. In this chapter we focus instead on countering the (potentially large) effects of arbitrary perturbations, the $\epsilon G$ term in Equation 4.5.

The normalized correlation coefficient $r_P$, introduced in Section 3.5, is a good starting point for investigating in a quantitative way the concept of robustness. The evaluation of $r_P$ requires that the signals to be compared be normalized to zero average and unit variance. The way this is achieved in practice is to use the sample estimates of the average and of the centered second-order moment, the variance. A little algebra shows that the quantities that these estimators are supposed to compute are not defined when signal noise is Cauchy distributed due to the long tails of the distribution causing divergences. While, in practice, the Cauchy distribution appears in its truncated form for which all moments can be computed, it is of interest to consider the location and scale estimators for its full form. It is possible to demonstrate that the parameters $\mu_C$ (location) and $\sigma_C$ (scale) of Equation 4.1 can be estimated as

$$\mu_C = \text{med}(X) \tag{4.6}$$

$$\sigma_C = \text{mad}(X) \tag{4.7}$$

where $\text{med}(X)$ and $\text{mad}(X)$ are respectively the median and median absolute deviation of the distribution (see Intermezzo 4.3). If we go back to the definition of normalized correlation in Equation 3.94 and move one of the sample points $(x_i, y_i)$ far away from the other ones we obtain

$$r_P \xrightarrow{x_i \to \infty,\, y_i = x_i} 1 \tag{4.8}$$

$$r_P \xrightarrow{x_i \to \infty,\, y_i = -x_i} -1 \tag{4.9}$$

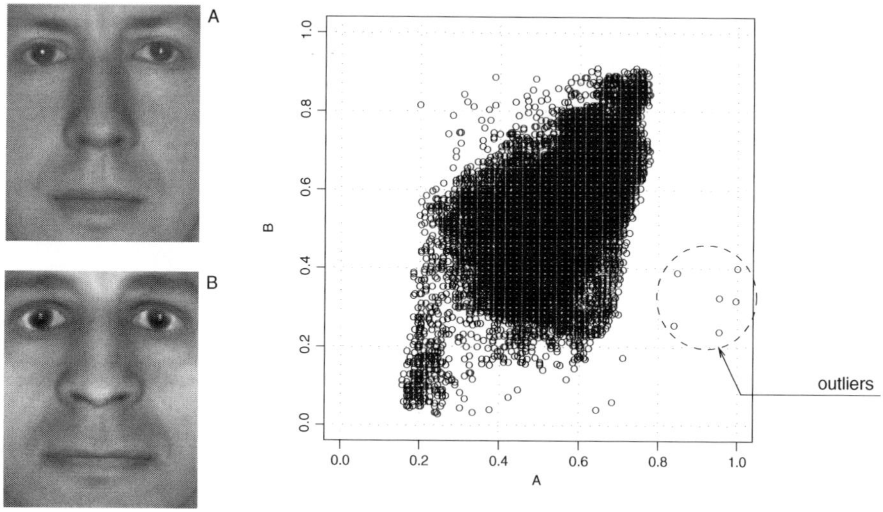


**Figure 4.3.** The plot shows the joint distribution of the corresponding pixels of the geometrically normalized images (eyes are in the same position) of two different persons. The points in the lower right corner are due to the different position of specularities of the eyes.

so that even a single point grossly out of place can drive the value of the coefficient to the extremal values independently of the value of all the other points. More generally, it is possible to make $r_P$ assume any predefined value within $[-1, 1]$ by moving a single point $(x_i, y_i)$: a single outlier can arbitrarily bias the value of the normalized correlation coefficient. This is a very simple example of what the non-robustness of an estimator implies. Several pieces of $r_P$ break down under the effect of the outlying point. The sample average and the sample variance used for the normalization both diverge. But the situation is even worse: even assuming that somehow they would not be influenced by the outlier, the effect of the latter would surface when the sum of the products of the normalized values is computed, the normalization of $r_P$ being lost.

As a preliminary step towards curing the robustness deficit of $r_P$, let us formalize what we mean by location and scale estimates for a set of (scalar) observations $\{x_i\}_{i=1}^N$. A location estimate is a statistic $L_N$ that is equivariant under shift transformations of the data:

$$L_N(a + x_1, \ldots, a + x_N) = a + L_N(x_1, \ldots, x_N), \ \forall a \in \mathbb{R}. \tag{4.10}$$

A scale estimate is any positive statistic $S_N$ that is equivariant under scale transformations of the data:

$$S_N(bx_1, \ldots, bx_N) = bS_N(x_1, \ldots, x_N), \ \forall b \in \mathbb{R}^+. \tag{4.11}$$

Sometimes a dispersion estimator is used in place of a scale estimator

$$S_N(bx_1 + a, \ldots, bx_N + a) = bS_N(x_1, \ldots, x_N). \tag{4.12}$$

The value of location and scale of a set of values is then not uniquely determined: it depends on the specific estimator chosen. We are free to choose among the available estimators the one that best suits a specific problem from a computational and/or stability perspective.

**Intermezzo 4.3.** The median and the median absolute deviation

Let $X = \{x_1, x_2, \ldots, x_N\}$ be a random sample, sort the data in ascending order from the smallest to the largest and write them as $x_{(1)}, x_{(2)}, \ldots, x_{(N)}$. The $x_{(i)}$ are called order statistics and $x_{(i)}$ is the $i$th-order statistic.

**Definition 4.1.1.** *The sample median* $\text{med}(X)$ *is defined as*

$$\text{med}(X) = x_{((N+1)/2)} \quad \textit{if } N \textit{ is odd}$$

$$\text{med}(X) = \frac{x_{(N/2)} + x_{((N/2)+1)}}{2} \quad \textit{if } N \textit{ is even.} \tag{4.13}$$

**Definition 4.1.2.** *The sample median absolute deviation is defined as*

$$\text{mad}(X) = \text{med}(\{|x_i - \text{med}(X)|\}_{i=1}^{N}). \tag{4.14}$$

From the definitions we see that at least half of the samples are lower than $\text{med}(N)$ and at least half of the samples are within a distance $\text{mad}(N)$ of $\text{med}(N)$. The med and mad estimators are respectively location and scale (dispersion) estimators that are useful for many distributions. As a matter of fact, the location and scale parameters of a location–scale parametric distribution can often be expressed as $c_\mu \,\text{med}(X)$ and $c_\sigma \,\text{mad}(X)$. For the normal distribution $N(\mu, \sigma^2)$ we have

$$c_\mu = 1 \tag{4.15}$$

$$c_\sigma = 1.483. \tag{4.16}$$

The computation of the median of a set of values does not require a complete sorting of the values and, accordingly, algorithms exist with complexity (slightly) lower than $O(N \log N)$, the one of full sorting algorithms. There are, however, two major possibilities for improving on this complexity in image processing tasks. The first one is based on the fact that in many applications the number of possible different pixel values is limited and small: only 256 values are possible for an 8-bit image. The computation of the median can then rely on value histograms and reduces to $O(N)$ or $O(r^2)$ if instead of using the number of pixels we use the radius $r$ of a circle with the same number of pixels in the image. The other source of improvement is related to the fact that the computation of the median, be it for image smoothing or for template normalization, must often be computed for a window located at each pixel position. This allows us to significantly amortize the computational cost as moving from one pixel to the neighboring one does not require a full recomputation. The set of pixels covered by the sliding window changes only slightly when moving from one position to a nearby one. When scanning the image in left to right, top to bottom order, a column is discarded and a new one is inserted: this leads easily to an amortized per pixel complexity of $O(r)$ and, not so easily, to $O(\log r)$ and even $O(1)$ amortized complexities (Perreault and Hebert 2007; Weiss 2006).

According to Equation 4.10, the sample mean is a location estimator while the sample standard deviation, the square root of the variance, is a scale estimator as defined by Equation 4.11. The Pearson correlation coefficient $r_{\text{P}}$ corresponds to the covariance of two zero-mean, unit-variance random variables $X'$ and $Y'$

$$r_{\text{P}} = \text{cov}(X'Y'). \tag{4.17}$$

However, for square integrable random variables, of not necessarily zero mean and/or unit variance, we have that

$$\text{cov}(XY) = \frac{1}{4ab}[\sigma^2(aX + bY) - \sigma^2(aX - bY)] \tag{4.18}$$

which allows us to express the normalized correlation coefficient in terms of two scale estimators

$$r_P = \frac{1}{4}[\sigma^2(X' + Y') - \sigma^2(X' - Y')] \tag{4.19}$$

$$= \frac{\sigma^2(X' + Y') - \sigma^2(X' - Y')}{\sigma^2(X' + Y') + \sigma^2(X' - Y')} \tag{4.20}$$

where we have exploited, in order to get Equation 4.20, the unit-variance normalization of the two random variables. If we substitute $\sigma$ by a different scale estimator $S$ we obtain an alternative correlation coefficient

$$r_S = \frac{S^2(X' + Y') - S^2(X' - Y')}{S^2(X' + Y') + S^2(X' - Y')} \in [-1, 1]. \tag{4.21}$$

We can now choose a scale estimator $S$ that does not suffer from the robustness deficit characteristic of $\sigma$. As we will see, there are many possibilities and a simple one, the median absolute deviation or mad, is provided by Equation 4.14. The companion location estimator is provided by the median, whose definition is recalled in Equation 4.13.

Making the value of a single data point, the outlier, arbitrarily large has limited impact on the median and median absolute deviation location and scale estimators. The position of the samples within the order statistics would change by at most one and the median value would be nearly unchanged. In a similar way the computation of mad($X$) would be nearly unaffected by the large distance of the outlier from the median. The correlation coefficient

$$r_{\text{mad}} = \frac{\text{mad}^2(X + Y) - \text{mad}^2(X - Y)}{\text{mad}^2(X + Y) + \text{mad}^2(X - Y)} \tag{4.22}$$

is then less sensitive than $r_P$ to arbitrary perturbations of the datasets from which it is computed.

Let us formalize the concept of robustness as applied to a statistical estimator. A key concept is that of the sensitivity curve of an estimator $T$ based on a sample $\{\boldsymbol{x}_i\}_{i=1}^N$ of finite size $N$ from the model distribution $F$:

$$\text{SC}_N(F; \boldsymbol{x}) = N\left\{T\left[\left(\left(1 - \frac{1}{N}\right)F_{N-1} + \frac{1}{N}\Delta_{\boldsymbol{x}}\right) - T(F_{N-1})\right]\right\} \tag{4.23}$$

where $F_N$ is the empirical distribution of $\{x_i\}_{i=1}^N$ and $\Delta_{\boldsymbol{x}}$ is a single observation at $\boldsymbol{x}$. In many cases the limit for $N \to \infty$ exists and is equal to $\psi(\cdot)$, the influence function of the estimator. The sensitivity curve is mainly a heuristic tool with an important intuitive interpretation: it describes the effect of a small contamination at the point $\boldsymbol{x}$ on the estimate, standardized by the mass of the contamination. Several robustness measures can be derived from the sensitivity curve of an estimator:

- The gross error sensitivity

$$\gamma^*(T, F) = \sup_{\boldsymbol{x}} |\psi(\boldsymbol{x}; T, F)| \tag{4.24}$$

which measures the worst approximate influence that a small amount of contamination of fixed size can have on the value of the estimator. If $\gamma^*(T, F) < \infty$ the estimator is said to be B-robust (from bias-robust).

- The local shift sensitivity

$$\lambda^*(T, F) = \sup_{x \neq y} \frac{|\psi(\boldsymbol{y}; T, F) - \psi(\boldsymbol{x}; T, F)|}{\|\boldsymbol{y} - \boldsymbol{x}\|} \tag{4.25}$$

which measures the effect of shifting an observation slightly from point $\boldsymbol{x}$ to point $\boldsymbol{y}$ and is the smallest Lipschitz constant that $\psi$ obeys.

- The rejection point

$$\rho^* = \inf\{r > 0; \psi(\boldsymbol{x}; T, F) = 0 \text{ when } |\boldsymbol{x}| > r\} \tag{4.26}$$

which, when finite, represents the point at which observations are completely rejected.

It is also possible to consider the robustness of the variance of an estimator: the analog of the influence function is called the change-of-variance function. The associated robustness measure is $\kappa^*$, the change-of-variance sensitivity: when it is finite, the estimator is said to be V-robust (from variance-robust). Variance robustness is important as it affects the estimation of confidence intervals.

While the above robustness measures are essentially local, a very important global measure is the breakdown point which describes up to what distance from the model distribution $F$ the estimator still gives some relevant information.

**Definition 4.1.3.** *The finite sample breakdown point $\epsilon^*$ of the estimator $T_N$ at the sample $x_1, \ldots, x_N$ is given by*

$$\epsilon^*(T_N; \boldsymbol{x}) = \frac{1}{N} \max\left\{k; \max_{i_1,\ldots,i_k} \sup_{y_1,\ldots,y_k} T_N(z) < \Omega_M, \min_{i_1,\ldots,i_k} \sup_{y_1,\ldots,y_k} T_N(z) > \Omega_m\right\} \tag{4.27}$$

*where $\Omega_m$, $\Omega_M$ represent boundary values of the parameter to be estimated and the sample $z = (z_1, \ldots, z_N)$ is obtained by replacing the k data points $(x_{i_1}, \ldots, x_{i_k})$ by the arbitrary values $(y_1, \ldots, y_k)$.*

Examples of $\{\Omega_m, \Omega_M\}$ are $\{-\infty, +\infty\}$ for location estimators, $\{0, +\infty\}$ for scale estimators, and $\{-1, 1\}$ for correlation measures. This leads to the following technical definition of robustness.

**Definition 4.1.4.** *An estimator is said to be robust if it has a strictly positive breakdown point $\epsilon^* > 0$.*

The breakdown point represents the largest fraction of data points that can be moved arbitrarily without completely removing the information provided by the estimator. The most widely used location and scale estimators, namely the sample mean and sample standard deviation, are both not robust (and not B-robust). On the other side, the median and the median absolute deviation are both robust and they achieve the maximal possible breakdown point (for equivariant estimators) $\epsilon^* = 0.5$. The normalized correlation coefficient is not robust, as we already remarked, and its influence function can be computed explicitly when the target distribution is the bivariate normal of Equation 3.102 with correlation value $\rho$:

$$\psi(r_{\mathrm{P}}; N(x_1, x_2; 0, 1, \rho)) = x_1 x_2 - \rho \frac{x_1^2 + x_2^2}{2} \tag{4.28}$$

and is reported in Figure 4.6. As the influence function $\psi(r_{\mathrm{P}}; N(x_1, x_2; 0, 1, \rho))$ is unbounded, $r_{\mathrm{B}}$ is not B-robust.

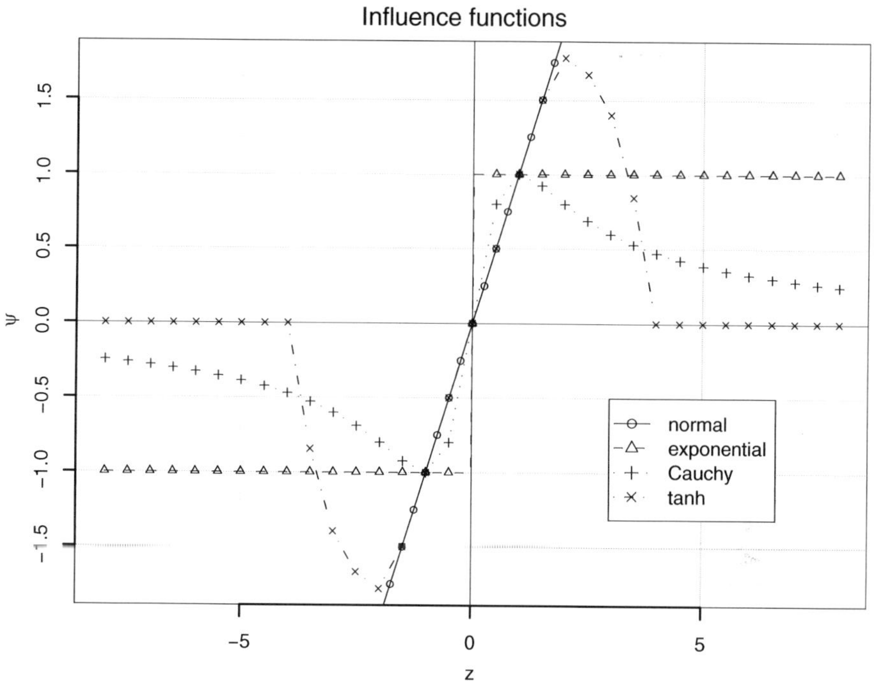


**Figure 4.4.** The influence function associated to some commonly used distributions and the one defining the tanh estimator.

## 4.2. M-estimators

Let us consider the problem of deciding whether a reference pattern $R$ corrupted by Gaussian noise is present at a given location within an image $I$. Based on the assumed model of noise, the probability that the pattern is actually present, let it be hypothesis $H$, is given by

$$P(H) \propto \prod_i \exp\left[-\frac{(I(i) - R(i))^2}{2\sigma^2}\right] \tag{4.29}$$

where $i$ runs over the positions of the pixels. Let us now modify the task to that of determining a geometric transformation $G_{\boldsymbol{\theta}}$ characterized by a vector of parameters $\boldsymbol{\theta}$ that aligns image $R$ to image $I$ by minimizing an appropriate distance $d$ between image $I$ and the transformed version $G_{\boldsymbol{\theta}}(R)$. Let us further assume that, at each image position, the variable $I - G_{\boldsymbol{\theta}}(R)$ is distributed according to $N(0, \sigma^2)$. Choosing the transformation parameters $\hat{\boldsymbol{\theta}}$ so that the $L_2$ distance between $I$ and $R$ is minimized is the same as choosing them to maximize the probability of the transformed image. There is then an equivalence between choosing a particular pixel additive distance for the alignment and the probability according to which the differences between the images to be aligned are distributed. If the differences are not normally distributed, minimization of the $L_2$ distance will in general provide different results from a maximum likelihood approach. It is apparent that whenever the maximum likelihood approach can be applied, it should be preferred to the minimization of a distance that has no probabilistic grounding but has merely a descriptive characterization. Let us formalize the connection between distance and likelihood. Denoting the error at each pixel with $z_i = (I - G_{\boldsymbol{\theta}}(R))(i)$ and letting

$$P(I - G_{\boldsymbol{\theta}}(R)(i)) = f_0(z_i) \tag{4.30}$$

we have, under the assumption that errors at different pixels are independent, that

$$\hat{\boldsymbol{\theta}} = \text{argmax} \prod_i f_0(z_i) \tag{4.31}$$

$$= \text{argmin} \sum_i [-\ln f_0(z_i)] \tag{4.32}$$

$$= \text{argmin} \sum_i \rho_{f_0}(z_i) \tag{4.33}$$

where

$$\rho_{f_0}(z) = -\ln(f_0(z)). \tag{4.34}$$

Function $\rho_{f_0}(\cdot)$ plays the same role as the squaring operation in the case of least squares estimation. We can then introduce an alternative additive (pseudo) distance function $d_f$ to compare images

$$d_f^2(I_1, I_2) = \sum_i \rho_f(I_1(i) - I_2(i)) \tag{4.35}$$

using it to replace the usual $L_2$ distance. The estimators described by Equation 4.33 are called M-estimators from their relation to maximum likelihood estimation. If $\rho$ has a derivative

$$\psi(z) = \frac{d\rho(z)}{dz} \tag{4.36}$$

$$= -\frac{f_0'(z)}{f_0(z)} = \Lambda(z) \tag{4.37}$$

Equation 4.37 establishes a direct link with the underlying model distribution $f_0$ (which is used to obtained the maximum likelihood estimate). The following equation holds at the estimated parameters:

$$\sum_i \psi(z_i) \frac{\partial G_\theta}{\partial \theta_k} = 0 \quad k = 1, \ldots, n \tag{4.38}$$

where the differential impact of each parameter is weighted by the factor $\psi(z_i)$. It is not by accident that we used the same symbol to denote the influence function and the derivative of $\rho$: it is possible to prove that the influence function of an M-estimator is proportional to $d\rho/dz$. The influence functions associated to the normal, exponential, and Cauchy distributions are shown in Figure 4.4. This result allows us to establish whether the corresponding estimator is robust (its $\psi$-function must be bounded) or not (its influence function is unbounded).

Before casting the computation of correlation in this framework, we can familiarize ourselves with M-estimators by a couple of related, simple problems: the robust estimation of the location and scale parameters of a distribution. The first one is that of approximating an image with a single, constant value:

$$G_\theta = \theta. \tag{4.39}$$

We want to determine the value of $\theta$ that minimizes the overall distance

$$\hat{\theta} = \underset{\theta}{\operatorname{argmin}} \sum_i \rho_f(y_i - \theta) \tag{4.40}$$

so that Equation 4.38 becomes

$$\sum_i \psi(z_i) = \sum_i \psi(y_i - \theta) = 0. \tag{4.41}$$

We can also consider $\theta$ as a location estimator since it satisfies Equation 4.10 (the M-estimator approach to scale estimation is discussed in Intermezzo 4.4). If we take $\rho(z) = z^2$, we have $\psi(z) \propto z$ and the solution is the arithmetic average of the values: the larger the value, the greater its contribution (as reflected by $\psi$). Things become more interesting when we change $\rho$. If we assume $\rho(z) = |z|$ we have that $\psi(z) = \operatorname{sign}(z)$: the maximum likelihood estimate of $\theta$ corresponds to the median of the values (that minimizes the mean absolute deviation). In this case the contribution of each point is constant and it is limited to its sign: a large value does not significantly change the final estimate. In the case of the Cauchy distribution $C(x; 0, \sigma_C)$ we have that

$$\rho_C(z) = \ln\left(1 + \frac{z^2}{\sigma_C^2}\right) - \ln \pi\sigma_C \tag{4.42}$$

$$\psi_C(z) = \left(1 + \frac{z^2}{\sigma_C^2}\right)^{-1} 2\frac{z}{\sigma_C^2}. \tag{4.43}$$

**Intermezzo 4.4.** Scale M-estimators

The problem of estimating the scale parameter of a distribution is not dissimilar from that of location estimation. In many cases of interest, including the Gaussian and Cauchy distributions, the probability density function can be rewritten as

$$f_0(x; \boldsymbol{\theta}) = \frac{1}{\theta_2} g\left(\frac{x - \theta_1}{\theta_2}\right). \tag{4.44}$$

The maximization of the dataset log-likelihood contains the following terms:

$$\frac{\partial}{\partial\theta_2}\left[\ln\left(\frac{1}{\theta_2} g\left(\frac{x - \theta_1}{\theta_2}\right)\right)\right] = -\frac{1}{\theta_2} - \frac{g'}{g}\frac{x}{\theta_2^2} = \frac{1}{\theta_2}\left[-1 - \frac{f_0'(z)}{f_0(z)} z\right] \tag{4.45}$$

from which we have that the MLE influence function $\chi$ for a scale problem is

$$\chi_{\text{MLE}}(z) = -1 - z\frac{f'(z)}{f(z)}. \tag{4.46}$$

In the case of a Gaussian distribution we have that $\chi(z) = z^2 - 1$ and we recover the variance as the usual scale estimate. As before, different probability distributions result in different estimators and a robust one is given by the median of absolute deviations defined in Equation 4.14. The influence function of this estimator is bounded and its breakdown point is equal to 0.5.

Two important things should be noted: the influence of each point decreases when the value becomes very large and the probability distribution depends on a scale factor $\sigma_C$. If the scale factor is not estimated from the data, the resulting location estimate is not equivariant with respect to data scaling: the reason is that the value of $z$ would scale with the data but $z/\sigma_C$ would not. The dependency on a scale factor is particularly important for estimators whose $\psi$-function decreases to zero when the argument value goes to infinity. For these types of estimators, called redescending M-estimators, the scale factor determines the value from which data start to be considered as unreliable, and, consequently, from which their weight on the estimator should be decreased, possibly nullified. The idea of obtaining robustness using weighted estimates is quite powerful per se and it also allows us to understand an important limitation of M-estimators. In order to downweight the contribution of points that are unlikely under the assumed model distribution, we should already have an idea of the parameters of the distribution, but this is the problem that we need to solve. The conundrum can be solved by means of an iterative, reweighting procedure that starts from a robust estimation that does not itself depend on a scale parameter but, on the contrary, is able to provide it. For many symmetric data distributions whose probability decays to zero with increasing absolute value, the median and mad estimators are able to provide a meaningful, maximally robust estimate of the distribution location and scale parameters. Starting from these estimates, we may refine the weight of data points according to their probability. Let us formalize the approach for a location estimator $L_N$:

$$L_N(x_1, \ldots, x_N) = \frac{\sum_{i=1}^N w_i x_i}{\sum_{i=1}^N w_i} \tag{4.47}$$

where the weights depend on the observations through

$$w_i = w\left(\frac{x_i - L_N}{S_N}\right) \tag{4.48}$$

where $S_N$ is a scale estimate. The location estimator $L_N$ satisfies

$$L_N = \left[\sum_{i=1}^N w\left(\frac{x_i - L_N}{S_N}\right)x_i\right] \Big/ \left[\sum_{i=1}^N w\left(\frac{x_i - L_N}{S_N}\right)\right] \tag{4.49}$$

which can be solved by iterating to convergence

$$L_N^{t+1} = \left[\sum_{i=1}^N w\left(\frac{x_i - L_N^t}{S_N}\right)x_i\right] \Big/ \left[\sum_{i=1}^N w\left(\frac{x_i - L_N^t}{S_N}\right)\right] \tag{4.50}$$

starting from

$$L_N^0 = \operatorname{med}(x_1, \ldots, x_N)$$
$$S_N = \operatorname{mad}(x_1, \ldots, x_N).$$

Equation 4.49 can be rewritten as

$$\left[\sum_{i=1}^N w\left(\frac{x_i - L_N}{S_N}\right)(x_i - L_N)\right] \Big/ \left[\sum_{i=1}^N w\left(\frac{x_i - L_N}{S_N}\right)\right] = 0 \tag{4.51}$$

leading to

$$\sum_{i=1}^{N} \psi(x_i - L_N) = 0 \tag{4.52}$$

with $\psi(z) = zw(z)$: reweighted estimators are M-estimators. An iterative algorithm to solve Equation 4.41 can be obtained by applying Newton's method. It can be proved that the estimates resulting from the first iteration, known as one-step M-estimators, exhibit the same properties (robustness, variance, etc.) of their fully iterated versions

$$L_N = L_N^0 + S_N \left[\sum_{i=1}^{N} \psi\left(\frac{x_i - L_N^0}{S_N}\right) x_i\right] \Big/ \left[\sum_{i=1}^{N} \psi'\left(\frac{x_i - L_N^0}{S_N}\right)\right] \tag{4.53}$$

where we have made explicit the dependence of $\psi(\cdot)$ on a scale factor.

Among the many available location and scale estimators (see also Intermezzo 4.4), the so-called tanh estimators stand out for being both B-robust and optimally $V$-robust (see Intermezzo 4.5) among redescending estimators: they have the lowest variance under the constraint $\kappa^* < \kappa$. It is possible to use them in Equation 4.21, obtaining $r_{\text{tanh}}$ (also identified by $R$ in the graphs of the present chapter), a robust version of the Pearson correlation coefficient.

**Intermezzo 4.5.** Tanh estimators

Among the redescending estimators of location and scale, the tanh estimators are at the same time B-robust and optimally V-robust. The location tanh estimator is defined by

$$\psi_{\text{tanh}}(x) = \begin{cases} x & 0 \le |x| \le p \\ \sqrt{A(k-1)}\tanh\left[\frac{1}{2}\sqrt{(k-1)\frac{B^2}{A}}(c - |x|)\right]\text{sign}(x) & p \le |x| \le c \\ 0 & c \le |x| \end{cases} \tag{4.54}$$

$$p = \sqrt{A(k-1)}\tanh\left[\frac{1}{2}\sqrt{(k-1)\frac{B^2}{A}}(c-p)\right], \quad 0 < p < c. \tag{4.55}$$

Figure 4.4 presents a plot of the estimator for the following (typical) parameter set:

$$(c = 4,\ k = 5,\ A = 0.857\,044,\ B = 0.911\,135,\ p = 1.803\,134).$$

The scale tanh estimator at a model distribution $f(x)$ is defined by

$$\chi_{\text{tanh}}(x) = \begin{cases} -x\dfrac{f'(x)}{f(x)} - 1 + a & 0 \le |x| \le p \\ \sqrt{A_1(k-1)}\tanh\left[\frac{1}{2}\sqrt{(k-1)\frac{B_1^2}{A_1}}\ \ln(r/|x|)\right] & p \le |x| \le c \\ 0 & c \le |x| \end{cases} \tag{4.56}$$

$$p\Lambda(p) - 1 + a = \sqrt{A_1(k-1)}\tanh\left[\frac{1}{2}\sqrt{(k-1)\frac{B_1^2}{A_1}}\ \ln(r/p)\right], \quad 0 < p < c. \tag{4.57}$$

A typical parameter set is

$$(c = 4,\ k = 8,\ A = 0.765\,560,\ B = 1.026\,474,\ p = 1.684\,394).$$

These estimators are also very efficient (for the normal distribution), typically reaching, depending on the parameters, more than 90% (60%) of the Rao–Cramer bound for location (scale).

The results presented so far allow us to normalize our images in a technically robust way before proceeding to the computation of a similarity measure. Using a robust offset and scale normalization we can compute a robust correlation coefficient using Equation 4.21.

We have seen in Section 3.5 that estimating the match probability using the normalized correlation coefficient is related to a least squares regression problem. A similar approach can be used for different matching measures, such as the one reported in Equation 4.35: the solution of the associated regression problem can be used to estimate a match probability under the assumed pattern distribution. Let us now formalize the complete problem of image registration in the framework of (robust) maximum likelihood estimation. We want to estimate $\theta$, $\mu$ such that the likelihood of $I - \theta(R - \mu)$ is maximal assuming a Cauchy distribution for it. Unfortunately the solution of the corresponding Equation 4.38 is not available in closed form. We need to rely on an iterative refinement starting from a robust estimate of the parameters:

$$\theta_0 = \frac{\text{mad}(I)}{\text{mad}(R)} \tag{4.58}$$

$$\mu_0 = \text{med}(I) - \text{med}(R). \tag{4.59}$$

The equations to estimate the parameters are

$$z_i = y_i - \theta(x_i - \mu) \tag{4.60}$$

$$0 = \sum_i \psi_C(z_i)(\mu - x_i) \tag{4.61}$$

$$0 = \sum_i \psi_C(z_i)\theta \tag{4.62}$$

leading to

$$\theta = \left[\sum_i \frac{(x_i - \mu)^2}{1 + z_i^2/\sigma_C^2}\right]^{-1} \sum_i \frac{y_i(x_i - \mu)}{1 + z_i^2/\sigma_C^2} \tag{4.63}$$

$$\mu = \left[\sum_i \frac{(\theta x_i - y_i)}{1 + z_i^2/\sigma_C^2}\right]\left(\sum_i \frac{\theta}{1 + z_i^2/\sigma_C^2}\right)^{-1}. \tag{4.64}$$

The scale parameter $\sigma_C$ should be estimated beforehand in a robust way. We iterate the computations of Equations 4.63–4.64 till the estimate is stabilized. Once the normalizing parameters $\theta$ and $\mu$ are computed we can compute a matching probability using Equation 4.30.

This approach can be generalized. In fact, the above formulation does not provide an affinely invariant matching measure: the scale parameter of the distribution is not estimated together with the transformation parameters, and the regression problem itself is asymmetric. However, it is possible to proceed in a fully affine-invariant way by specifying only the probability density function describing the templates. Let us represent the values of two corresponding pixels from image $I$ and from the reference template $R$ as $\boldsymbol{x} = (x_I, x_R) \in \mathbb{R}^2$ and let $f_0(\boldsymbol{x}) = f_0(|\boldsymbol{x}|)$, a spherically symmetric probability density governing the distribution of $\boldsymbol{x}$. Under a general non-degenerate affine transformation $\boldsymbol{x} \to \boldsymbol{z} = V(\boldsymbol{x} - \boldsymbol{t})$

$$f_0(\boldsymbol{x}; \boldsymbol{t}, V) = |\det V| f_0(|V(\boldsymbol{x} - \boldsymbol{t})|). \tag{4.65}$$

We need to estimate the matrix $V$ and the vector $\boldsymbol{t}$ maximizing the log-likelihood of the dataset

$$(\hat{\boldsymbol{t}}, \hat{V}) = \underset{\boldsymbol{t}, V}{\operatorname{argmax}} \left[ N \log(\det V) + \sum_{i=1}^{N} \log f_0(|V(\boldsymbol{x}_i - \boldsymbol{t})|) \right]. \tag{4.66}$$

The maximization over all non-degenerate affine transformations makes the result independent of the (relative) affine transformations of the two patterns. Setting the derivatives to zero with respect to $\boldsymbol{t}$ and $V$ leads to

$$\hat{\boldsymbol{t}} = \frac{\sum_{i=1}^{N} w(r_i)\, \boldsymbol{x}_i}{\sum_{i=1}^{N} w(r_i)} \tag{4.67}$$

$$(\hat{V}^T \hat{V})^{-1} = \sum_{i=1}^{N} w(r_i)(\boldsymbol{x}_i - \boldsymbol{t})(\boldsymbol{x}_i - \boldsymbol{t})^T \tag{4.68}$$

$$r_i = \|V(\boldsymbol{x}_i - \boldsymbol{t})\| \tag{4.69}$$

$$w(r) = \frac{f_0'(r)}{r f_0(r)} \tag{4.70}$$

from which we see that $(\hat{V}^T \hat{V})^{-1} = \Sigma^{f_0}$ can be interpreted as a pseudo-covariance matrix. We can then compute the maximum likelihood correlation coefficient $r_{f_0}$

$$r_{f_0} = \frac{\Sigma_{12}^{f_0}}{\sqrt{\Sigma_{11}^{f_0} \Sigma_{22}^{f_0}}} \tag{4.71}$$

that gives us an affine-invariant similarity measure associated to $f_0$: it is the natural generalization of the Pearson coefficient to a distribution $f_0$ and can be computed for any pseudo-covariance matrix. The distribution $f_0$ need not represent the actual distribution of patterns and can be chosen for convenience to gain specific robustness properties. If we choose

$$f_0 = \frac{1}{2\pi} e^{-r^2/2}$$

we recover the usual sample estimates and the Pearson correlation coefficient.

The above equations define the transformation parameters in an implicit way: the existence and uniqueness of the solution to the joint estimation problem has been proven only for particular cases. The equations can be solved with an iterative procedure computing $\boldsymbol{t}_0$ as the coordinate-wise sample median of $\{z_i\}$ and $\hat{V}_0$ as the square root of the inverse of the $z$ sample covariance matrix. As the latter is not robust, the presence of outliers can bias it; in such cases it is useful to employ a robust estimator of the covariance matrix such as the one presented in Section 4.4. The left-hand sides of Equations 4.67–4.68 provide the updated estimates using the values from the previous iteration for the right-hand side computations. Iteration can be stopped when

$$\|\hat{V}_{n+1}^{-1} \hat{V}_n - I\| < \delta_V \tag{4.72}$$

$$\|\hat{V}_{n+1}(\hat{\boldsymbol{t}}_{n+1} - \hat{\boldsymbol{t}}_n)\| < \delta_t \tag{4.73}$$

where $\delta_V$ and $\delta_t$ are user-selected thresholds.

## 4.3. $L_1$ Similarity Measures

The probabilistic approach presented in Chapter 3 and further pursued in the previous section is appealing from a theoretical point of view but (often) suffers from significant computational complexity. In many cases we may approach template matching from a more pragmatic point of view, relying on any measure that can reliably assess the similarity of two different patterns. The resulting values can then be fed as inputs into a classification function that can be as simple as thresholding, like the cases considered so far, or a more complex device such as a support vector machine to be described in Chapter 12. This section follows this more pragmatic approach focusing on the design of simple similarity measures that possess better robustness properties than the $L_2$ norm.

Starting from the possibility of normalizing in a robust way the signals to be compared, we can introduce two different similarity measures based on the $L_1$ norm:

$$g(\mathbf{x}', \mathbf{y}') = 1 - \frac{\sum_i |x_i' - y_i'|}{\sum_j (|x_j'| + |y_j'|)} \tag{4.74}$$

$$l(\mathbf{x}', \mathbf{y}') = \frac{1}{N} \sum_i \left(1 - \frac{|x_i' - y_i'|}{|x_i'| + |y_i'|}\right) \tag{4.75}$$

where, without loss of generality, $\mathbf{x}'$ and $\mathbf{y}'$ are assumed to be normalized vectors with zero location and unit scale. The normalization should be obtained with robust estimators of location and scale. This means that we cannot assume $\|\mathbf{x}'\| = 1$ but that the location and scale of the two signals are matched by the following transformation:

$$\mathbf{x} \to \mathbf{x}' = \frac{\mathbf{x} - L_N(\mathbf{x})}{S_N(\mathbf{x})}. \tag{4.76}$$

The similarity measures $g(\cdot, \cdot)$ and $l(\cdot, \cdot)$ satisfy the following relations:

$$g(\mathbf{x}', \mathbf{y}'), l(\mathbf{x}', \mathbf{y}') \in [0, 1] \tag{4.77}$$

$$g(\mathbf{x}', \mathbf{y}') = 1, l(\mathbf{x}', \mathbf{y}') = 1 \iff \mathbf{x}' = \mathbf{y}' \tag{4.78}$$

$$g(\mathbf{x}', \mathbf{y}') = 0, l(\mathbf{x}', \mathbf{y}') = 0 \iff \mathbf{x}' = -\mathbf{y}' \tag{4.79}$$

closely resembling those of the standard correlation coefficient. In many template matching cases we do not need to compare directly different similarity measures: only the results obtained with them are compared. Sometimes, however, as in the present case where we focus on the characteristics of stability of different similarity measures, it is convenient to calibrate the measures so that the resulting values are directly comparable. While a simple scaling and translation can map the ranges of $g$ and $l$ to those of the Pearson correlation coefficient, a detailed comparison of the characteristics requires a more sophisticated mapping. In order to align the different similarity measures we map them so that they estimate the correlation parameter $\rho$ of a bivariate normal distribution. The similarity measure $g$ is an estimator of

$$E_g = E\left\{1 - \frac{|x - y|}{E\{|x| + |y|\}}\right\} = \phi_g(\rho) \tag{4.80}$$

while $l$ is an estimator of

$$E_l = E\left\{1 - \frac{|x - y|}{|x| + |y|}\right\} = \phi_l(\rho) \tag{4.81}$$

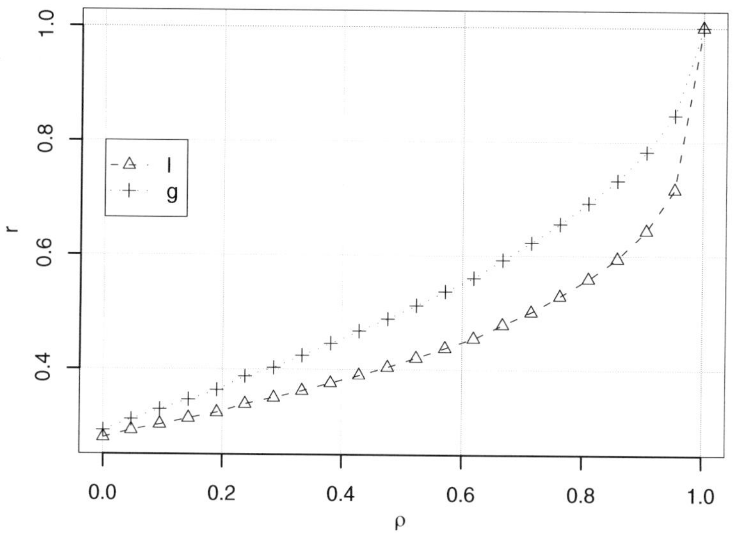


**Figure 4.5.** The $L_1$ similarity measures described in the text can be transformed into estimators of the correlation parameter of a bivariate normal distribution.

where $E\{\cdot\}$ has the usual meaning of expected value. The mappings $\phi_g$ and $\phi_l$ can be inverted and $\rho$ can be computed from $E_g$ and $E_l$:

$$\rho = \phi_g^{-1}(E_g) \tag{4.82}$$

and similarly for $\phi_l$. When the size of the sample is sufficiently large, $g$ and $l$ are good approximations of the expected values $E_g$ and $E_l$ and can be used as estimators of $\rho$:

$$r_G = \phi_g^{-1}(g), \quad r_L = \phi_l^{-1}(l). \tag{4.83}$$

The mappings $\phi_g$ and $\phi_l$ can be computed numerically under the binormal assumption and are shown in Figure 4.5. While $r$, $r_G$, and $r_L$ are consistent estimators of $\rho$, i.e. they are asymptotically unbiased, they are all biased for finite samples. This is not a problem when using them as similarity measures, but is a shortcoming when they are used as estimators for the bivariate normal. This defect can be reduced with a procedure called jackknife that assumes the following expansion for the bias of an estimator $\phi(\theta)$ based on a sample of size $N$:

$$\text{bias}\{\phi(\theta)\} = E\{\phi(\theta) - \theta\} = \sum_{k=1}^{\infty} \frac{a_k}{N^k}. \tag{4.84}$$

If we denote by $\phi_{N\backslash i}$ the estimate based on the sample of size $N-1$ obtained by neglecting the $i$th observation, the jackknifed estimator $\phi_{\text{jk},N}$

$$\phi_{\text{jk},N} = N\phi_N - \frac{N-1}{N}\sum_i \phi_{N\backslash i} \tag{4.85}$$

has no first-order term, hence a reduced bias. The so-called jackknife pseudo-values

$$\phi_{N,i} = N\phi_N - (N-1)\phi_{N\backslash i} \tag{4.86}$$

can be used to estimate the variance of the estimator

$$\mathrm{var}_N(\phi) = \frac{1}{N-1}\sum_{i=1}^{N}(\phi_{N,i} - \phi_N)^2. \tag{4.87}$$

The jackknife pseudo-values are proportional to the sensitivity curve of the estimator and can then be used to diagnose outlying observations. Not all estimators can be jackknifed successfully, $r_{\tanh}$ being one example due to its finite rejection point.

Having obtained a set of aligned estimators, we can compare some of their characteristics: the breakdown point, the sensitivity curve, their efficiency (i.e. their variance at the model distribution). The breakdown points for a sample of size $N$ are

$$\epsilon_r^* = 0 \tag{4.88}$$

$$\epsilon_{r_{\tanh}}^* = 1/2 \tag{4.89}$$

$$\epsilon_{r_G}^* = 0 \tag{4.90}$$

$$\epsilon_{r_L}^* = (N-1)/N \tag{4.91}$$

meaning that the only robust estimators of the group are $r_{\tanh}$ and $r_L$. The breakdown point for $r_L$ is somewhat surprising and points out a possibly controversial aspect of its nature whose discussion is beyond the scope of this book (see Section 4.5 for references). The sensitivity curves of the estimators are shown in Figure 4.6. Inspection of the sensitivity curves $\mathrm{SC}_N$ of the estimators allows us to compare the stability of the estimate to a small contamination of the model distribution (in this case the bivariate normal). Both $r_{\tanh}$ and $r_L$ have bounded sensitivity curves with a similar range even if $r_{\tanh}$ values are usually lower than those of $r_L$, and $r_G$ has an unbounded curve, as $r$, but with significantly lower values.

The robustness of the estimators can be appreciated in the comparison of patterns corrupted by noise as shown in Figure 4.7 and when they are used to discriminate patterns as in a face recognition system, a case that we now describe in some detail. Let us consider a set of frontal face images that have been previously normalized so that all eyes are in predefined positions (see Chapter 14 for details). The regions corresponding to the eyes, nose, and mouth are extracted from each image and are used for the comparison. Each unknown face is compared in turn to each face in the database, assessing the similarity of corresponding facial features by means of a correlation coefficient, generating a final integrated score using a weighted geometric average (see Section 14.3 for details). Following a sliding window approach, each region from the database was allowed to move over the corresponding region of the image to be classified, allowing for limited misalignments. Each person in the database is represented by more than one image and the unknown face is identified as the person providing the highest integrated score. The $r_P$ and $r_{\tanh}$ coefficients have the same performance, lower than that of the $r_G$ and $r_L$ coefficients (see Table 4.1). The lower performance of $r_{\tanh}$, in spite of its robustness, is due to its finite rejection point which, for the task at hand, throws away discriminating information. The high sensitivity of $r_P$ to even small contaminations of the templates, such as those due to defective geometric normalization of the templates, is the main reason for its reduced performance. The $L_1$-related estimators perform nearly equally well, clearly outperforming $r_P$ and $r_{\tanh}$. A useful indicator of discrimination that can be easily computed is the average value of the $R_{mM}$ ratio, defined as the similarity of the face to be recognized with the database entry of the

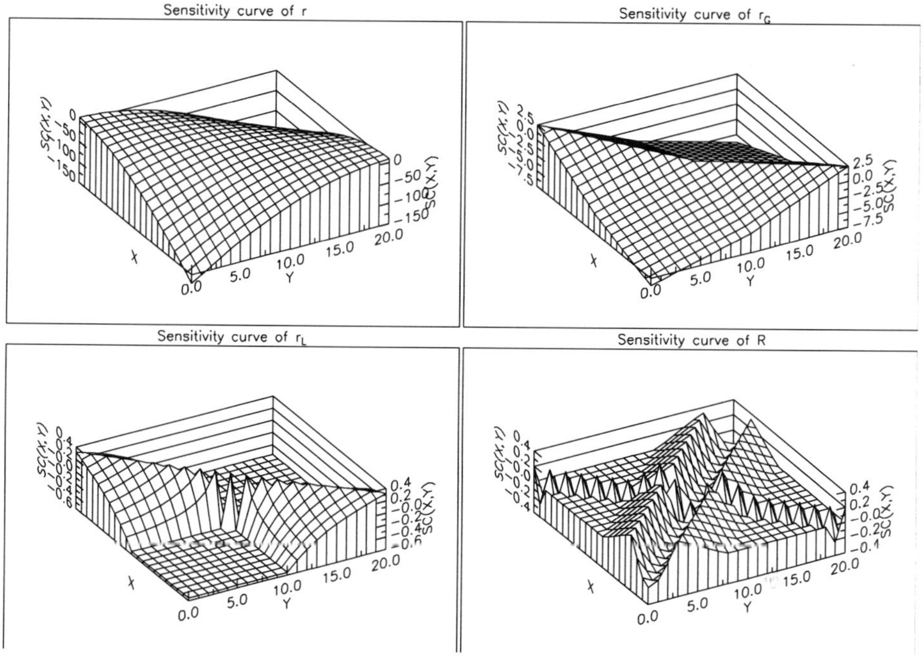


**Figure 4.6.** The sensitivity curves of the estimators of correlation based on the $L_1$ similarity measures compared to the sensitivity curve of the Pearson correlation coefficient (upper left plot). The estimator denoted by $R$ is $r_{\mathrm{tanh}}$. Source: Brunelli and Messelodi (1995).

**Table 4.1.** Computational complexity, correct recognition percentage, and separation ratio of the correlation estimators in a face recognition task. Source: Brunelli and Messelodi (1995).

| Estimator | Complexity | Recognition | $R_{mM}$ |
|---|---|---|---|
| $r_{\mathrm{P}}$ | 1 | 81 | 1.19 |
| $r_{\mathrm{tanh}}$ | 15 | 81 | 1.22 |
| $r_G$ | 1.4 | 88 | 1.12 |
| $r_L$ | 1.6 | 87 | 1.18 |
| $g$ | 1.4 | 87 | 1.29 |
| $l$ | 1.6 | 87 | 1.32 |

corresponding user divided by the highest similarity among the remaining entries. In the task of face recognition we may rely on the unmapped versions $g$ and $l$ as we are not estimating the correlation of a bivariate normal. Interestingly $g$ and $l$ achieve the same performance of their mapped counterpart but exhibit a significantly higher $R_{mM}$ ratio. This is particularly useful in face verification tasks, where a similarity threshold is introduced to reject an identification as unreliable. The reason for the higher $R_{mM}$ ratio can be found by inspecting the relation linking $g$ and $l$ to $\rho$ (see Figure 4.5). When $\rho \approx 1$ the dynamics of $g$ and $l$ are higher, somehow providing higher resolution and enhanced discriminability. The computational complexities of the four estimators normalized to that of $r_{\mathrm{P}}$ vary significantly, giving the $L_1$-based similarity estimators an additional advantage over the robustified version.

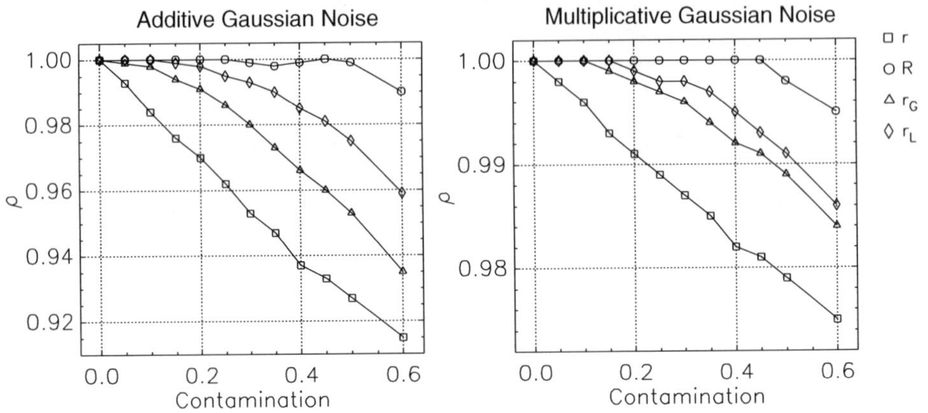


**Figure 4.7.** The sensitivity of some correlation estimators to additive and multiplicative Gaussian noise. Source: Brunelli and Messelodi (1995).

## 4.4. Robust Estimation of Covariance Matrices

In the previous sections we considered the problem of robust estimation of the location and scale parameters of a set of scalar data. A related and very important problem is that of estimating in a robust way the covariance matrix of a set of vector observations. As we saw in Chapter 3, the covariance matrix is the key structure in probabilistic detection of patterns corrupted by additive Gaussian noise, and, as we will see in Chapter 8, it is the starting point of the most basic dimensionality reduction algorithm, principal component analysis. The sensitivity to outliers of the sample variance, the usual scale estimator, carries over unchanged to the estimation of covariance matrices. Let us consider the case of $N$ samples $\boldsymbol{x}_i \in \mathbb{R}^{n_d}$. The covariance matrix $\Sigma$

$$\Sigma = \frac{1}{N-1} \sum_{i=1}^{N} (\boldsymbol{x}_i - \boldsymbol{\mu})(\boldsymbol{x}_i - \boldsymbol{\mu})^T \tag{4.92}$$

is then an $n_d \times n_d$ matrix. The robust estimation of $\Sigma$ requires the removal, or downweight, of outlying values, but the situation differs substantially in the scalar and vector cases. In particular, will a single outlying coordinate value in a sample cause the downweight, or straightforward removal, of the whole point? Or will we cope with it at the coordinate level?

As in many pattern recognition applications the number of available samples is scarce, we focus on the second class of approaches, and more specifically on the robust estimator proposed by Gnanadesikan and Kettenring (1972) whose structure is simple. The starting point is the observation that the covariance of two random variables can be written as

$$\Sigma_{jk} = \frac{1}{4}[\sigma^2(X_j + X_k) - \sigma^2(X_j - X_k)] \tag{4.93}$$

where $\sigma^2(X)$ denotes the variance (dispersion) estimator of the random variable $X$. If $\sigma$ is a robust estimator, we obtain a robust estimate of each separate covariance $\Sigma_{jk}$. Direct application of this approach is plagued by two problems:

1. The resulting matrix $\Sigma$ is not necessarily positive definite as a covariance matrix should be.

2. The estimator $\Sigma$ is not affine equivariant

$$\Sigma(A\boldsymbol{X}+\boldsymbol{b}) \neq A\Sigma(\boldsymbol{X})+\boldsymbol{b} \tag{4.94}$$

where $A$ is a transformation matrix and $\boldsymbol{b}$ is an offset vector.

Of the two problems, the first one is the most important as many algorithms rely on the covariance matrix being positive definite. A simple solution is based on the observation that the eigenvalues of the covariance matrix are the variances along the respective eigenvectors (see also Section 8.1 for a detailed discussion of this topic). We can then orthogonalize the estimator obtained from Equation 4.93, enforce positive definiteness, and back project by reversing the orthogonalizing rotation. Let $X_j$, $j=1,\ldots,n_d$, represent our random variables, i.e. the coordinates of the random vectors $\boldsymbol{X}$, and $\sigma_S$ a robust dispersion estimator (such as the median absolute deviation or a more efficient one):

1. Let $D=\mathrm{diag}[\sigma_S(X_j)]|_{j=1,\ldots,n_d}$ and the scaled samples $\boldsymbol{y}_i=D^{-1}\boldsymbol{x}_i$, where $i=1,\ldots,N$.

2. Compute a robustified covariance matrix $\Sigma_S$

$$\Sigma_{S,jk}=\frac{1}{4}[\sigma_S^2(Y_j+Y_k)^2-\sigma_S^2(Y_j-Y_k)^2]. \tag{4.95}$$

   Due to the scaling operation, $\Sigma_{S,jk}$ can actually be considered as a correlation matrix. However, the normalization used in Equation 4.93 does not guarantee that $\Sigma_{S,jk}\in[-1,1]$.

3. Decompose $\Sigma_S$ using the singular value decomposition (see Section 8.1 for details) as $\Sigma_S=E\Lambda E^T$ where $\Lambda=\mathrm{diag}(\lambda_1^2,\ldots,\lambda_{nd}^2)$ is the diagonal matrix of the eigenvalues of $\Sigma_S$ and $E$ is a matrix whose columns represent the eigenvectors of $\Sigma_S$.

4. Project the scaled samples onto the orthogonalized space

$$\boldsymbol{z}_i=E^T\boldsymbol{y}_i \tag{4.96}$$

   and let $A=DE$: the estimator of $\Sigma$ is then given by

$$\Sigma=A\Lambda_z A^T \tag{4.97}$$

   where $\Lambda_z=\mathrm{diag}(\sigma_S^2(Z_1),\ldots,\sigma_S^2(Z_{n_d}))$. The location estimator corresponding to the robust covariance matrix can be obtained similarly:

$$\boldsymbol{\mu}=A\boldsymbol{\nu} \tag{4.98}$$

   where $\boldsymbol{\nu}=(\mu(Z_1),\ldots,\mu(Z_{n_d}))$, $\mu(\cdot)$ being a robust location estimator.

The estimate can be refined by iterating steps 3–4 to convergence with the substitution $\Sigma\rightarrow\Sigma_S$. Approximate affine equivariance can be obtained by employing a reweighting step. Using Equation 4.96 we can associate an 'outlyingness' score (weight) to each data point

$$\mathrm{AO}_i=\sum_j\left[\frac{z_{ij}-\mu(Z_j)}{\sigma(Z_j)}\right]^2. \tag{4.99}$$

We can then compute the reweighted estimates

$$\boldsymbol{\mu}_w = \frac{\sum_i w_i \boldsymbol{x}_i}{\sum_j w_j} \tag{4.100}$$

$$\Sigma_w = \frac{\sum_i w_i (\boldsymbol{x}_i - \boldsymbol{\mu}_w)(\boldsymbol{x}_i - \boldsymbol{\mu}_w)^T}{\sum_j w_j} \tag{4.101}$$

where $w_i = w(\mathrm{AO}_i)$ takes care of downweighting the contribution of outliers.

## 4.5. Bibliographical Remarks

The introductory remarks on the Cauchy distribution as a good model for image differences are based on Sebe *et al.* (2000). The paper by Stewart (1999) motivates the use of robust estimation techniques in computer vision with several examples. Two fundamental books on robust statistics are Huber (1981) and Hampel *et al.* (1986); this chapter emphasizes the approach based on the influence function of the latter. The influence function of the Pearson correlation coefficient can be found in Croux and Dehon (2008). The book by Rey (1978) describes the jackknife procedure in detail. The limitations of the concept of breakdown point are discussed by Davies and Gather (2005): the rejoinder explicitly addresses the case of correlation. The computation of a generalized correlation coefficient from a pseudo-covariance matrix is based on Huber (1981) and Kim and Fessler (2004).

Cauchy estimators are considered in Mizera and Muller (2002) and Muller (2004). The hyperbolic (tanh) estimators and their optimality are discussed in Hampel *et al.* (1986). The $L_1$ similarity measures and their usage as estimators of correlation are discussed by Brunelli and Messelodi (1995). The same paper also considers their application to face recognition. An extensive survey of robust similarity measures is reported by Chambon and Crouzil (2003). The orthogonalized version of the Gnanadesikan–Kettenring estimator (Gnanadesikan and Kettenring 1972) is due to Maronna and Zamar (2002) and an extension to the case of missing data is considered by Copt and Victoria-Feser (2004).

## References


Brunelli R and Messelodi S 1995 Robust estimation of correlation with applications to computer vision. *Pattern Recognition* **28**, 833–841.

Chambon S and Crouzil A 2003 Dense matching using correlation: new measures that are robust near occlusions. *Proceedings of the British Machine Vision Conference (BMVC'03)*, vol. 1, pp. 143–152.

Copt S and Victoria-Feser M 2004 Fast algorithms for computing high breakdown covariance matrices with missing data. *Theory and Applications of Recent Robust Methods*. Birkhäuser, pp. 71–82.

Croux C and Dehon C 2008 Robustness versus efficiency for nonparametric correlation measures. Université Libre de Bruxelles, Ecares, ECARES Working Papers, 2008_002.

Davies P and Gather U 2005 Breakdown and groups (with rejoinder). *Annals of Statistics* **33**, 977–1035.

Gnanadesikan R and Kettenring J 1972 Robust estimates, residuals, and outlier detection with multiresponse data. *Biometrics* **28**, 81–124.

Hampel F, Rousseeuw P, Ronchetti E and Stahel W 1986 *Robust Statistics: The Approach Based on Influence Functions*. John Wiley & Sons, Ltd.

Huber P 1981 *Robust Statistics*. John Wiley & Sons, Inc.

Kim J and Fessler J 2004 Intensity-based image registration using robust correlation coefficients. *IEEE Transactions on Medical Imaging* **23**, 1430–1444.

Maronna R and Zamar R 2002 Robust estimates of location and dispersion for high-dimensional datasets. *Technometrics* **44**, 307–317.

Mizera I and Muller C 2002 Breakdown points of Cauchy regression-scale estimators. *Statistics & Probability Letters* **57**, 79–89.

Muller C 2004 Redescending M-estimators in regression analysis, cluster analysis and image analysis. *Discussiones Mathematicae – Probability and Statistics* **24**, 59–75.

Perreault S and Hebert P 2007 Median filtering in constant time. *IEEE Transactions on Image Processing* **16**, 2389–2394.

Rey W 1978 *Robust Statistical Methods*, Lecture Notes in Mathematics, vol. 690. Springer.

Sebe N, Lew MS, Huijsmans DP 2000 Toward improved ranking metrics. *IEEE Transactions on Pattern Analysis and Machine Intelligence* **22**, 1132–1143.

Stewart C 1999 Robust parameter estimation in computer vision. *SIAM Review* **41**, 513–537.

Weiss B 2006 Fast median and bilateral filtering. *ACM Transactions on Graphics* **25**, 519–526.

# 5 ORDINAL MATCHING MEASURES

Fading in music: that the comparison
May stand more proper, my eye shall be the stream.

*The Merchant of Venice*
WILLIAM SHAKESPEARE

Linear correspondence measures like correlation and the sum of squared differences between intensity distributions have been shown to be fragile. Similarity measures based on the relative ordering of intensity values have demonstrable robustness both to monotonic image mappings and to the presence of outliers.

## 5.1. Ordinal Correlation Measures

We have seen in Chapter 2 that different sensing processes map photon count information in specific ways to provide the final images used for template matching. A simple example is given by gamma correction and automatic gain adjustment (see Equation 2.48) that map photon counts in a nonlinear way. Another example is provided by scanned films of different characteristics, again resulting in a nonlinear map of the original information (see Equation 2.39). An important aspect of these nonlinear mappings is that they affect the complete image in a homogeneous way: all pixels are subject to the same functional mapping. Another very important aspect of these transformations is that they are usually monotonic. As the techniques introduced in the previous chapters to compare images are very sensitive to global nonlinear transformations, it is important to find ways to cope with this common, but unwanted, source of variability. We present two different techniques, addressing the problem from two very different perspectives. The first one, called histogram equalization (see Intermezzo 5.1), attempts to undo the mapping by transforming the intensity values so that their distribution matches that of a reference pattern. The resulting patterns can then be compared using the techniques already introduced.

The second approach is also based on the transformation of the original values, but instead of transforming the original pixel value into a normalized intensity, it associates to each value its rank among the population of values representing the original image sorted in ascending order. More formally, let us sort a set of values $\{x_i\}$ so that

$$x_{i_1} \leq x_{i_2} \leq \cdots \leq x_{i_N}. \tag{5.1}$$

The rank $\pi(i)$ of an element is defined by $\pi(i_j) = j$ and the ranking of the set is fully characterized by a permutation of the set of integers $[1, \ldots, N]$ (see Figure 5.1).

*Template Matching Techniques in Computer Vision: Theory and Practice* Roberto Brunelli

**Intermezzo 5.1.** Histogram equalization

Histogram equalization employs a monotonic, nonlinear mapping which reassigns the intensity values of pixels in the input image so that the output image contains a uniform, or otherwise specified, distribution of intensities.

Let us first address the problem of mapping to a uniform distribution $U$. Let $r$ be a variable whose pdf is $P_r$ (the histogram of the image to undergo the mapping). We want to find a transformation $s = F_{P_r \to U}(r) : r \in [0, 1] \to s \in [0, 1]$, $r = F_U^{-1}(s)$ such that the pdf of $s$ is uniform. The solution is based on the cumulative distribution of $P_r$:

$$s = F_U(r) = \int_0^r P_r(w)\,dw. \tag{5.2}$$

In fact, recalling that $P_s(s) = P_r(r)dr/ds$, we have that

$$\frac{ds}{dr} = P_r(r) \tag{5.3}$$

$$P_s(s) = P_r(r)\frac{1}{P_r(r)} = 1, \quad \text{the uniform distribution.} \tag{5.4}$$

The solution for the general case, $P_s(s) = T(s)$ with $T$ given, exploits the invertibility of the mappings:

$$s = F_{T \to U}^{-1}(F_{H \to U}(r)) \tag{5.5}$$

where $H$, $U$, and $T$ denote respectively the original, uniform, and target pdfs. The fact that image intensity values are discrete does not pose any major difficulty, and the discrete counterparts of Equation 5.2 can be used:

$$F_{T \to U} = \frac{\sum_{j=0}^{i} T(j)}{\sum_{j=0}^{n} T(j)} \tag{5.6}$$

$$F_{H \to U} = \frac{\sum_{j=0}^{i} H(j)}{\sum_{j=0}^{n} H(j)} \tag{5.7}$$

assuming that the intensity values are restricted to the interval $[0, n]$. The target probability distribution need not be associated with a particular image: it can be specified out of convenience. In the latter case, if it is chosen to be a Gaussian distribution, the joint distribution of the intensity values of two normalized images would approximate a bivariate normal distribution leading to some advantages in the estimation of the significance of correlation values.

Three considerations are in order:

- A monotonic, increasing transformation does not modify the ranking of the values: the rank transformation is invariant to any monotonic, increasing transformation of the original values.
- The distribution of the transformed values, i.e. the ranks, is known and simple; in the case of no ties, i.e. no values are equal, it is a uniform distribution.
- The impact of the arbitrary modification of $k$ values on the ranking values is bounded by $k$.

We are then able to achieve the invariance we were looking for without the need to compute any parameter from the data, or assuming any specific kind of mapping. Furthermore, the fact that the distribution of the transformed values does not depend on that of the original data opens the way to meaningful significance tests of the statistics computed on rank-transformed data. As mapping values to ranks throws away a significant amount of information, we may wonder whether the resulting data are rich enough to support pattern discrimination

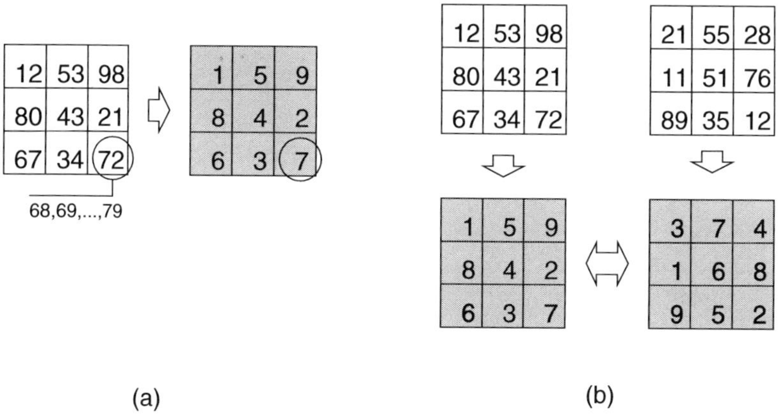


**Figure 5.1.** The first step in the computation of an ordinal correlation measure is the transformation (a) of the original intensity values into the corresponding ranks in the set sorted by ascending values. Besides invariance to monotonic transformation, the rank transform features limited sensitivity to noise: the rank associated to the lower right pixel (7) does not change if the pixel value changes in the interval [68, 79]. In order to compare two patterns, they must be first transformed into the corresponding ranks (b) and the resulting values compared using a distance function or a similarity measure such as the correlation coefficient.

in practical cases. A quantitative indicator of the amount of information preserved is the relative efficiency of statistical estimators based on rank information with respect to the optimal one in the case of non-ordinal estimators. A qualitative indicator is the possibility of reconstructing a pattern from a rank-based representation. This is not possible if we also discard spatial information, but local rank-based descriptions of an image, to be introduced in the next section, contain enough information to provide a recognizable reconstruction of a pattern (see Intermezzo 5.2). While the non-parametricity of rank-based estimators can already be considered a form of robustness, their invariance to monotonic transformations of the data coupled with the bounded impact on ranks of the value of any single sample leads to the same form of robustness addressed in the previous chapter.

**Intermezzo 5.2.** Reconstructing patterns from a rank-based description

A measure of the descriptive power of an image transform is its ability to support the reconstruction of the original image. The description of the value of a single image pixel by means of its relative ranking with respect to its neighbors can be proven experimentally to support such reconstruction using a simple algorithm.

Let $W(\boldsymbol{x}, 1)$ be the $3 \times 3$ square neighborhood of a pixel at $\boldsymbol{x}$ and let $\theta_{\pm}$ be the symmetric Heaviside function whose value is $-1$ for negative arguments and 1 otherwise. Let image $I$ be unknown except for the local ordinal relations at each pixel:

$$\{\theta_{\pm}(I(\boldsymbol{x}) - I(\boldsymbol{x}'))\}_{\boldsymbol{x}' \in W(\boldsymbol{x},1) \setminus \boldsymbol{x}}.$$

An approximation $R$ of $I$ can be obtained by iterating to equilibrium the update rule

$$R_{t+1}(\boldsymbol{x}) = R_t(\boldsymbol{x}) + \alpha \sum_{\boldsymbol{x}' \in W(\boldsymbol{x},1) \setminus \boldsymbol{x}} (\theta_{\pm}(R_t(\boldsymbol{x}) - R_t(\boldsymbol{x}')) - \theta_{\pm}(I(\boldsymbol{x}) - I(\boldsymbol{x}'))). \tag{5.8}$$

While the quality of the reconstruction varies with the class of patterns, it is particularly good for face images (Sadr *et al.* 2002).

As our final goal is to compare rank-transformed patterns, in order to assess their similarity we need to introduce a distance $d_\pi(\boldsymbol{\pi}_x, \boldsymbol{\pi}_y)$ between their corresponding rankings. If we assume that a maximum value for $d_\pi$ exists over all possible ranking permutations, it is possible to normalize it so that it resembles the Pearson correlation coefficient

$$r_\pi = \left[1 - \frac{2d_\pi(\boldsymbol{\pi}_x, \boldsymbol{\pi}_y)}{\max d_\pi}\right] \in [-1, 1]. \tag{5.9}$$

Different choices of $d_\pi$ lead to different ordinal correlation coefficients and the next sections address three important cases. In the case of uncorrelated patterns we may compute the distribution of $r_\pi$ explicitly and derive a significance test for the null hypothesis of uncorrelated patterns. This possibility relies on the fact that under the null hypothesis we need to compute the distribution of $r_\pi$ over all possible permutations of the set $\{1, \ldots, N\}$ chosen uniformly. From a pattern matching perspective, a significance test of the computed correlation value allows us to reject a match with a given confidence if the value of the test is below the corresponding threshold. This is a much more reliable procedure than using an empirically selected threshold.

### 5.1.1. SPEARMAN RANK CORRELATION

Let us consider a set of data $\{x_i\}$, e.g. the intensity values of the pixels of a single channel image, and let $\{\pi_{xi}\}$ be the set of corresponding ranks. Similarly, let $\{\pi_{yi}\}$ be the ranks of the data from another set $\{y_i\}$. The Spearman correlation coefficient $r_S$ is defined as the linear correlation coefficient of the ranks:

$$r_S = \frac{\sum_i(\pi_{xi} - \bar{\pi}_x)(\pi_{yi} - \bar{\pi}_y)}{\sqrt{\sum_i(\pi_{xi} - \bar{\pi}_x)^2}\sqrt{\sum_i(\pi_{yi} - \bar{\pi}_y)^2}} \tag{5.10}$$

where $\bar{\pi}_y = \bar{\pi}_x = (N + 1)/2$. When there are no ties, i.e. no values from a set are equal, $r_S$ is related to the Euclidean distance of the ranks by the simple relation

$$r_S = 1 - \frac{6d_{\pi S}}{N^3 - N} \tag{5.11}$$

$$d_{\pi S} = \sum_i (\pi(x_i) - \pi(y_i))^2 \tag{5.12}$$

showing that the Spearman rank correlation coefficient is a special case of Equation 5.9. Under the null hypothesis of uncorrelated data, and for $N > 20$, the distribution of the quantity

$$t = r_S\sqrt{\frac{N-2}{1-r_S^2}} \tag{5.13}$$

is approximately Student with $N - 2$ degrees of freedom. This result corresponds to the one reported in Equation 3.111 for the Pearson correlation coefficient where $n_d$, representing the pattern size, corresponds to $N$, the present dimension of the dataset. It is then possible to determine a threshold based on a required confidence value. The resemblance of $r_S$ to $r_P$ may lead to the (wrong) conclusion that $r_S$ is not robust. The reason this is not true is that an arbitrary change in a single value $x_k$ (or $y_l$) does not modify $\bar{\pi}_x$ (or $\bar{\pi}_y$) and $|\Delta\pi| \leq 1$. This can be better appreciated by means of the sensitivity curve of $r_S$ at the bivariate normal distribution $b(x, y; \rho)$

$$\psi_S(x, y; b(0, 1, \rho)) = -3\left(\frac{6}{\pi} \arcsin\left(\frac{\rho}{2}\right)\right) - 9 + 12[P_b^X(x)P_b^Y(y) + E_b\{P_b^X(X)J(Y \geq y)\} + E_b\{P_b^Y(Y)J(X \geq x)\}] \tag{5.14}$$

where $J(\cdot)$ stands for the indicator function, $P_b^X(X)$ represents the marginal distribution (at the bivariate normal) of the random variable $X$ and similarly for $Y$. It is possible to normalize $r_S$ so that its expected value at the bivariate normal matches the expected value of $r_P$, hence of $\rho$,

$$r'_S = 2 \sin\left(\frac{\pi}{6} r_S\right) \tag{5.15}$$

whose influence function is

$$\psi'_S(x, y, b(0, 1, \rho)) = \left(\frac{\pi}{3} \text{sign}(\rho) \sqrt{1 - \frac{\rho^2}{4}}\right) \psi_S(x, y; b(0, 1, \rho)). \tag{5.16}$$

The asymptotic variance of $r_S$ in this specific estimation task can also be computed in closed form and the corresponding efficiency is found to be always higher than 70% of that of $r_P$, which represents the optimal value.

### 5.1.2. KENDALL CORRELATION

It turns out that it is even possible to dispense with the usage of ranks, relying solely on their relative ordering: higher, lower, or same in rank. In this case we do not need to compute the ranks as the relative ordering of the ranks is the same as that of the corresponding values. The Kendall correlation coefficient $r_K$, often denoted by $\tau$, is based on this idea and requires consideration of all $N(N-1)/2$ different couples of data points from each set. The actual definition is based on Equation 5.9 with the following definition of distance:

$$d_K = \sum_{i<j} \frac{1}{2}[1 - \text{sign}(\pi_{xi} - \pi_{xj}) \text{ sign}(\pi_{yi} - \pi_{yj})] \tag{5.17}$$

$$= \sum_{i<j} \frac{1}{2}[1 - \text{sign}(x_i - x_j) \text{ sign}(y_i - y_j)]. \tag{5.18}$$

The Kendall coefficient can also be seen as the value of the Pearson coefficient computed between two binary vectors of size $N(N-1)/2$, each element of which corresponds to one possible data pair $(i, j)$ (from a single set) and is given the value $\text{sign}(x_i - x_j)$. An immediate consequence of the definition is that the computation of $r_K$ is expensive, being $O(N^2)$. Under the null hypothesis of uncorrelated data, $r_K$ is approximately normally distributed even for small samples, with zero expectation value and

$$\sigma_K^2 = \frac{4N + 10}{9N(N-1)}. \tag{5.19}$$

The Kendall coefficient is robust: an arbitrary perturbation of a single value in the original datasets has a bounded impact on $r_K$. Its sensitivity curve can be computed explicitly at the bivariate normal and it is found to be

$$\psi_K(x, y; b(0, 1, \rho)) = 2\left\{2P_b[(X - x)(Y - y) > 0] - 1 - \frac{2}{\pi} \arcsin(\rho)\right\}. \tag{5.20}$$

The Kendall coefficient can be normalized to match $r_P$ by appropriate scaling

$$r'_K = \sin\left(\frac{\pi}{2} r_K\right) \tag{5.21}$$

and its influence function is then

$$\psi'_K(x, y; b(0, 1, \rho)) = \left(\frac{\pi}{2}\text{sign}(\rho)\sqrt{1 - \rho^2}\right)\psi_K(x, y; b(0, 1, \rho)). \tag{5.22}$$

The asymptotic variance of $r_K$ in this specific estimation task can also be computed in closed form and the corresponding efficiency is always higher than that of $r_S$ and always higher than 80% of that of $r_P$, which represents the optimal value. The high efficiency achieved by the Kendall and Spearman coefficients (in the normal bivariate case) is somehow striking, considering the fact that a lot of information is discarded (only rank information is used) and that they are B-robust operators. Table 5.2 reports some data showing that ordinal correlation measures are indeed more robust to illumination variations than the Pearson correlation coefficient.

### 5.1.3. BHAT–NAYAR CORRELATION

The Spearman and Kendall rank correlation coefficients are robust to random, arbitrary modification of the original values but are still sensitive to biased outliers that perturb the overall ranking structure in a coherent way. In an attempt to increase the robustness to these situations another measure of association was introduced in the past. The idea is based on ranking one of the sets with respect to the other (see Figure 5.2). The result is a vector $(s^1, s^2, \ldots)$ whose $i$th element is the rank of the $y$ element corresponding to the $x$ element whose rank is $i$:

$$s^i = \pi_y(k), \quad k = \pi_x^{-1}(i). \tag{5.23}$$

Some attention must be paid to the case of ties, i.e. pixels having the same value. A practical solution is to break ties exploiting the (fixed) scan order of the pixels: pixels with the same value receive progressively increasing ranks based on their scan order position. In the case of perfect ranking correspondence, namely perfect positive correlation, we have $s^i = i$.

We may quantify how far $s^i$ is from perfect correlation by computing the number of elements that are out of order:

$$d_m^i = \sum_{j=1}^{i} J(s^j > i) = i - \sum_{j=1}^{i} J(s^j \leq i) \tag{5.24}$$

where $J(\cdot)$ is the indicator function. The resulting vector $\boldsymbol{d}_m$ is invariant to arbitrary relabeling of data because it includes a normalizing permutation, the one given by Equation 5.23.

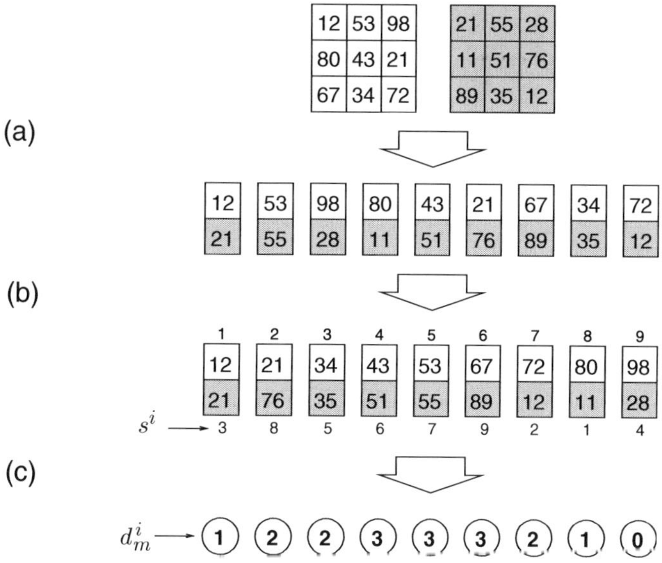


**Figure 5.2.** The steps for the computation of the Bhat–Nayar coefficient (see Equations 5.23 and 5.24).

The relative ordering established by Equation 5.23 is the key of the robustness to biased outliers. To understand this, let us consider a perfect correlation situation with the exception of the first three values of $x$, which, for some reason, have a perfect reverse correlation: they have the lowest values instead of the highest. This means that three correspondences are spoilt but the others are correct: if a set of values is biased, the impact will be limited to them (as a fraction) because the relative ordering of the remaining ones will remain correct. Furthermore, as the notation suggests, it can be proven that $d^i_m$ is symmetric: it does not depend on the choice of the pivot dataset ($x$ or $y$) used in Equation 5.23. It can be shown that $\max d^i_m = \lfloor N/2 \rfloor$ so that

$$r_{\mathrm{BN}} = 1 - \frac{2 \max_{i=1}^{N} d^i_m}{\lfloor N/2 \rfloor} \tag{5.25}$$

defines a properly normalized correlation coefficient (see Equation 5.9). The Bhat–Nayar coefficient of Equation 5.25 is a simplified version of another coefficient $r_g$

$$r_g = \frac{\max_{i=1}^{N} d^i_{-m} - \max_{i=1}^{N} d^i_m}{\lceil N/2 \rceil} \tag{5.26}$$

where the negative subscript of $d_{-m}$ indicates that the pivot set must be first subject to a reverse permutation. The $r_{\mathrm{BN}}$ coefficient inherits its robustness properties from the same source of the previous ones, usage of rank information, and turns out to be more resistant to structurally organized (biased) outliers. The computation of $r_{\mathrm{BN}}$ can be performed efficiently using the following formula:

$$d^{i+1}_m = d^i_m - \left( \sum_{j=1}^{i} J(s^j = i+1) \right) + J(s^{i+1} > i+1) \tag{5.27}$$

so that only the last term on the right-hand side (RHS) needs to be computed if $\sum_{j=1}^{i} J(s^j = i+1)$ was memorized during the computation of $d_m^i$. If the intensity values are represented by integers in the range $[0, 2^k - 1]$ the sorting operations required for ranking can be performed with counting sort of complexity $c_1 O(N + 2^k)$, instead of comparison sort, whose complexity is $c_2 O(N \log N)$. Under the assumption $c_1 = c_2$, the former is more efficient for $N \geq 64$ (this corresponds to $8 \times 8$ windows or larger). The influence function can be computed numerically. A shortcoming of $r_{\mathrm{BN}}$ lies in its reduced efficiency, and discriminatory power, resulting from synthesizing the information in $\boldsymbol{d}_m$ with its maximum value. As a result $|r_{\mathrm{BN}}|$ may assume only $1 + \lfloor N/2 \rfloor$ discrete values. A finer grain can be obtained complementing $r_{\mathrm{BN}}$ with information from all the elements of $\boldsymbol{d}_m$ to get a new coefficient $r'_{\mathrm{BN}}$

$$r'_{\mathrm{BN}} = 1 - \left( \frac{\max_{i=1}^{N} d_m^i}{\lfloor N/2 \rfloor} + \frac{\sum_{i=1}^{N} d_m^i}{\lfloor N^2/4 \rfloor} \right). \tag{5.28}$$

The computational complexity of $r'_{\mathrm{BN}}$ is essentially the same as that of $r_{\mathrm{BN}}$ as all quantities are already computed when the latter is determined. The computational complexity of the Bhat–Nayar coefficient is much better than that of the Kendall coefficient, $O(N^2)$, being similar to that of $r_{\mathrm{S}}$ and greater than that of $r_P$ and of the simpler $L_2$ distance matching.

## 5.2. Non-parametric Local Transforms

The ordinal approaches considered so far are characterized by a single rank transformation applied to the complete pattern. There is a different set of transformations that transform the value of each pixel (or of a subset of pixels) using only local relative rank information in the spirit of the Kendall coefficient. The new pixel values are then insensitive to local monotonic transformations, and can be used to introduce image matching operators that are more resistant to outliers than the Pearson coefficient. The descriptive power of these local ordinal transformations remains significant, as argued in Intermezzo 5.2, and the computation of pattern similarity based on them is of wide applicability.

### 5.2.1. THE CENSUS AND RANK TRANSFORMS

The relative rank (value) information from the neighborhood of a pixel can be used to define two local image transforms, the census and the rank transforms, that inherit some of the robustness of the ordinal correlation measures presented in the previous sections. Let us consider a pixel $I(\boldsymbol{x})$ and its neighborhood of $W(\boldsymbol{x}, l)$ of size $l$. Denoting by $\otimes$ the operation of concatenation, the census transform is defined as

$$C(\boldsymbol{x}) = \bigotimes_{\boldsymbol{x}' \in W(\boldsymbol{x}, l) \setminus \boldsymbol{x}} \theta_0(I(\boldsymbol{x}) - I(\boldsymbol{x}')) \tag{5.29}$$

where $\theta_0(\cdot)$ is the Heaviside function satisfying $\theta_0(0) = 0$ and the neighborhood is scanned in a predefined order (see Figure 5.3 and Figure 5.6). If we assume a square $3 \times 3$ neighborhood the result of the transformation is a string of 8 bits describing the ordinal relations of the central pixel with its neighbors. The corresponding description for a larger, $5 \times 5$ neighborhood would require 24 bits. The census transform does not scale well with window size as it requires 1 bit for each additional pixel considered. However, if the transform must

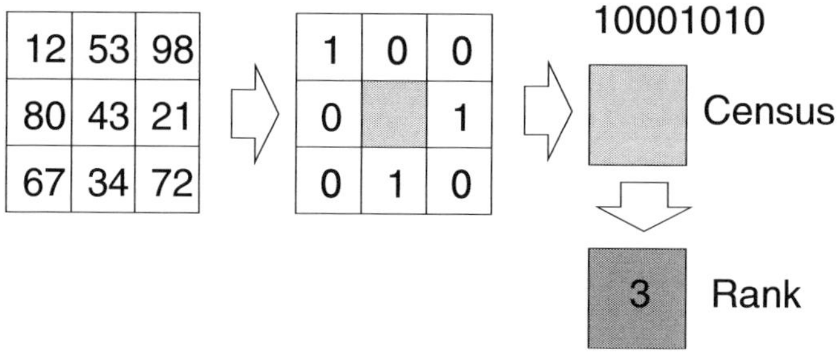


**Figure 5.3.** The census and rank transforms are based on local ordinal information. The census transform preserves more information and the rank transform can be computed from it.

be computed for a block of pixels, the typical case in template matching applications, some of the bits are redundant: as shown in Figure 5.4a, the number of comparisons required equals half the number of pixels in the neighborhood considered. Subsampling allows the use of larger windows: the case reported in Figure 5.4b requires only 8 bits to represent a $9 \times 9$ window that would otherwise require 80 bits, resulting in severe efficiency loss due to the hardware limitations of current processors. The transformed pixels can be compared by means of distances defined for bit vectors such as $D_T$ (Tanimoto), $D_{DK}$ (Dixon–Koehler) or $D_H$ (Hamming):

$$D_H(\boldsymbol{a}, \boldsymbol{b}) = \frac{1}{N} \sum_{i=1}^{N} ((a_i + b_i) \mod 2) \tag{5.30}$$

$$D_T(\boldsymbol{a}, \boldsymbol{b}) = \begin{cases} 1 & \text{if } \boldsymbol{a} = \boldsymbol{b} = 0 \\ 1 - \dfrac{\boldsymbol{a} \cdot \boldsymbol{b}}{\boldsymbol{a} \cdot \boldsymbol{a} + \boldsymbol{b} \cdot \boldsymbol{b} - \boldsymbol{a} \cdot \boldsymbol{b}} & \text{otherwise} \end{cases} \tag{5.31}$$

$$D_{DK}(\boldsymbol{a}, \boldsymbol{b}) = D_H(\boldsymbol{a}, \boldsymbol{b}) D_T(\boldsymbol{a}, \boldsymbol{b}). \tag{5.32}$$

The Tanimoto distance, which is equivalent to the Jaccard distance, favors matches done on bits with value 1 while the Hamming distance is influenced only by the number of non-matching bits. The Dixon–Koehler distance is a compromise of the two (see Table 5.1).

The census transform can be modified to enhance its capability of describing the local neighborhood of a pixel, but losing part of its robustness:

$$C(\boldsymbol{x}) = \bigotimes_{\boldsymbol{x}' \in W(\boldsymbol{x}, l)} \theta(I(\boldsymbol{x}) - \bar{I}) \tag{5.33}$$

where $\bar{I}$ represents the average intensity over the window $W(\boldsymbol{x}, l)$. This provides a vector of 9 bits per pixel, as the comparison is performed towards the average value of the neighborhood and exploiting the test with the central pixel as well. The modified transform presents the same invariance of the original one to monotonic variations of the pixel values but it shows greater sensitivity to salt-and-pepper-like noise. If the value of a single pixel is modified by noise so that is becomes white (black), the estimate of the local average will be severely biased towards high (low) values. As a consequence, the results of the tests will be modified towards 0 (or 1), providing an off-mark value. In the original census transform

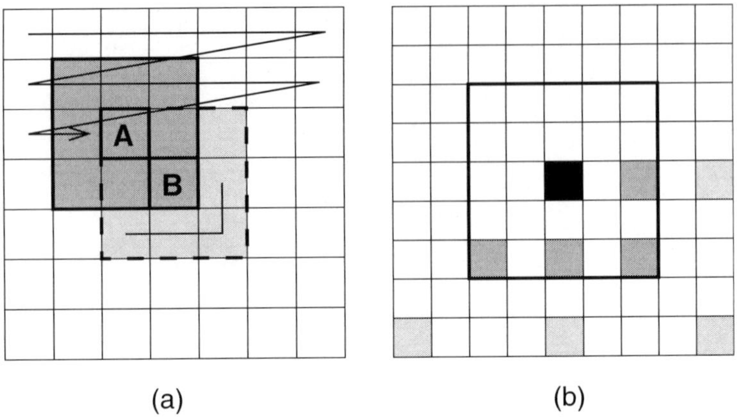


**Figure 5.4.** Only half of the comparisons required by Equation 5.29 are actually needed to capture all the information. As shown in (a), pixel B needs to be compared only to the pixels following it (in left to right, top to bottom scan order) because the information conveyed by the comparison to the pixels preceding it has already been incorporated in the corresponding transform values. As an example, when the transform of B is computed, the relative ranking of pixels A and B has already been considered when computing the transform at A. Larger windows can be used by adopting a subsampling strategy (b) which, besides being more efficient, is often more informative due to the larger spacing between the compared pixels: the correlation between them is reduced and the relative ordering more stable.

**Table 5.1.** A few practical examples of the distances defined in the text for the comparison of binary vectors.

| $a$ | $b$ | $D_H$ | $D_T$ | $D_{DK}$ |
|---|---|---|---|---|
| 101010 | 101010 | 0.00 | 0.00 | 0.00 |
| 101010 | 111111 | 0.50 | 0.50 | 0.25 |
| 101010 | 000000 | 0.50 | 1.00 | 0.50 |
| 011001 | 101001 | 0.33 | 0.50 | 0.17 |
| 011101 | 101101 | 0.33 | 0.40 | 0.13 |
| 011101 | 100010 | 1.00 | 1.00 | 1.00 |

the modification of a single, non-central, pixel has a bounded impact on the description of the neighborhood: a single bit is changed. Robustness can be recovered at the expense of computation by substituting the average value with a robust estimate of it such as the median (see Intermezzo 4.3 for hints on the efficient computation of the median on a sliding image window).

The rank transform is defined as

$$R(\boldsymbol{x}) = \pi_{W(\boldsymbol{x},l)}(I(\boldsymbol{x})) \tag{5.34}$$

and it corresponds to the number of pixels in $W(\boldsymbol{x}, l)$ whose value is lower than $I(\boldsymbol{x})$ (see Figure 5.3 and Figure 5.6).

The rank transform is related to the Mann–Whitney–Wilcoxon rank-sum test commonly used for the comparison of (medical) treatments. Let us consider the pixels in $W(\boldsymbol{x}, l)$ as observations divided into two groups: $n_t$ correspond to treatment cases, $n_c$ to control cases. The test aims at deciding whether the observations on the treatment cases are significantly

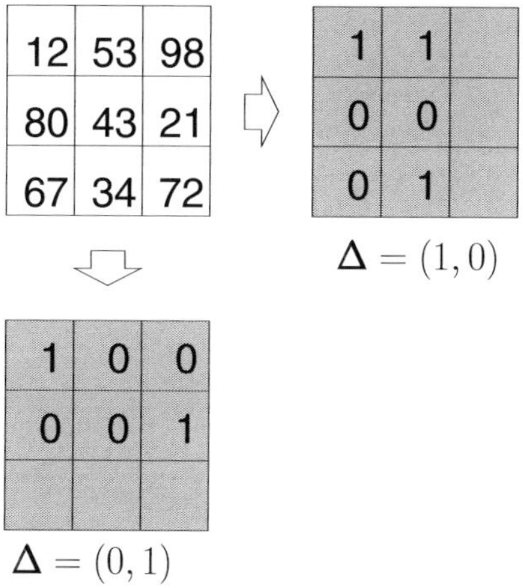


**Figure 5.5.** The incremental sign transform is essentially a rank transform where the neighborhood is limited to a single pixel. Multiple neighborhoods can be considered, averaging the resulting matching scores.

higher than those on the control. The statistic used is the sum of the ranks of treatment cases, $U = \sum_{i=1}^{n_t} \pi(i)$, and the test is effected by comparing it to a threshold value depending on the required significance. The rank transform corresponds then to a test with a single treatment case, the central pixel, and $n_c$ is the number of pixels to which the central pixel is compared. If $n_c$ is sufficiently large, the distribution of $U$ is approximately normal:

$$p(U) = \frac{1}{\sqrt{2\pi}\sigma_U} \exp\left[-\frac{(U-\mu_U)^2}{2\sigma_U^2}\right] \tag{5.35}$$

where

$$\mu_U = n_c/2 \tag{5.36}$$

$$\sigma_U = \sqrt{\frac{n_c(n_c+2)}{12}}. \tag{5.37}$$

The importance of the connection is that the binarization of the rank transform has a well-defined statistical meaning: a pixel is set to 1 only if its intensity is significantly higher than that of the neighboring pixels, where the significance is related to the critical Wilcoxon value imposed by the chosen threshold. The resulting binary map is more stable and it outperforms the integer-valued rank transform in face detection tasks, approaching that of the more descriptive census transform. Figure 5.6 presents some examples of the census and rank transforms while Table 5.2 quantifies their sensitivity to illumination variations by means of several correlation coefficients.

### 5.2.2. INCREMENTAL SIGN CORRELATION

All of the transforms considered previously are more computationally expensive than simple correlation. However, the good results obtained with a binarized rank transform suggest that

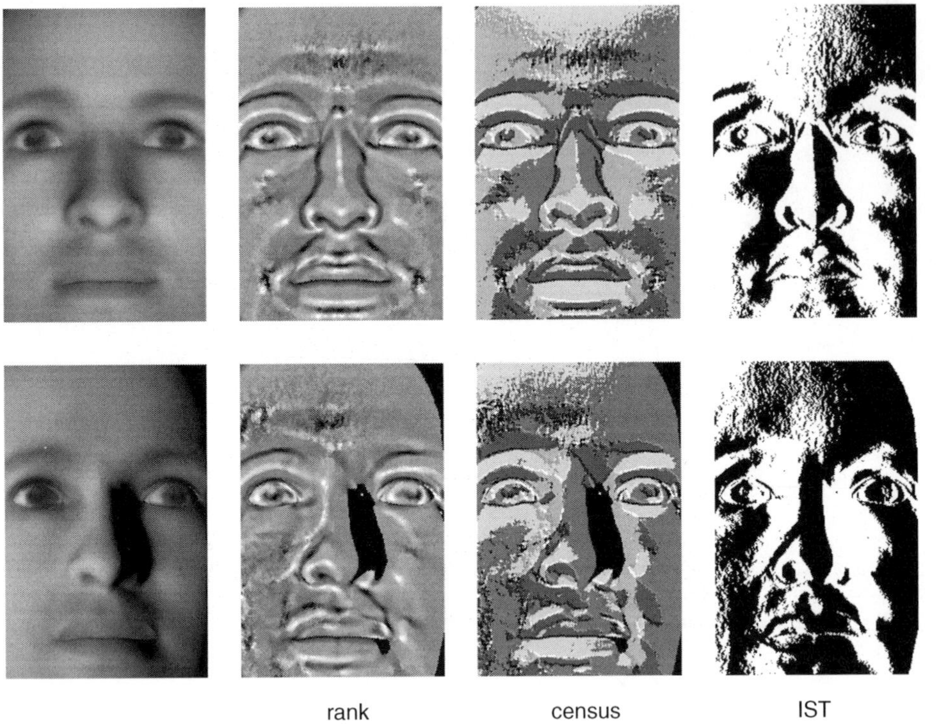


**Figure 5.6.** From left to right: a sample image and its rank, census, and incremental sign transforms. The two rows correspond to two different illumination conditions.

**Table 5.2.** The correlation coefficient of the image pairs presented in Figure 5.6 neglecting the flat, deep shadow area.

| Transform | Pearson | Spearman | Kendall |
|---|---|---|---|
| None | 0.31 | 0.56 | 0.42 |
| Rank | 0.60 | 0.44 | 0.35 |
| Census | 0.58 | 0.56 | 0.41 |
| IST | 0.36 | 0.28 | 0.19 |

even very simple transforms based on the relative rank of neighboring pixels may be effective. The incremental sign transform (IST, see Figure 5.5 and Figure 5.6) is such a transform, based on the relative ranking of two pixels whose positions differ by $\boldsymbol{\Delta}$:

$$B_{\boldsymbol{\Delta}}(x, y) = \begin{cases} 1 & I(x, y) < I(x + \Delta_x, y + \Delta_y) \\ 0 & \text{otherwise.} \end{cases} \tag{5.38}$$

Transformed images are then bit strings that can be compared by means of binary distance such as $D_T$ (Tanimoto), $D_{DK}$ (Dixon–Koehler) or $D_H$ (Hamming) defined in Equations 5.30–5.32, the last being the most widely used. Local illumination effects described by monotonic changes, such as shadowing, have limited impact on the transformed image: only the pixels at the boundary of the change are affected by the perturbation. Multiple pixel relations $\mathbf{\Delta}_i$ can be used and the overall similarity can be computed by averaging the resulting values:

$$d_{\text{ist}}^k(I_1, I_2) = \frac{1}{k} \sum_{i=1}^{k} D_H(B_{\mathbf{\Delta}_i}(I_1), B_{\mathbf{\Delta}_i}(I_2)). \tag{5.39}$$

The choice of $\mathbf{\Delta}$, which in the original formulation is assumed to be $(0, 1)$ or $(1, 0)$, is of some importance for the same reasons outlined in the subsampled version of the census transform (see Figure 5.4). The correlation of adjacent pixels is high and the expected difference (statistically) low: even moderate amounts of noise may cause $B_{\mathbf{\Delta}}(x, y)$ to flip, impairing the comparison process. The IST can be adapted to the case of variable patterns, modeling sign probability (see Intermezzo 5.3) and it can be used to create adaptive masks for the computation of a selective Pearson correlation coefficient (see Intermezzo 5.4).

**Intermezzo 5.3.** Managing variability with the incremental sign transform

Let us consider the problem of detecting a pattern $T$ belonging to a class exhibiting significant variations among its instances, such as faces. Furthermore, the image of each instance may be corrupted by noise. The resulting pattern variability may generate variability of the IST that can be quantified by computing the probability of the values 0/1 for each pixel within the pattern:

$$P_T(v, x, y) = \frac{1}{N} \sum_{i=1}^{N} \delta(v, B_{\mathbf{\Delta}}(T_i(x, y))) \tag{5.40}$$

where the sum is extended to all patterns within the class. The set should also include patterns perturbed by noise that can be synthetically generated if the model of the noise is known. Pixels where the information provided by the IST is stable will be characterized by $P \approx 1$, while pixels whose local ranking is uncorrelated among the patterns of the class will have $P \approx 0.5$. The values of $P_T$ can be used to define a new distance as follows:

$$\mathrm{d}_{\text{ist},P} = \frac{1}{N} \sum_{i=1}^{N} (1 - P_T(B_{\mathbf{\Delta}}(I; x_i, y_i), x_i, y_i)) \tag{5.41}$$

where $I$ denotes that the transform is computed over the image where the pattern must be found. The knowledge of the patterns is compiled in $P_T$ and it is the reason why we do not compute the distance between two transformed images (Mita *et al.* 2006).

## 5.3. Bibliographical Remarks

The use of rank transforms and of the associated correlation measures has a long history in statistics. The robustness and efficiency of the Spearman and Kendall coefficients are considered in Croux and Dehon (2008).The Bhat–Nayar coefficient is described in detail by Bhat and Nayar (1998) and is based on ideas originally presented by Gideon and Hollister (1987). The modified version, featuring increased discriminatory power, is described in Scherer *et al.* (1999).

The census and rank transforms were introduced by Zabih and Woodfill (1994) and many interesting remarks can be found in the paper by Cyganek (2006), where the ideas are extended to color spaces. The connection of the rank transform to the Wilcoxon test is remarked upon by Smeraldi (2003) who exploits it to build a novel kind of image transform, the ranklet. Schonfeld (2000) investigates the problem of template matching using morphological methods based on order statistics filter.

**Intermezzo 5.4.** The selective Pearson coefficient

The computation of the Pearson coefficient, like that of the other correlation coefficients considered, is completely oblivious to the position of the corresponding values within the comparison window. The IST can be used to remove some of the pixels from the computation of the Pearson (or other) coefficients when the value of the transform does not match that of the reference patterns:

$$r_P = \frac{\sum_{i=1}^{N-1} c_i (T_i - \hat{T})(I_i - \hat{I})}{\sqrt{\sum_{i=1}^{N-1} c_i (T_i - \hat{T})^2} \sqrt{\sum_{i=1}^{N-1} c_i (I_i - \hat{I})^2}} \tag{5.42}$$

$$c_i = \begin{cases} 1 - |B_{\mathbf{\Delta}}(T; i) - B_{\mathbf{\Delta}}(I; i)| & \text{even } i \\ c_{i-1} & \text{odd } i \end{cases} \tag{5.43}$$

where, for simplicity, $i$ identifies pixels within image patches serialized in left to right, top to bottom order, and $\mathbf{\Delta} = (1, 0)$. The reason for replicating the $c$ values in Equation 5.43 is that the values would not be independent anyway, as the corresponding pixel pairs overlap. A typical situation where the use of the masked correlation coefficient is advantageous is in the presence of occlusions. The standard computation of the Pearson coefficient would include all pixels, including those belonging to the occluding object. On the contrary, the masked coefficient neglects all the pixels that are locally inconsistent with the sign transform. A maximum allowed percentage of occlusion should be imposed to prevent comparisons with too small a sample. Let us note that in the case of variations due to local illumination effects, the mask would keep nearly all of the pixels due to the invariance of the transform, but the subsequent application of the Pearson coefficient would provide a wrong matching score due to the inability of the latter to account for such effects (Kaneko *et al.* 2003).

The idea underlying the incremental sign transform was introduced by Kaneko *et al.* (2002) and its use in masking the correlation coefficient is described by Kaneko *et al.* (2003), while its probabilistic variant is presented by Mita *et al.* (2006).

## References


Bhat D and Nayar S 1998 Ordinal measures for image correspondence. *IEEE Transactions on Pattern Analysis and Machine Intelligence* **20**, 415–423.

Croux C and Dehon C 2008 Robustness versus efficiency for nonparametric correlation measures. Université Libre de Bruxelles, Ecares, ECARES Working Papers, 2008_002.

Cyganek B 2006 Matching of the multi-channel images with improved nonparametric transformations and weighted binary distance measures. *Proceedings of the 11th International Workshop on Combinatorial Image Analysis (IWCIA 2006)*, Lecture Notes in Computer Science, vol. 4040, pp. 74–88. Springer.

Gideon R and Hollister R 1987 A rank correlation coefficient resistant to outliers. *Journal of the American Statistical Association* **82**, 656–666.

Kaneko S, Murase I and Igarashi S 2002 Robust image registration by increment sign correlation. *Pattern Recognition* **35**, 2223–2234.

Kaneko S, Satoh Y and Igarashi S 2003 Using selective correlation coefficient for robust image registration. *Pattern Recognition* **36**, 1165–1173.

Mita T, Kaneko T and Hori O 2006 A probabilistic approach to fast and robust template matching and its application to object categorization. *Proceedings of the 18th IAPR International Conference on Pattern Recognition (ICPR'06)*, vol. 2, pp. 597–601.

Sadr J, Mukherjee S, Thoresz K and Sinha P 2002 The fidelity of local ordinal encoding. *Proceedings of Advances in Neural Information Processing Systems*, vol. 14, pp. 1279–1286.

Scherer S, Werth P and Pinz A 1999 The discriminatory power of ordinal measures – towards a new coefficient. *Proceedings of the IEEE Conference on Computer Vision and Pattern Recognition (CVPR'99)*, vol. 1, pp. 1076–1081.

Schonfeld D 2000 On the relation of order-statistics filters and template matching: optimal morphological pattern recognition. *IEEE Transactions on Image Processing* **9**, 945–949.

Smeraldi F 2003 A nonparametric approach to face detection using ranklets. *Proceedings of the 4th International Conference on Audio- and Video-Based Biometric Person Authentication*, Lecture Notes in Computer Science, vol. 2688, pp. 351–359. Springer.

Zabih R and Woodfill J 1994 Non-parametric local transforms for computing visual correspondence. *Proceedings of the 3rd European Conference on Computer Vision (ECCV'94)*, Lecture Notes in Computer Science, vol. 801, pp. 151–158. Springer.

# 6 MATCHING VARIABLE PATTERNS

Therein do men from children nothing differ.

*Much Ado About Nothing*
WILLIAM SHAKESPEARE

While finding a single, well-defined shape is useful, finding instances of a class of shapes can be even more useful. Intraclass variability poses new problems for template matching and several interesting solutions are available. This chapter focuses on the use of projection operators on a one-dimensional space to solve the task. The use of projection operators on multidimensional spaces will be covered in Chapter 8.

## 6.1. Multiclass Synthetic Discriminant Functions

In Chapter 3 we solved the problem of finding the optimal filter by which a single, deterministic signal potentially corrupted by noise can be detected. An important extension is the detection of a signal belonging to a given distribution of signals. As an example, we may consider the problem of locating the eyes within an image of a face. In many practical cases we do not know the identity of the person represented in the image so that we may not select the corresponding matching signal, i.e. an image of the eyes of that specific person. However, a set of different eyes may be available, possibly including the correct ones. We would like to design a single correlation filter to optimally detect all eyes stored in the available catalogue (and hopefully more). We will follow an approach similar to the one considered in Section 3.4, optimizing a signal to noise ratio, but formulating the problem in a discrete setting. Let us consider a set of $N$ discrete signals $\{\boldsymbol{\phi}_i\}_{i=1,\ldots,N}$ belonging to an $n_d$-dimensional space. The signals could be natively one dimensional, e.g. time series, or multidimensional, e.g. images. In the latter case, we consider them as linearized by concatenation of the constituent data. We want to find a $d$-dimensional vector $\hat{\boldsymbol{h}}$ that maximizes the following signal to noise ratio:

$$E_{\phi\eta}\{r^2_{\mathrm{SNR}}\} = \frac{1}{N}\frac{\sum_{i=1}^{N} |\boldsymbol{h}^T\boldsymbol{\phi}_i|^2}{E\{\boldsymbol{h}^T\boldsymbol{\eta}\}^2} \tag{6.1}$$

where $\boldsymbol{\eta} \in \mathbb{R}^{n_d}$ represents the noise corrupting our signal. The denominator is then the expected value of the correlation of the filter with noise while the numerator represents the average correlation over the set of available signals. We may make all summations explicit obtaining

$$\hat{\boldsymbol{h}} = \operatorname*{argmax}_{\boldsymbol{h}} \frac{1}{N} \frac{\sum_{q=1}^{N} (\sum_{i=1}^{n_d} h_i \phi_{qi})(\sum_{j=1}^{n_d} h_j \phi_{qj})}{\sum_{i=1}^{n_d} \sum_{j=1}^{n_d} (h_i \Sigma_{ij}^{\eta} h_j)} \tag{6.2}$$

where the effect of noise $\eta$ is fully characterized by its covariance $\Sigma_{ij}^{\eta}$. If the noise covariance is proportional to the identity matrix, i.e. $\Sigma_{ij}^{\eta} = \sigma_{\eta}^2 \delta_{ij}$, the noise is said to be white (its Fourier power spectrum is flat). By rearranging the summation order in the numerator and introducing the covariance values $\sigma_{ij}^{\phi}$ for the set of signals we may rewrite the previous equation as

$$\hat{\boldsymbol{h}} = \operatorname*{argmax}_{\boldsymbol{h}} \frac{\sum_{i=1}^{n_d} \sum_{j=1}^{n_d} h_i \Sigma_{ij}^{\phi} h_j}{\sigma_{\eta}^2 |\boldsymbol{h}|^2}. \tag{6.3}$$

If the assumption of white noise does not hold, we may apply a linear whitening transformation to the signals and work in the transformed domain. The denominator of the RHS in Equation 6.3 represents the energy of the filter and we can fix it to 1 without loss of generality

$$|\boldsymbol{h}|^2 = 1. \tag{6.4}$$

The optimization of Equation 6.3 requires the maximization of the numerator subject to the energy constraint on the filter of Equation 6.4. As the covariance matrix $\Sigma^{\phi}$ is real and symmetric, by the spectral theorem we may find an orthonormal basis $\{\boldsymbol{\psi}_i\}$ of $\mathbb{R}^{n_d}$ composed of eigenvectors of $\Sigma^{\phi}$ with real eigenvalues $\lambda_i^2$ and we may express $\boldsymbol{h}$ as a linear combination of these eigenvectors:

$$\boldsymbol{h} = \sum_{i=1}^{n_d} (\boldsymbol{h}^T \boldsymbol{\psi}_i) \boldsymbol{\psi}_i = \sum_{i=1}^{n_d} c_i \boldsymbol{\psi}_i \tag{6.5}$$

where the expansion coefficients $c_i$ are given by the scalar product with the corresponding eigenvector. Exploiting the orthonormality of the eigenvectors, we may write the numerator of Equation 6.3 as

$$\sum_{i=1}^{n_d} \sum_{j=1}^{n_d} h_i \Sigma_{ij}^{\Phi} h_j = \left( \sum_{i=1}^{n_d} c_i \boldsymbol{\psi}_i \right) \left( \sum_{j=1}^{n_d} \lambda_j^2 c_j \boldsymbol{\psi}_j \right) = \sum_{i=1}^{n_d} \lambda_i^2 c_i^2 \tag{6.6}$$

finally obtaining

$$\hat{\boldsymbol{h}} = \operatorname*{argmax}_{\sum_i c_i^2 = 1} \sum_i \lambda_i^2 c_i^2. \tag{6.7}$$

If we sort the eigenvalues so that $\lambda_1^2 \geq \lambda_2^2 \geq \cdots \geq \lambda_k^2 \geq \cdots$, we have

$$\sum_i \lambda_i^2 c_i^2 \leq \lambda_1^2 \sum_i c_i^2 = \lambda_1^2 \tag{6.8}$$

so that the maximum possible value, namely $\lambda_1^2$, is achieved when $c_1 = 1$: the optimal filter function is given by the dominant eigenvector.

While correlation filters are optimal for the discrimination of patterns from white noise, they have three major limitations:

1. The output of the correlation peak degrades rapidly with geometric image distortions.
2. They cannot be used for multiclass pattern recognition.
3. The peak is often broad, making its detection difficult.

It has been noted that one can obtain better performance from a multiple correlator (i.e. one computing the correlation with several templates) by forming a linear combination of the resulting outputs instead of, for example, taking the maximum value. The filter synthesis technique known as synthetic discriminant functions (SDFs) starts from this observation and builds a filter as a linear combination of matched spatial filters (MSFs) for different patterns. The coefficients of the linear combination are chosen to satisfy a set of constraints on the filter output, specified as a predefined output value for each of the patterns used for the synthesis of the filter. By forcing the filter output to different values for different patterns, multiclass pattern recognition can be achieved. Let $\boldsymbol{u} = (u_1, \ldots, u_N)^T$ be a vector representing the required filter output for each of the images:

$$\boldsymbol{\phi}_q \otimes \boldsymbol{h} = u_q, \quad q = 1, \ldots, N. \tag{6.9}$$

The filter $\boldsymbol{h}$ can be expressed as a linear combination of the images $\boldsymbol{\phi}_q$

$$\boldsymbol{h}(x) = \sum_{q=1}^{N} b_q \boldsymbol{\phi}_q(x) \tag{6.10}$$

because any additional contribution from the space orthogonal to the images would yield a zero contribution when correlating with the image set since correlation is proportional to projection. If we denote by $X$ the matrix whose columns represent the images, by enforcing the constraints we obtain the following set of equations:

$$\boldsymbol{b} = (X^T X)^{-1} \boldsymbol{u} \tag{6.11}$$

which can be solved if the images are linearly independent. The resulting filter is appropriate for pattern recognition applications in which the input object can be a member of several classes and different distorted versions of the same object (or different objects) can be expected within each class. If $M$ is the number of classes, $N_c$ the number of different patterns within each class $c$, and $N$ the overall number of patterns, $M$ filters can be built by solving

$$\boldsymbol{b}_c = (X^T X)^{-1} \boldsymbol{\delta}_c, \quad c = 1, \ldots, M \tag{6.12}$$

where

$$\delta_{ck} = \begin{cases} 1 & \sum_{j=1}^{c-1} N_j < k < \sum_{j=1}^{c} N_j \\ 0 & \text{otherwise} \end{cases} \tag{6.13}$$

$k = 1, \ldots, N$, and image $\phi_k$ belongs to class $c$ if $\delta_{ck} = 1$. Discrimination of different classes can be obtained also using a single filter and imposing different output values. However, the performance of such a filter is expected to be inferior to that of a set of class-specific filters due to the high number of constraints imposed on filter outputs. While this approach makes it easy to obtain predefined values on a given set of patterns it does not allow us to control off-peak response: the constraints only apply to the perfectly aligned patterns. This can prevent reliable classification when the number of constraints becomes large. The effect of filter clutter can also appear in the construction of a filter giving a fixed response over a set of images belonging to the same class. The problem can be alleviated by the introduction of least squares SDFs (lsSDFs). These filters are computed using only a subset of the training images and the coefficient of the linear combination is chosen to minimize the error of the filter output on all of the available images. Let $Y$ be the matrix whose columns are the subset of images used for the construction of the filter:

$$\boldsymbol{h} = \sum_{r=1}^{n} b_r \boldsymbol{\phi}_r \tag{6.14}$$

where $n < N$. Using matrix notation we may rewrite Equation 6.14 as

$$(X^T Y)\boldsymbol{b} = \boldsymbol{u} \tag{6.15}$$

obtaining an overdetermined system as we have more constraints $u_i$ than free parameters $b_r$. As it is not possible to satisfy exactly all the constraints, we may search for a solution that minimizes the violation of the constraints as quantified by the square error

$$\Delta^2 = [X^T Y\boldsymbol{b} - \boldsymbol{u}]^T [X^T Y\boldsymbol{b} - \boldsymbol{u}]. \tag{6.16}$$

The expansion vector $\boldsymbol{b}$ is found by setting the derivative of $\Delta^2$ to zero with respect to $\boldsymbol{b}$

$$\frac{\partial \Delta^2}{\partial \boldsymbol{b}} = \frac{\partial(\boldsymbol{b}^T R^T R\boldsymbol{b} - \boldsymbol{b}^T R^T \boldsymbol{u} - \boldsymbol{u}^T R\boldsymbol{b} + \boldsymbol{u}^T \boldsymbol{u})}{\partial \boldsymbol{b}} = 0 \tag{6.17}$$

where $R = X^T Y$, giving

$$R^T R\, \boldsymbol{b} = R^T \boldsymbol{u} \tag{6.18}$$

and the expansion vector is finally given by

$$\boldsymbol{b} = (R^T R)^{-1} R^T \boldsymbol{u}. \tag{6.19}$$

In this case the matrix $R = X^T Y$ is rectangular and the least squares estimation of $\boldsymbol{b}$ relies on the computation of the pseudo-inverse of $R$:

$$R^+ = (R^T R)^{-1} R^T. \tag{6.20}$$

The dimension of the matrix to be inverted is $n \times n$ where $n$ is the number of images used to build the filter and not the (greater) number $N$ of training images. The subset of training images used to actually build the filter can be chosen in several ways. A first solution is to choose them at random while a more informed choice can be based on clustering all available images, the number of clusters matching the required number of images used for filter construction (see Intermezzo 12.1 for a simple clustering algorithm). For each cluster, the available image closest to its center is selected for filter construction. Reducing the number of templates used in the construction of the filter reduces the problem of filter clutter. Least squares SDFs produce a matched filter whose appearance markedly differs from that

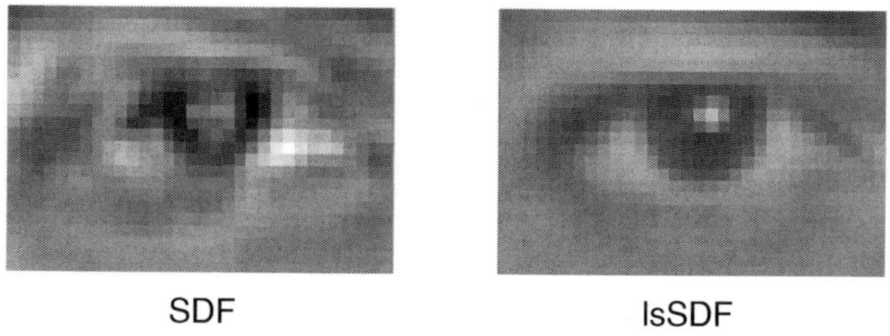


**Figure 6.1.** The MSF resulting from using 20 building images in the SDF (left) and 2 in the least squares SDF (right) when using the same set of training images. The difference in contrast of the two images is related to the magnitude of the MSFs. The performance of the two filters is similar. Source: Brunelli and Poggio (1997).

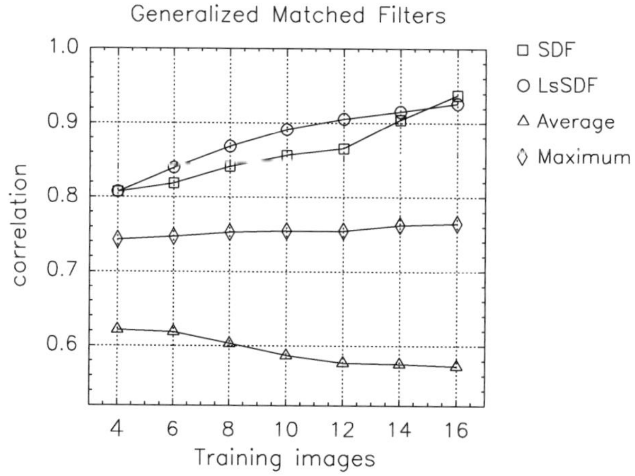


**Figure 6.2.** An increasing percentage of a set of 30 eye images is used to build an SDF, an average SDF, or a set of prototype MSFs from which the highest score is extracted. The least squares SDF uses only four building templates and the performance is computed over a disjoint set of 30 images. Source: Brunelli and Poggio (1997).

obtained using all available images (see Figure 6.1) and provide good performance using a small number of templates (see Figure 6.2).

Yet another variant of SDFs is given by symbolic encoded filters. Let us assume that the number of classes is of the form $M = 2^k$ so that we may associate to each of them a $k$-digit binary number. We can build a set of $k$ filters whose outputs for images of class $c$ are 0 or 1 depending on the value of the corresponding digit in the binary encoding of $c$. In order to use these filters for classification, the response of the bank of $k$ filters applied to a pattern to be classified is thresholded and the resulting $k$ bits used to compose the binary number identifying the pattern class. Even in this case a least squares SDF approach can be employed.

The MSF obtained from a projection SDF algorithm can achieve distortion invariance and retain shift invariance. However, the resulting filter cannot prevent large side lobe levels from occurring in the correlation plane for the case of false or true targets. The next section describes a general approach by which off-peak response can be controlled with different optimization criteria.

## 6.2. Advanced Synthetic Discriminant Functions

The signal to noise ratio maximized by the MSF is limited to the correlation peak: it does not take into account the off-peak response and the resulting filters often exhibit a sustained response well apart from the location of the central peak. This effect is usually amplified in the case of SDF when many constraints are imposed on the output of the filter. In order to locate the correlation peak reliably, it should be very localized. However, it can be expected that the greater the localization of the filter response (approaching a $\delta$ function), the more sensitive the filter to slight deviations from the patterns used in its synthesis. This suggests that the best response of the filter should not really be a $\delta$ function, but rather a shape, like a Gaussian, whose dispersion can be tuned to the characteristics of the pattern space. In this section we will review the synthesis of such filters in the frequency domain. Let us assume for the moment that there is no noise and that the patterns have dimension $n_d$. The correlation of the $i$th pattern with filter $\boldsymbol{h}$ is represented by

$$z_i(l) = (\phi_i \otimes h)(l), \quad l = 0, \ldots, n_d - 1. \tag{6.21}$$

In the following, Fourier transform quantities are identified by a subscripted $\omega$. The filter is required to produce an output $u_i$ for each training image:

$$z_i(0) = u_i. \tag{6.22}$$

Exploiting the Parseval identity, which essentially states that the Fourier transform (with symmetric normalization factors) preserves the scalar product, we may rewrite the previous equation as

$$X_\omega^\dagger \boldsymbol{h}_\omega = \boldsymbol{u} \tag{6.23}$$

where $(\cdot)^\dagger$ denotes the complex conjugate transpose. Working in the Fourier domain is convenient because of the correlation theorem (see Intermezzo 3.2), which states that the circular correlation of two patterns is given by the scalar product of their Fourier transforms. It is important to note that circular correlation does not provide, in general, the same results as linear correlation. However, we are interested in controlling the off-peak response of the autocorrelation of a signal, a function that peaks when the signal is aligned to itself. The structure of the autocorrelation function is usually well preserved by circular correlation so that we may exploit the convenience offered by the correlation theorem. Using Parseval's theorem, the energy $E_i$ of the $i$th circular correlation plane is given by

$$E_i = \sum_{l=0}^{n_d-1} |z_i(l)|^2 = \sum_{k=0}^{n_d-1} |z_{\omega,i}(k)|^2 \tag{6.24}$$

$$= \sum_{k=0}^{n_d-1} |h_\omega(k)|^2 |\phi_{\omega,i}(k)|^2. \tag{6.25}$$

When the signal is perturbed with noise $\eta$, the output of the filter will also be corrupted:

$$z_i(0) = \phi_i(0) \otimes h(0) + \eta(0) \otimes h(0). \tag{6.26}$$

Under the assumption of zero-mean noise, the variance of the filter output due to noise is

$$E_\lambda = \sum_{k=0}^{n_d-1} |h_\omega(k)|^2 s_\omega(k) \tag{6.27}$$

where $s_\omega(k)$ is the noise spectral energy. The goal is to obtain a filter whose average correlation energy over the different training images and noise is as low as possible while at the same time meeting the constraints on filter outputs conventionally imposed at position 0. A first possibility is to minimize

$$E_a = \sum_i (E_i + E_\lambda) \tag{6.28}$$

$$= \sum_i \sum_k |h_\omega(k)|^2 (|\phi_{\omega,i}(k)|^2 + s_\omega(k)) \tag{6.29}$$

subject to the constraints of Equation 6.23. However, minimizing the average energy (or filter variance due to noise) does not minimize each term of Equation 6.28, corresponding to the correlation energy with the $i$th pattern or to noise variance. A more stringent bound can be obtained by considering the spectral envelope of the different terms in Equation 6.29:

$$E_M = N \sum_k |h_\omega(k)|^2 \max(|\phi_{\omega,1}(k)|^2, \ldots, |\phi_{\omega,N}(k)|^2, s_\omega(k)) \tag{6.30}$$

where the factor $N$ takes care of the summation over $i$ in Equation 6.29. If we introduce the diagonal matrix $T_{\omega,kk} = N \max(|\phi_{\omega,1}(k)|^2, \ldots, |\phi_{\omega,N}(k)|^2, s_\omega(k))$, the synthesis of filter $\boldsymbol{h}_\omega$ can be summarized as

$$\hat{\boldsymbol{h}}_\omega = \underset{\boldsymbol{h}_\omega}{\operatorname{argmin}}\ E_M = \underset{\boldsymbol{h}_\omega}{\operatorname{argmin}}\ \boldsymbol{h}_\omega^\dagger T_\omega \boldsymbol{h}_\omega$$

$$X_\omega^\dagger h_\omega = \boldsymbol{u}. \tag{6.31}$$

The problem is very similar to the kinds of problems that are solved by means of Lagrangian multipliers, with one important difference: the independent variable $\boldsymbol{h}$ is complex. In this case the Lagrangian multiplier technique can be used only if we make the Lagrangian real: the gradient can then be computed with respect to the complex conjugate quantity, $\boldsymbol{h}_\omega^*$, while keeping $\boldsymbol{h}_\omega$ constant. In order to make the Lagrangian real we add to the function to be minimized the real part of the constraint:

$$\mathcal{E}(\boldsymbol{h}_\omega, \boldsymbol{\lambda}) = \boldsymbol{h}_\omega^\dagger T_\omega \boldsymbol{h}_\omega - \sum_{i=1}^{N} [\lambda_i^* (\boldsymbol{x}_{\omega,i}^\dagger \boldsymbol{h}_\omega - u_i) + \lambda_i (\boldsymbol{x}_{\omega,i}^T \boldsymbol{h}_\omega^* - u_i)] \tag{6.32}$$

where $\boldsymbol{\lambda} = (\lambda_1, \ldots, \lambda_N)$ are the Lagrangian multipliers and the term within the square brackets is the real part of the constraints. The solution is obtained by setting the gradient of $\mathcal{E}(\boldsymbol{h}_\omega, \boldsymbol{\lambda})$ to zero with respect to $\boldsymbol{h}_\omega^*$:

$$T_\omega \boldsymbol{h}_\omega - \sum_i \lambda_i \boldsymbol{x}_{\omega,i} = T_\omega \boldsymbol{h}_\omega - X_\omega \boldsymbol{\lambda} = 0. \tag{6.33}$$

Substituting $\boldsymbol{h}_\omega = T_\omega^{-1} X_\omega \boldsymbol{\lambda}$ in the constraint we obtain

$$X_\omega^\dagger T_\omega^{-1} X_\omega \boldsymbol{\lambda} = \boldsymbol{u}$$

hence

$$\boldsymbol{\lambda} = (X_\omega^\dagger T_\omega^{-1} X_\omega)^{-1} \boldsymbol{u}$$

from which the solution can be found

$$\boldsymbol{h}_\omega = T_\omega^{-1} X_\omega (X_\omega^\dagger T_\omega^{-1} X_\omega)^{-1} \boldsymbol{u}. \tag{6.34}$$

The use of the spectral envelope has the effect of reducing the emphasis given by the filter to the high-frequency content of the signal. It is important to note that the resulting filter in Equation 6.34 can be seen as a cascade of a whitening filter $P = T^{-1/2}$, which can be considered as a projection matrix, and a conventional SDF based on the transformed data (see Equation 6.19). In fact,

$$\boldsymbol{h}_\omega = P_\omega (P_\omega X_\omega)(X_\omega^\dagger P_\omega P_\omega X_\omega)^{-1} \boldsymbol{u} \tag{6.35}$$

which, introducing the transformed pattern matrix $\breve{X} = PX$ and exploiting the fact that $P$ is diagonal and real so that $P = P^\dagger$, can be rewritten as

$$\boldsymbol{h}_\omega = P_\omega \breve{X}_\omega (\breve{X}_\omega^\dagger \breve{X}_\omega)^{-1} \boldsymbol{u} \tag{6.36}$$

and, introducing $\breve{\boldsymbol{h}}_\omega = \breve{X}_\omega (\breve{X}_\omega^\dagger \breve{X})_\omega^{-1}$,

$$\boldsymbol{h}_\omega = P_\omega \breve{\boldsymbol{h}}_\omega \boldsymbol{u}. \tag{6.37}$$

In this case the whitened spectrum is the envelope of the spectra of the real noise and of the training images. A least squares approach may again be preferred when a large number of examples are used in the construction of the filter. In this case all available images are used to estimate $T$ but only a subset of them is used to build the corresponding SDF. It is possible to use white noise of tunable energy to model intraclass variability:

$$E_M(\alpha) = \sum_k |\boldsymbol{h}_\omega(k)|^2 \max(|\phi_{\omega,1}(k)|^2, \ldots, |\phi_{\omega,N}(k)|^2, \alpha). \tag{6.38}$$

Adding white noise limits the emphasis given to high frequencies, reducing the sharpness of the correlation peak and increasing the tolerance to small variations of the templates (see Figure 6.3). An additional function of white noise is that of ensuring the invertibility of matrix $T$ when the spectral envelope is null for some value of $k$. An alternative way to obtain a sharp correlation peak is by means of nonlinear image transformations emphasizing high frequencies. Another way to control intraclass performance is that of modeling the shape of the correlation peak. As already mentioned, the shape of the correlation peak is expected to be important both for its detection and for the requirements imposed on the filter which can impair its ability to correlate well with patterns even slightly different from the ones used in the construction of the filter. Let us denote by $f_\omega(k)$ the required shape. The shape of the peak can be constrained by minimizing the squared deviations of its output from $f_\omega(k)$:

$$E_S = \sum_{i=1}^{N} \sum_{k=0}^{n_d - 1} |\boldsymbol{h}_\omega(k)^* \boldsymbol{\phi}_\omega(k) - f_\omega(k)|^2 \tag{6.39}$$

where, for instance, $f(x) = \exp(-x^2/2\sigma^2)$ is a Gaussian amplitude function. By switching to matrix notation, the resulting energy can be expressed as

$$E_S = \boldsymbol{h}_\omega^\dagger D_\omega \boldsymbol{h}_\omega + \boldsymbol{f}_\omega^\dagger \boldsymbol{f}_\omega - \boldsymbol{h}_\omega^\dagger A_\omega \boldsymbol{f}_\omega - \boldsymbol{f}_\omega^\dagger A_\omega \boldsymbol{h}_\omega \tag{6.40}$$

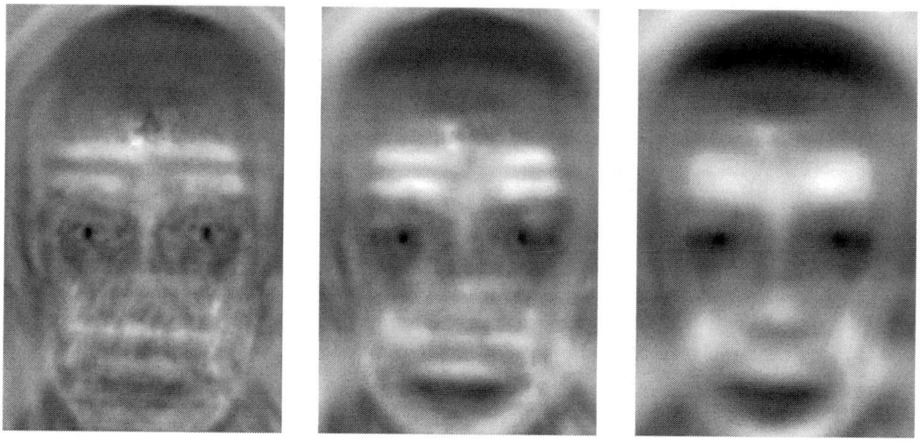

**Figure 6.3.** Using an increasing amount of added white noise reduces the emphasis given to high frequencies and the resulting response approaches that of a standard MSF. The pictures present the result of correlation with an SDF based on the spectral envelope of 10 training images and different amounts of white noise (left: $\alpha = 1$; middle: $\alpha = 5$) compared to the output of normalized cross-correlation (right) using one of the images used for SDF synthesis, without any spectral enhancement (the darker the image, the higher the matching value). Source: Brunelli and Poggio (1997).

where $A_\omega$ is a diagonal matrix whose elements are the sum of the components of $\boldsymbol{\phi}_{\omega,i}$ and $D_\omega$ is a diagonal matrix whose elements are the sum of the squares of the components of $\boldsymbol{\phi}_{\omega,i}$. The first term on the RHS of Equation 6.40 corresponds to the average correlation energy of the different patterns. Again, we suggest the use of the spectral envelope $T_\omega$ instead of $D_\omega$, employed in the original approach, thereby minimizing the following energy:

$$E'_S = \boldsymbol{h}^\dagger_\omega T_\omega \boldsymbol{h}_\omega + \boldsymbol{f}^\dagger_\omega \boldsymbol{f}_\omega - \boldsymbol{h}^\dagger_\omega A_\omega \boldsymbol{f}_\omega - \boldsymbol{f}^\dagger_\omega A_\omega \boldsymbol{h}_\omega \geq E_S. \tag{6.41}$$

The minimization of $E'_S$ subject to the constraints of Equation 6.23 can be done again using the Lagrange multiplier approach and is found to be

$$\boldsymbol{h}_\omega = T_\omega^{-1} X_\omega (X_\omega^\dagger T_\omega^{-1} X_\omega)^{-1} \boldsymbol{u} + T_\omega^{-1} A_\omega \boldsymbol{f}_\omega - T_\omega^{-1} X_\omega (X_\omega^\dagger T_\omega^{-1} X_\omega)^{-1} X_\omega^\dagger T_\omega^{-1} A_\omega \boldsymbol{f}_\omega. \tag{6.42}$$

These filters provide a controlled, sharp correlation peak subject to the constraints on the filter output and the required correlation peak shape. Working in the Fourier domain eases the task of computing this class of filters, which can then be transformed back to the spatial domain where standard correlation can be applied if needed, remembering, however, that their design is based on circular correlation.

## 6.3. Non-orthogonal Image Expansion

SDFs are not the only way to obtain sharp correlation peaks: non-orthogonal image expansion provides an alternative that turns out to be closely related. Matching by expansion is based on expanding the signal with respect to basis functions (BFs) all of which are translated versions of the template to be detected. Such an expansion is feasible if the BFs are linearly independent and complete. It can be proven that self-similar BFs of compact support are independent and complete under very weak conditions. Let us suppose that we need to approximate a discrete $d$-dimensional signal $g(x)$ by a linear combination of BFs $\phi_i(x)$:

$$g'(x) = \sum_{i=0}^{n_d-1} c_i \phi_i(x) \tag{6.43}$$

where $\phi_i(x)$ now represents the $i$th circulated translation of $\phi$. The coefficients are estimated by minimizing the square error of the approximation $|g - g'|^2$. The approximation error is orthogonal to the BFs so that we need to solve the following system of equations:

$$\begin{cases} \sum_{j=0}^{n_d-1} \langle \phi_0, \phi_j \rangle c_j &= \langle g, \phi_0 \rangle \\ \vdots & \\ \sum_{j=0}^{n_d-1} \langle \phi_{n_d-1}, \phi_j \rangle c_j &= \langle g, \phi_{n_d-1} \rangle. \end{cases} \tag{6.44}$$

If we denote by $\Phi^T = [\boldsymbol{\phi}_1, \dots]$ the matrix whose columns are the BFs, we can use the following compact notation:

$$\Phi \Phi^T \boldsymbol{c} = \Phi \boldsymbol{g}. \tag{6.45}$$

If the set of BFs is linearly independent the equations give a unique solution for the expansion coefficients and these expansion coefficients are given by

$$\boldsymbol{c} = (\Phi^T)^{-1} \boldsymbol{g}. \tag{6.46}$$

If the $j$th BF used for the expansion is the template we need to locate, we see that, at the template location, $c_i = \delta_{ij}$ as we only need the template itself to reconstruct its instance within the signal. Therefore, at the template location, the expansion vector $\boldsymbol{c}$ maximizes the following signal to noise ratio

$$\text{SNR}_{\text{exp}} = \log\left(\frac{c_j^2}{\sum_{i \neq j} c_i^2}\right) \tag{6.47}$$

whenever the $j$th BF represents the template. The resulting expansion matching is essentially equivalent to the advanced SDF approach if we choose as BFs the circulated versions of the template we need to locate:

$$\Phi = \begin{pmatrix} \phi(0) & \phi(1) & \cdots & \phi(n_d - 1) \\ \phi(1) & \phi(2) & \cdots & \phi(0) \\ \vdots & \vdots & \cdots & \vdots \\ \phi(n_d - 1) & \phi(0) & \cdots & \phi(n_d - 2) \end{pmatrix}. \tag{6.48}$$

The circulated templates provide a set of independent vectors provided that the DFT of the signal has no null entries. In fact, if we consider an advanced SDF minimizing the average circulating correlation energy (instead of the spectral envelope $T$), with a single constraint fixing its response to 1 at the template, it is not difficult to show that the energy of the circulating correlation is 1 for the correct shift of the template and 0 otherwise: it is the same situation considered by expansion matching. The elements of the diagonal matrix $T$ correspond to the magnitude of the DFT coefficients of the signal and invertibility of $T$, required for the solution, is granted if there are no null DFT entries. This result leads to the idea of phase correlation, which is investigated in some detail in Section 10.2.2. The connection can be clearly seen if we work out the advanced SDF directly in the spatial domain, assuming without loss of generality that $\boldsymbol{g} = \boldsymbol{\phi}_0 = \boldsymbol{\phi}$ is the template that we want to expand

$$\hat{\boldsymbol{h}} = \operatorname{argmin}\left(\frac{\sum_i c_i^2}{c_1^2}\right) = \operatorname{argmin}\left(\frac{\boldsymbol{h}^T \Phi^T \Phi \boldsymbol{h}}{(\boldsymbol{\phi}^T \boldsymbol{h})^2}\right). \tag{6.49}$$

By equating to zero the derivative with respect to $\boldsymbol{h}$ we find after some algebra that

$$\hat{\boldsymbol{h}} = \left(\frac{\boldsymbol{h}^T \Phi^T \Phi \boldsymbol{h}}{\boldsymbol{h}^T \boldsymbol{\phi}}\right)(\Phi^T \Phi)^{-1}\boldsymbol{\phi} \propto (\Phi^T \Phi)^{-1}\boldsymbol{\phi} \tag{6.50}$$

from which we have

$$\Phi\,\hat{\boldsymbol{h}} \propto (\Phi^T)^{-1}\boldsymbol{\phi} = \boldsymbol{c}. \tag{6.51}$$

The vector of the expansion coefficients of the template can then be obtained by means of correlation with the minimum average correlation SDF. The idea of expansion matching is also closely related to correlation SDFs where multiple shifted templates are explicitly used to shape the correlation peak. Let us consider a set of templates obtained by shifting the original pattern (possibly with circulation) on the regular grid defined by image coordinates. We can build a filter requiring that its correlation with the original pattern be 1 when there is no shift and 0 for every non-null shift. This corresponds to a filter whose response is given by $\boldsymbol{c} = (0, \ldots, 0, 1, 0, \ldots, 0)$ as pointed out above.

## 6.4. Bibliographical Remarks

The optimality of the principal component of a set of templates as a correlation filter is addressed in Kumar *et al.* (1982). The approach to multi-object distortion-invariant matching based on synthetic discriminant functions was developed over several years and is described in a series of papers starting from Casasent (1984).

Control of correlation plane sidelobe response by means of multiple constraints for shifted templates is addressed by Casasent and Chang (1986). Minimum variance SDFs controlling the correlation peak only were introduced by Kumar (1986) and improved by the introduction of minimum average correlation energy filters (Mahalanobis *et al.* 1987) whose performance is analyzed in Mahalanobis and Casasent (1991). A space formulation of the latter is presented in Sudharsanan *et al.* (1990). Shaping the response peak in the Fourier domain is explored in Casasent *et al.* (1991) and compared to previous approaches (Casasent and Ravichandran 1992; Kumar *et al.* 1992). The use of the spectral envelope of the training images is introduced in Ravichandran and Casasent (1992). Reduction of filter clutter by means of least squares SDFs is covered in Brunelli and Poggio (1997).

The approach based on non-orthogonal expansion was proposed by Ben-Arie and Rao (1993) who also present the conditions under which the set of a circulated version of a template provides a complete basis. The connection with advanced SDF is explored in Ben-Arie and Rao (1994) and Brunelli and Poggio (1997).

## References

Ben-Arie J and Rao K 1993 A novel approach for template matching by non-orthogonal image expansion. *IEEE Transaction on Circuits and Systems for Video Technology* **3**, 71–84.

Ben-Arie J and Rao K 1994 Optimal template matching by nonorthogonal image expansion using restoration. *Machine Vision and Applications* **7**, 69–81.

Brunelli R and Poggio T 1997 Template matching: matched spatial filters and beyond. *Pattern Recognition* **30**, 751–768.

Casasent D 1984 Unified synthetic discriminant function computational formulation. *Applied Optics* **23**, 1620–1627.

Casasent D and Chang WT 1986 Correlation synthetic discriminant functions. *Applied Optics* **25**, 2343–2350.

Casasent D and Ravichandran G 1992 Advanced distortion-invariant minimum average correlation energy (MACE) filters. *Applied Optics* **31**, 1109–1116.

Casasent D, Ravichandran G and Bollapragada S 1991 Gaussian-minimum average correlation energy filters. *Applied Optics* **30**, 5176–5181.

Kumar B 1986 Minimum-variance synthetic discriminant functions. *Journal of the Optical Society of America A* **3**, 1579–1584.

Kumar B, Casasent D and Murakami H 1982 Principal-component imagery for statistical pattern recognition correlators. *Optical Engineering* **21**, 43–47.

Kumar B, Mahalanobis A, Song S, Sims S and Epperson J 1992 Minimum squared error synthetic discriminant functions. *Optical Engineering* **31**, 915–922.

Mahalanobis A and Casasent D 1991 Performance evaluation of minimum average correlation energy filters. *Applied Optics* **30**, 561–572.

Mahalanobis A, Kumar B and Casasent D 1987 Minimum average correlation energy filters. *Applied Optics* **26**(17), 3633–3640.

Ravichandran G and Casasent D 1992 Minimum noise and correlation energy optical correlation filter. *Applied Optics* **31**, 1823–1833.

Sudharsanan S, Mahalanobis A and Sundareshan M 1990 Unified framework for the synthesis of synthetic discriminant functions with reduced noise variance and sharp correlation structure. *Optical Engineering* **29**, 1021–1028.

# 7 MATCHING LINEAR STRUCTURE: THE HOUGH TRANSFORM

> The forms of things unknown, the poet's pen
> Turns them to shapes and gives to airy nothing
> A local habitation and a name.
>
> *A Midsummer Night's Dream*
> WILLIAM SHAKESPEARE

Finding simple shapes, such as lines and circles, in images may look like a simple task but computational issues coupled with noise and occlusions require some not so naive solutions. In spite of the apparent diversity of lines and areas, it turns out that common approaches to the detection of linear structures can be seen as an efficient implementation of matched filters. The chapter describes how to compute salient image discontinuities and how simple shapes embedded in the resulting map can be located with the Radon/Hough transform.

## 7.1. Getting Shapes: Edge Detection

The image templates analyzed in the previous chapters are characterized by the fact that they correspond to two-dimensional image patches. As a direct consequence of this, they have been described in essentially two ways: by giving the values of all the pixels belonging to the template region or by giving a (possibly smaller) set of coefficients specifying an approximating linear combination of reference templates. While this approach is effective and general, we may wonder whether a better representation exists supporting more efficient computations while potentially being as general as a pixel-based description. An image representation that makes explicit features of general interest, such as the boundaries of physical structures, would be particularly useful in the analysis of image content. Among the principles guiding the quest for such visual encoding, we are mainly interested in:

- *concision*, providing a compressed image representation hopefully supporting faster computation;
- *reliability*, in particular with respect to the effects of the sensing process;
- *explicitness*, basing the encoding on features directly related to interesting image content.

A representation based on significant image discontinuities, edges, is interesting from a template matching point of view: many vision tasks are related to the detection and description of shapes as purely geometrical structures identified by their boundaries. As the outlines of the shapes, curves in two-dimension (2D) and surfaces in three-dimension (3D), have a lower dimension than the corresponding interiors, we may expect a concise description and computational savings. We have seen in Chapter 2 that the sensing process is characterized by two intrinsic limitations: diffraction and photon noise. The consequence of the former is that only a limited depth of field can be achieved and, in general, parts of the image will be affected by focus blur. The effect of photon noise is that our image data will always be corrupted by intensity-dependent Gaussian noise. Even without considering penumbral effects due to extended light sources and objects with rounded edges, edges will often appear not as clean step discontinuities but as noisy, smoothed steps. We have seen that out of focus imaging results in point sources being spread over a circular region, an effect that can be modeled using a pill-box filter, while in focus imaging may be limited by Fraunhofer diffraction that can be modeled by a convolution with an Airy kernel. The latter can be approximated very well by a Gaussian kernel, which also provides an analytic approximation to the pill-box filter. We can then formalize our model of an edge as the convolution of a step edge, with the discontinuity in the $x$ direction, by a Gaussian kernel

$$e(x, y) = \left( \frac{1}{\sqrt{2\pi\sigma_e^2}} e^{-(x^2+y^2)/(2\sigma_e^2)} \right) * ((I_1 - I_0)\theta(x)) + I_0 + \eta(x, y) \tag{7.1}$$

$$= \frac{I_1 - I_0}{2} \text{erf}\left( \frac{x}{\sqrt{2}\sigma_e} + 1 \right) + I_0 + \eta(x, y) \tag{7.2}$$

where $\theta(x)$ is the Heaviside function, the values $I_0$, $I_1$ represent the lowest and highest intensity values at the two sides of the transition, $\sigma_e$ identifies the scale over which the transition happens, and $\eta(x, y)$ represents noise. A first step towards edge detection is the computation of the image gradient at each point along any direction: the 'edgeness' of a pixel is the maximum absolute value of the directional derivatives. A Gaussian regularized estimate of the gradient can be obtained using a basis of filters $\{g_{x1}, g_{y1}\}$ whose properly combined response gives the magnitude of the gradient $G_1^\theta$ for any direction $\theta$:

$$g_{x1}(x, y, \sigma_1) = -\frac{x}{2\pi\sigma_1^4} e^{-(x^2+y^2)/(2\sigma_1^2)}, \quad G_{x1} = g_{x1} * I \tag{7.3}$$

$$g_{y1}(x, y, \sigma_1) = -\frac{y}{2\pi\sigma_1^4} e^{-(x^2+y^2)/(2\sigma_1^2)}, \quad G_{y1} = g_{y1} * I \tag{7.4}$$

$$G_1^\theta(x, y, \sigma_1) = \cos\theta\, G_{x1}(x, y, \sigma_1) + \sin\theta\, G_{y1}(x, y, \sigma_1) \tag{7.5}$$

from which the maximum gradient intensity and the direction $\theta_M$ at which it is obtained may be derived:

$$G_1^{\theta_M}(x, y, \sigma_1) = \sqrt{G_{x1}^2(x, y, \sigma_1) + G_{y1}^2(x, y, \sigma_1)} \tag{7.6}$$

$$\theta_M = \arctan\left(G_{y1}(x, y, \sigma_1)/G_{x1}(x, y, \sigma_1)\right). \tag{7.7}$$

As the presence of noise may cause $G_1^{\theta_M}(x, y, \sigma_1) > 0$ even when no edge is present, we need to determine when the response of the filter is not due to noise. If the distribution of the filter response over noise is known, we may determine a gradient value threshold $c_1$ over which we have confidence $1 - \alpha$ in the value not being due to noise. If we fix the maximum number, say $k$, of type I errors over an image of a given number of pixels $N$, we can determine an appropriate confidence value to be used at each pixel:

$$N\alpha_I = k \tag{7.8}$$

as the confidence is the area of the (right) tail of the distribution of given statistics from $\alpha_I$ upwards. This means that we are able to choose a value of the statistics such that the probability of exceeding it under the null hypothesis is equal to $k/N$. The distribution of $\mathrm{G}_1^{\theta\mathrm{M}}$ over noise can be obtained in closed form only for a few cases, Gaussian noise being one of them. While photon noise can be approximated with a Gaussian distribution, its dependence on the intensity value prevents an analytic derivation. A working solution is to locally approximate photon noise with a Gaussian distribution based on the local average image value using a convolution with a Gaussian filter whose spread matches that of the first derivative filter ($\sigma_1$).

**Intermezzo 7.1.** A simple way to estimate noise variance

Let us assume that our sensor is affected by independent and identically distributed random noise whose standard deviation we must estimate. A simple way to obtain the estimate is based on the following result:

**Proposition 7.1.1.** *The standard deviation of the result of a linear operator $\mathfrak{L} : \mathbb{R}^n \to \mathbb{R}$ applied to a set of independent and identically distributed random variables of standard deviation $s_n$ is the product of the $L_2$ norm of the linear transformation and the standard deviation of the random variables: $s_{\mathfrak{L}} = \|\mathfrak{L}\|_2 s_n$.*

In order to estimate the noise of the sensor, we can acquire a defocused image of a plain flat surface and apply a simple high-pass filter such as $(1/\sqrt{2}, -1/\sqrt{2})$ whose $L_2$ norm is unitary. By the previous proposition, the statistics of the filter output directly provide an estimate of the noise standard deviation, as the filter is zero mean besides being unit norm.

**Intermezzo 7.2.** The distribution of a function of a random variable

It is often necessary to compute the probability density function of a function of a random variable whose pdf is known. The general solution of the problem relies on the following result:

**Proposition 7.1.2.** *Let $x$ be a random variable and $y = g(x)$ a function of it. If we denote by $\{x_i\}$ the real roots of the equation*

$$y = g(x) \tag{7.9}$$

*the pdf $p_y(y)$ is given by*

$$p_y(y) = \sum_i p_x(x_i)\left|\frac{dg(x_i)}{dx}\right|. \tag{7.10}$$

*When $g(\cdot)$ is invertible and its inverse differentiable, the expression simplifies to*

$$p_y(y) = p_x(g^{-1}(y))\left|\frac{dg^{-1}(y)}{dy}\right|. \tag{7.11}$$

For each point in the image we may now compute the gradient at different scales and choose the minimum scale for which the magnitude of the gradient turns out to be significant:

$$\hat{\sigma}_1 = \inf\{\sigma_1 : \mathrm{G}_1^{\theta\mathrm{M}}(x, y, \sigma_1) > c_1(\sigma_1)\}. \tag{7.12}$$

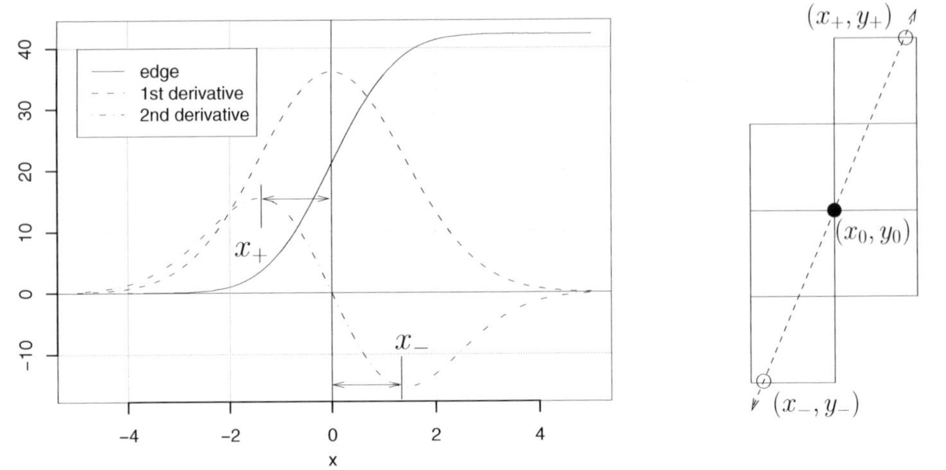


**Figure 7.1.** The profile of a smoothed edge and the corresponding values of the first and second derivatives. The points used in detecting reliable edges are shown on the right where the dashed line represents the direction maximizing gradient intensity.

This choice ensures reliable edge detection from two different perspectives: noise, because we required a statistically significant response, and image structure, because we stick to the minimum scale that minimizes the effect of larger image structures (gathered by the convolution operator) on edge computation. However, this filtering procedure does not necessarily provide unique gradient extrema in the proximity of edges; ensuring such uniqueness is not a local problem.

If we look at Figure 7.1 we see that the response of the second derivative is not local, exhibiting a well-defined structure consisting of two peaks (of different sign) enclosing the smooth edge and a central zero crossing. It is then possible to extend the reasoning on the reliability of the magnitude of the gradient to the reliable detection of a combined structure built from the responses of $G_1$ and of the second derivative $G_2$. Using the steerable filters approach,

$$g_{x2}(x, y, \sigma_2) = \frac{1}{2\pi\sigma_2^4}\left(\frac{x^2}{\sigma_2^2} - 1\right)e^{-(x^2+y^2)/(2\sigma_2^2)} \tag{7.13}$$

$$g_{y2}(x, y, \sigma_2) = \frac{1}{2\pi\sigma_2^4}\left(\frac{y^2}{\sigma_2^2} - 1\right)e^{-(x^2+y^2)/(2\sigma_2^2)} \tag{7.14}$$

$$g_{xy2}(x, y, \sigma_2) = \frac{xy}{2\pi\sigma_2^6}e^{-(x^2+y^2)/(2\sigma_2^2)} \tag{7.15}$$

$$\begin{aligned} G_2^{\theta}(x, y, \sigma_2) = {} & \cos^2\theta\,(g_{x2} * I)(x, y, \sigma_2) \\ & - 2\cos\theta\sin\theta\,(g_{xy2} * I)(x, y, \sigma_2) \\ & + \sin^2\theta\,(g_{y2} * I)(x, y, \sigma_2). \end{aligned} \tag{7.16}$$

We choose as the final set of edges the significant zero crossings of $G_2^{\theta_M}(x, y, \sigma_2)$. Even for the second-derivative response we must choose a threshold $c_2(\sigma_2)$ to find the minimum scale $\sigma_2$ at which the response is significant. The actual computation of $c_1$ and $c_2$ depends on Proposition 7.1.2 (see Section 7.5 for references). The thresholds for the first and second derivatives can be expressed as a function of the scale of the filter and the required confidence at pixel level as follows:

$$c_1 = \frac{1}{2}\sqrt{-\frac{\ln(\alpha)}{\pi}\frac{\sigma_\eta}{\sigma_1^2}} \tag{7.17}$$

$$c_2 = \frac{1}{2}\sqrt{\frac{3}{2\pi}\frac{\sigma_\eta}{\sigma_2^3}}\,\mathrm{erf}^{-1}(1-\alpha) \tag{7.18}$$

where $\sigma_\eta$ is the standard deviation of noise. It is possible to show that the response of the first-derivative filter at a blurred edge decreases as $1/\sigma_1$. As the required confidence threshold $c_1$ decreases faster, as $1/\sigma_1^2$, the minimum reliable scale is well defined. The response of the second-order derivative operator at the peaks surrounding the zero crossing decreases, for a blurred edge, as $1/\sigma_2^2$, while $c_2$ decreases faster, as $1/\sigma_2^3$, so that the minimum reliable scale is a well-defined quantity even in this case (as there is a unique intersection of the curves).

Let $(x_0, y_0)$ be a candidate zero crossing for which the gradient was found to be significant. As the value of $G_2^{\theta_M}(x, y, \sigma_2)$ should be 0, we want it to be below threshold and, as it should separate a region of positive values from negative ones, we require it to be surrounded by two peaks at $(x_+, y_+)$ and $(x_-, y_-)$ along the gradient direction $\theta_M$ where the value of the $G_2$ operator is confidently (i.e. above threshold) positive $(x_+, y_+)$ and negative $(x_-, y_-)$:

$$G_2^{\theta_M}(x_+, y_+, \sigma_2) > c_2 \tag{7.19}$$

$$G_2^{\theta_M}(x_-, y_-, \sigma_2) < -c_2. \tag{7.20}$$

The distance between the two peaks is a measure of the slope of the edge and is important when the edge map is used to reconstruct the original image. Each edge pixel can then be characterized by its location, direction, scale, brightness, and contrast. It is found, experimentally, that this information is enough to reconstruct the original image in spite of the sparsity of the representation. The image values at the pixels where no edge information is available are obtained by diffusing the information at the edges using the heat equation. After diffusing the intensity information, the edge scale must be diffused as well, generating a dense scale map that can be used to defocus the intensity image with a space-variant smoothing kernel (see Figure 7.2). Besides providing a reliable map of edges and their orientation, this representation can be exploited for image compression showing a good capability of preserving important image structures even at high compression ratios without blurring them. Before switching to a description of how to search the resulting edge map for known shapes, let us remark that, apart from the significance level, no additional parameters must be specified.

## 7.2. The Radon Transform

Having computed an edge map of the image to be analyzed we now turn to the problem of spotting subsets of points representing instances of a specified class of shapes. Each member of the class is specified by a set of parameters which include its spatial position. The idea is

**Figure 7.2.** The pictures show the edge response, the reconstructed intensity, and the refined (reblurred) version using diffusion (lower right). Source: Elder (1999).

that of formalizing the detection process as a convolution operator whose maxima correspond to specific instances of shapes from the chosen family. As we will have to deal with several entities and domains, let us first introduce the necessary notation:

- $\mathbf{x}$ — the spatial coordinates;
- $I(\mathbf{x})$ — a $d$-dimensional image containing $n$-dimensional shapes;
- $\mathbf{q} = \{\mathbf{q}_s, \mathbf{x}_0\}$ — the parameters of the curve (intrinsic shape parameters $\mathbf{q}_s$ and shape location $\mathbf{x}_0$);
- $s(\mathbf{q})$ — an element of the class of shapes characterized by a vector of parameters $\mathbf{q}$;
- $\mathbf{s}(\mathbf{u}; \mathbf{q})$ — the image coordinates of a point belonging to shape $s(\mathbf{q})$, identified by the coordinates $\mathbf{u}$ on the shape itself;
- $\mathcal{K}(\mathbf{x}; \mathbf{q})$ — a set of $d - n$ constraints $\{\mathcal{K}_i(\mathbf{x}; \mathbf{q}) = 0\}_{i=1,\ldots,(d-n)}$ whose solutions identify the point belonging to the shape;
- $\mathrm{K}(\mathbf{q}; \mathbf{x})$ — a kernel (template) representing the projection of the curve identified by $\mathbf{q}$ in image space, therefore using spatial coordinates.

The description of a shape by means of a constraint function is not unique, even for a simple case like that of a line:

$$\mathcal{K}_{1a}(\mathbf{x}; (m, q)) = y - mx - q \tag{7.21}$$

$$\mathcal{K}_{1b}(\mathbf{x}; (\rho, \theta)) = x\cos\theta + y\sin\theta - \rho \tag{7.22}$$

where $\mathcal{K}_{1a}$ identifies a line with the usual slope/intercept equation while $\mathcal{K}_{1b}$ uses the *normal* parameterization based on the normal to the line passing through the origin (see Figure 7.3). Not all parameterizations are equally convenient from a computational point of view. In the

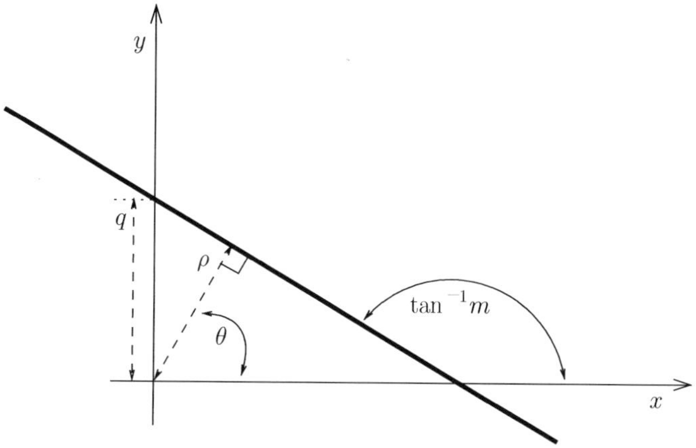


**Figure 7.3.** Two different parameterizations of a line: the $(m, q)$ or slope/intercept description and the $(\rho, \theta)$ or normal one.

case of lines, the parameter space of the slope/intercept representation is unbounded even in the case of finite images, due to lines parallel to the $y$ axis, while that of the normal representation is bounded. Given a parameterized class of shapes we see that each point in parameter space identifies a subset of points in image space. The Radon transform $\mathcal{R}$ characterizes a function (a curve) in image space by its integral projection over lines

$$\mathcal{R}(I;\, (\rho, \theta)) = \int_{\mathbb{R}} I(\rho \cos\theta - u \sin\theta,\, \rho \sin\theta + u \cos\theta)\, du. \tag{7.23}$$

If we consider $\theta$ fixed, we see that by varying $\rho$ we get the projection of image $I$ over a line orthogonal to the integration lines. The Fourier transform of this projection represents the Fourier transform of image $I$ over the line passing through the origin and with inclination $\theta$. If many projections are available, a good approximation to the complete Fourier transform of $I$ can be obtained by interpolation, and $I$ can be reconstructed; this important result is the basis of computed axial tomography. The line Radon transform can be generalized to an arbitrary shape given the corresponding constraints

$$\mathcal{R}_{s(\boldsymbol{q})}(I;\, \boldsymbol{q}) = \int_{\mathbb{R}^d} \delta(\mathcal{K}(\boldsymbol{x};\, \boldsymbol{q})) I(\boldsymbol{x})\, d\boldsymbol{x} \tag{7.24}$$

where $\delta(\cdot)$ is the Dirac delta function. Let us consider a binary image characterized by the presence of a single shape $s(\boldsymbol{q}_0)$: all points belonging to the shape have value 1, the others 0. The value of $\mathcal{R}_{s(\boldsymbol{q}_0)}(I;\, \boldsymbol{q}_0)$ is the $n$-dimensional volume of $s(\boldsymbol{q}_0)$. For $\boldsymbol{q} \neq \boldsymbol{q}_0$ the value is given by the volume of $s(\boldsymbol{q}_0) \cap s(\boldsymbol{q})$; in the case of lines, this would correspond to a single point for intersecting lines and would be null in the continuous case we are currently addressing. The definition of the Radon transform has the same form as a linear integral operator with kernel T

$$\mathfrak{L}_{\mathrm{T}} I(\boldsymbol{q}) = \int_{\mathbb{R}^d} \mathrm{T}(\boldsymbol{q};\, \boldsymbol{x}) I(\boldsymbol{x})\, d\boldsymbol{x} \tag{7.25}$$

where $\mathrm{T}(\boldsymbol{q}, \boldsymbol{x}) = \delta(\mathcal{K}(\boldsymbol{x};\, \boldsymbol{q}))$. The notation highlights the fact that the Radon transform computes the inner product of image $I$ with the template $\mathrm{T}_{\boldsymbol{q}}$: it performs template matching.

When we can make explicit the parameters corresponding to the location of the shape, the kernel has a shift-invariant structure

$$\mathrm{T}(\boldsymbol{q}_s, \boldsymbol{x}_0; \boldsymbol{x}) = \mathrm{T}(\boldsymbol{q}_s, \boldsymbol{x}_0 + \boldsymbol{d}; \boldsymbol{x} + \boldsymbol{d}) \quad \forall \boldsymbol{d} \tag{7.26}$$

so that the operator $\mathfrak{L}_\mathrm{T}$ can be expressed in terms of convolutions

$$\mathfrak{L}_\mathrm{T} I(\boldsymbol{q}_s, \boldsymbol{x}_0) = (M(\boldsymbol{q}_s) * I)(\boldsymbol{x}_0) \quad \text{with} \quad M(\boldsymbol{q}_s) = T(\boldsymbol{q}_s, \boldsymbol{0}; -\boldsymbol{x}) \tag{7.27}$$

where the notation $M(\boldsymbol{q}_s)$ emphasizes the relation with the matched filter approach presented in Section 3.4. Convolutions can be computed efficiently as multiplications in the Fourier domain (see Intermezzo 3.2). Let us now turn to the discretization of Equation 7.25. There are two different discretizations to be taken into account:

1. Given $\boldsymbol{q}$, under which conditions can we safely use a discretized version of $I$ and T and substitute the integral with a summation?
2. Under which conditions is it possible to sample $\mathfrak{L}_\mathrm{T} I(\boldsymbol{q})$ on a discrete set of points?

If the continuous functions T and $I$ are properly band limited, they can be expressed in terms of their discrete samples, provided they are dense enough to accommodate the highest bandwidth constraint. As the reconstruction is based upon an orthogonal base (the sinc functions), the value of the integral reduces to the sum of the product of the discrete values. While image $I$ is often band limited, e.g. due to the fact that we are working with a diffraction-limited optical system (see Chapter 2), kernel T usually is not. However, a band-limited version of the kernel can be easily obtained by convolving the original kernel with a Gaussian filter that has a band limit of $\sigma^{-1}/2$ with critical sample spacing $\sigma$.

Let us now turn to the sampling of parameter space by considering the Fourier transform on the $\boldsymbol{q}$ space

$$\mathcal{F}\{\mathfrak{L}_\mathrm{T} I\}(\tilde{\boldsymbol{q}}) = \mathcal{F}\left\{\int_{\mathbb{R}^d} \mathrm{T}(\boldsymbol{q}, \boldsymbol{x}) I(\boldsymbol{x})\, d\boldsymbol{x}\right\}(\tilde{\boldsymbol{q}}) \tag{7.28}$$

$$= \int_{\mathbb{R}^d} \mathcal{F}_{\boldsymbol{q}}\{\mathrm{T}(\boldsymbol{q}; \boldsymbol{x})\}(\tilde{\boldsymbol{q}}, \boldsymbol{x}) I(\boldsymbol{x})\, d\boldsymbol{x} \tag{7.29}$$

from which we derive that the band limit $\mathrm{up}^{\omega}_{\boldsymbol{q}}(\mathrm{T})$ holds for the transform as well, because the integral evaluates to zero whenever $\tilde{\boldsymbol{q}} > \mathrm{up}^{\omega}_{\boldsymbol{q}}(\mathrm{T})$. This means that we can sample the transform in parameter space without losing information as the continuous version can be reconstructed. Parameter space subcell precision estimation is then possible.

## 7.3. The Hough Transform: Line and Circle Detection

How can we exploit the previous results to spot shapes in a two-dimensional image? The first step is to generate a sparse map containing significant edges. If the edge map comprises a single, complete instance $s(\boldsymbol{q}_a)$ of our class of shapes, the corresponding value of the Radon transform at $\boldsymbol{q}_a$ will evaluate to the length of the curve while the values at $\boldsymbol{q} \neq \boldsymbol{q}_a$ will quantify $s(\boldsymbol{q}_a) \cap s(\boldsymbol{q})$. In order to determine $\boldsymbol{q}_a$ we need to compute the transform over the allowed parameter space, searching for values that are close to the maximum expected value $|s(\boldsymbol{q}_a)|$, the number of pixels in our shape. We have already remarked that one of the advantages of an edge-based image representation is its conciseness: what about exploiting the sparseness of the edge map to locate shapes? While a single edge point does not by itself identify a unique shape, it provides supporting evidence for a limited set of shapes, i.e. those passing through it. Let us consider the class of circles of given radius $r$ and variable center $\boldsymbol{a}$. Each edge point $\boldsymbol{x}$ is compatible with circles whose centers are given by $\boldsymbol{x} + (r\cos\theta, r\sin\theta)$ for some $\theta$. As the radius is fixed, the dimension of our parameter space is 2, corresponding to the number of coordinates identifying circle centers. We then quantize the two-dimensional parameter space creating what is called an accumulator array. For each edge point $\boldsymbol{x}$, we increment all cells intersecting the circle $\boldsymbol{x} + (r\cos\theta, r\sin\theta)$, sometimes called the locus of the given shape at $\boldsymbol{x}$: $\{\boldsymbol{a} : \boldsymbol{x} \in s(\boldsymbol{q}_a, \boldsymbol{a})\}$. At the end of the process, the cell containing the center of the circle present in the image will have received a vote from all the elements of its boundary while the other cells will have received fewer contributions.

The final state of the accumulator array is the same as what we would have obtained by computing the Radon transform but the computation required is considerably reduced. The above informal procedure conveys the essence of what is called the Hough transform, which is essentially a reformulation of the Radon transform from a different computational perspective:

**Radon** each point $\boldsymbol{q}$ in parameter space integrates the evidence from all afferent cells in image space: local maxima in parameter space for which the values of the Radon transform are above threshold identify shape instances;

**Hough** each feature point (edge pixel) in image space $\boldsymbol{x}$ distributes its support to all compatible cells in parameter space: the cells with locally maximal values (above threshold) correspond to shape instances.

Figure 7.4 presents a graphical comparison of the two approaches. Let us now formalize the definition of the Hough transform. Let $\{\boldsymbol{x}_i\}_{i=1,\ldots,k}$ be a set of points representing features of interest, e.g. our edge pixels. We can associate with this feature set the image function $I(\boldsymbol{x})$:

$$I(\boldsymbol{x}) = \sum_{i=1}^{k} \delta(\boldsymbol{x} - \boldsymbol{x}_i). \tag{7.30}$$

Let $\{\pi_i\}$ be a set of cells whose union covers the parameter space without overlapping. We can introduce the discretized version $\mathrm{K}_\pi$ of the template associated with our family of shapes:

$$\mathrm{K}_\pi(\boldsymbol{x}, i) = \begin{cases} 1 & \forall(\boldsymbol{x}, i) : \{\boldsymbol{q} : \mathrm{K}(\boldsymbol{q}, \boldsymbol{x}) = 0\} \cap \pi_i \neq 0 \\ 0 & \text{otherwise.} \end{cases} \tag{7.31}$$

The (discrete) Hough transform can then be formally written as

$$\mathcal{H}(\pi_i) = \sum_{j=1}^{k} K_\pi(\boldsymbol{x}_j, i) = \int \mathrm{K}_\pi(\boldsymbol{x}, i) I(\boldsymbol{x})\, d\boldsymbol{x} \tag{7.32}$$

essentially recovering Equation 7.25 in a discrete context, and shape instances correspond to the following cell set:

$$\left\{\pi_j : \operatorname*{argmax}_{\pi \in \mathfrak{N}(\pi_j)} \mathcal{H}(\pi) \text{ and } \mathcal{H}(\pi_j) > \theta\right\} \tag{7.33}$$

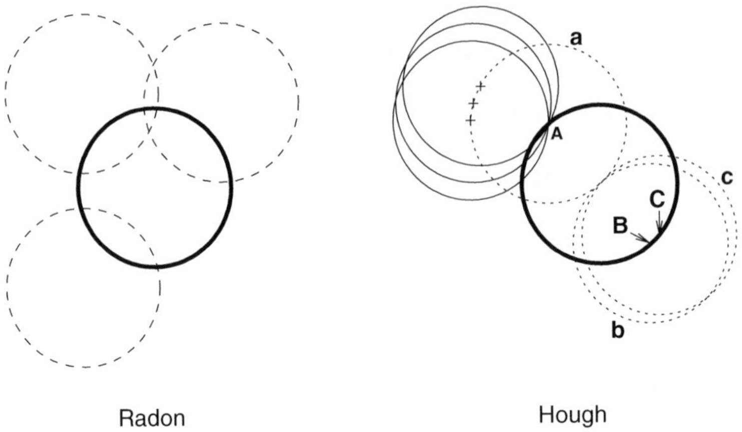


**Figure 7.4.** The Radon and Hough views of the shape matching problem. In the Radon approach (left), the supporting evidence for a shape with parameter $\boldsymbol{q}$ is collected by integrating over $s(\boldsymbol{q})$. In the Hough approach (right), each potentially supporting pixel (e.g. edge pixels $\boldsymbol{A}$, $\boldsymbol{B}$, $\boldsymbol{C}$) votes for all shapes to which it can potentially belong (all circles whose centers lie respectively on circles $\boldsymbol{a}$, $\boldsymbol{b}$, $\boldsymbol{c}$).

where $\mathfrak{N}(\pi_j)$ is a neighborhood of cell $\pi_j$ and $\theta$ a threshold corresponding to the expected counts from a valid shape instance. The detailed shape of the cells used in accumulator space is important: the definition presented in Equation 7.31 shows that, by changing the cell, the image space support of the kernel changes (see also Figure 7.5). Recalling that the kernel represents the template matched, we see that changing the shape of the cell amounts to changing the templates located by the Hough transform. The actual templates in image space vary not only when the parameterization is changed (say from the slope/intercept to the normal one for lines), but also within a single parameterization when the translational invariance of the continuous template might be lost in the discrete case.

In order to better appreciate the correspondence of the Hough transform with the idea of template matching and the reasons for its computational advantages, let us consider the case where the intrinsic shape parameters $\boldsymbol{q}_s$ are fixed and only the shape location changes (see Figure 7.6). This means that we restrict ourselves to translations of the shape considered and that the effective parameter space is the same as the image space, possibly represented at a different resolution. Let us work explicitly with the coordinates of the points belonging to a reference version of our template:

$$\boldsymbol{x}(\boldsymbol{q}_s) : \{\mathcal{K}_i(\boldsymbol{x}; \boldsymbol{q}) = 0\}_{i=1,\dots,(d-n)}. \tag{7.34}$$

A translation $\boldsymbol{\Delta}$ of these points identifies a translated version of our template:

$$\boldsymbol{y}(\boldsymbol{q}_s) = \boldsymbol{x}(\boldsymbol{q}_s) + \boldsymbol{\Delta}. \tag{7.35}$$

Let $\boldsymbol{b}$ be a point in the image plane. The set of all translated versions of our template passing through $\boldsymbol{b}$ for some $\boldsymbol{q}_s = \boldsymbol{q}'_s$ can be obtained in the following way:

$$\boldsymbol{y}(\boldsymbol{q}'_s) = \boldsymbol{b} = \boldsymbol{x}(\boldsymbol{q}'_s) + \boldsymbol{\Delta} \tag{7.36}$$

$$\boldsymbol{\Delta}(\boldsymbol{q}'_s, \boldsymbol{b}) = \boldsymbol{b} - \boldsymbol{x}(\boldsymbol{q}'_s) \tag{7.37}$$

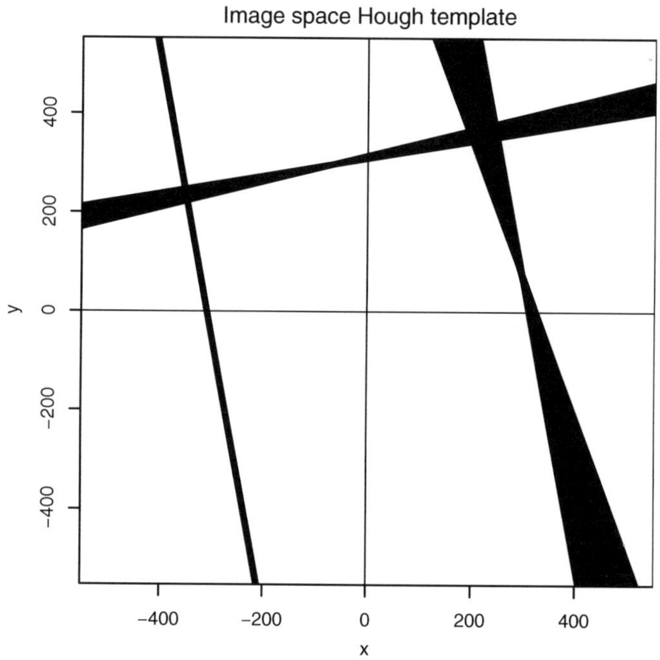


**Figure 7.5.** The shape of the cells defines the template in image space. This is very important as the equivariance of the transform to scaling and affine transformation depends on the shape of the cell. In the case of the normal parameterization, the transform is rotation invariant but not translation invariant if a rectangular cell is used. Invariance to translation is recovered if we use a line segment cell, the limit of the rectangular cell for $\Delta\theta \to 0$. The three examples in the figure are obtained with different angular resolutions of the rectangular cell: the thinner template corresponds to smaller $\Delta\theta$.

and, by replacing $\boldsymbol{x}(\boldsymbol{q}_s')$ by $\boldsymbol{x}(\boldsymbol{q}_s)$, we can obtain a translation of the reflection about the origin of $\boldsymbol{x}(\boldsymbol{q}_s)$. For a fixed value of $\boldsymbol{b}$, this can be plotted as a function of $\boldsymbol{q}_s$:

$$\boldsymbol{\Delta}(\boldsymbol{q}_s) = \boldsymbol{b} - \boldsymbol{x}(\boldsymbol{q}_s) \tag{7.38}$$

which is the equivalent of a matched filter (see Figure 7.6 and Equation 7.27). Referring back to the formal definition of the Hough transform, we observe that for each feature point $\boldsymbol{b}$ in the image, shape $\boldsymbol{\Delta}(\boldsymbol{q}_s)$ is considered and each of the accumulator cells it meets is incremented by 1. On the other side, the template matching approach considers all possible translations $\{\boldsymbol{\Delta}\}$ of the reference shape, computing for each of them the number of intersecting feature points. Let us now consider the situation from a dual point of view, namely that of accumulator space. If we focus on a point $\boldsymbol{\Delta}$ in accumulator space, we see from Equation 7.38 that it gets contributions from all image points $\boldsymbol{b}$ such that $\boldsymbol{b} - \boldsymbol{\Delta} = \boldsymbol{x}$ for some $\boldsymbol{x}$ belonging to the reference shape:

$$\mathcal{H}(\boldsymbol{\Delta}) = \sum_{\boldsymbol{b}} \mathrm{h}(\boldsymbol{b} - \boldsymbol{\Delta}) I(\boldsymbol{b}) \tag{7.39}$$

which we can rewrite as the convolution

$$\mathcal{H}(\boldsymbol{q}_x) = \sum_{\boldsymbol{x}} \mathrm{h}(\boldsymbol{x} - \boldsymbol{q}_x) I(\boldsymbol{x}). \tag{7.40}$$

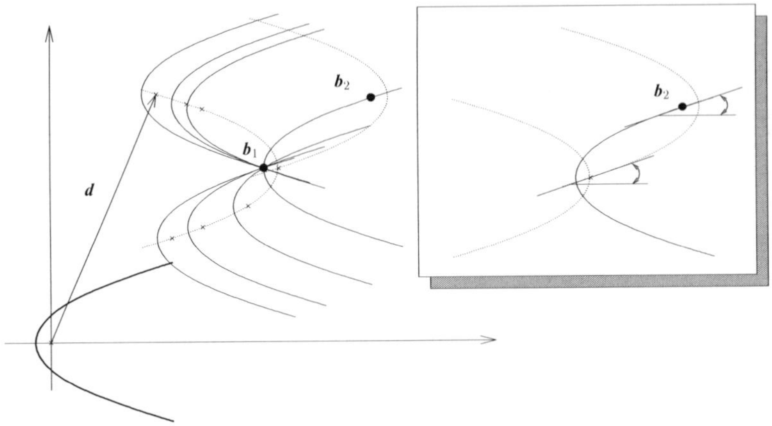


**Figure 7.6.** A pictorial presentation of the connections between the shape in image space and the corresponding template in accumulator space. In this case they coincide as we are only considering the translational parameter of the shape. Knowledge of curve orientation at edge point $\boldsymbol{b}_2$ allows us to identify a single corresponding point in parameter space, on the matched template that represents the shape locus.

A significant computational advantage presented by the Hough transform is that only feature points in the image are sources of computations: this must be confronted by the exhaustive scanning of image space characteristic of the standard template matching approach. An additional computational advantage derives from the analysis of the neighborhood of each image feature point. If a reliable estimate of the gradient direction exists at the edge point we may exploit its knowledge to limit the incrementation step in parameter space to those points that share the same orientation. As shown by the inset in Figure 7.6, the savings can be substantial: instead of updating the cells corresponding to the complete shape we may update a single point. Incorporating this type of knowledge in the standard template matching process is not possible. As Figure 7.6 shows, blindly relying on the gradient information to update a single cell in accumulator space may be dangerous: a small perturbation in gradient direction might result in a large displacement of the corresponding point along the curve. The correct procedure is then to take into account the uncertainty in the estimation of the direction and proceed to update all the points whose gradient falls within the uncertainty interval. As instances of the given shape appearing in the image are associated with peaks in accumulator space, we note that under realistic circumstances we cannot expect an impulse response as errors in the estimation of the parameters cause peak broadening. Failure to take into account the errors affecting the mapping from feature space to parameter space may result in no peaks at all: the voting from the feature points may be so dispersed that no accumulation results in the correct cells in parameter space. When errors are isotropic we can show that the procedure of distributing the contribution of a single feature point over multiple cells in parameter space is equivalent to a smoothing step in parameter space after a standard incrementation procedure has been carried out.

Let us focus on circle detection, and let us denote by $\Delta\phi(\boldsymbol{x})$ the uncertainty of angle $\phi(\boldsymbol{x})$ and by $\Delta(r)$ the uncertainty on the radius of the circles we are looking for. In order to cope with these uncertainties we increment all cells falling inside a square domain placed at the computed center $\pi_0$ and we weight each contribution according to a given point spread function $w$. The contents of the accumulator after considering the first point are

$$\mathcal{H}_s(\boldsymbol{q}) = w(\boldsymbol{q} - \boldsymbol{q}_0) \tag{7.41}$$

where $\boldsymbol{q} = (q_x, q_y, q_r)$ denotes the parameters of our circle and the subscript $s$ identifies smoothed accumulator space. If we denote by $C(\boldsymbol{q}_0)$ the number of points contributing to $\boldsymbol{q}_0$ we obtain

$$\mathcal{H}_s(\boldsymbol{q}) = C(\boldsymbol{q}_0) w(\boldsymbol{q} - \boldsymbol{q}_0) \tag{7.42}$$

and summing over all centers supported by image feature points

$$\mathcal{H}_s(\boldsymbol{q}) = \sum_{\boldsymbol{q}_0} C(\boldsymbol{q}_0) w(\boldsymbol{q} - \boldsymbol{q}_0). \tag{7.43}$$

Recalling that by definition $C(\boldsymbol{q}_0) = \mathcal{H}(\boldsymbol{q}_0)$, we may write

$$\mathcal{H}_s(\boldsymbol{q}) = \sum_{\boldsymbol{q}_0} \mathcal{H}(\boldsymbol{q}_0) w(\boldsymbol{q} - \boldsymbol{q}_0) \tag{7.44}$$

$$= \mathcal{H} * w \tag{7.45}$$

which is the convolution (in parameter space) with the point spread function used to weight the contribution of feature points. The derivation also shows that the cure for errors is formally equivalent to the employment of a band-limiting kernel $w$ applied to the original image (this follows from Equation 7.40 and the properties of convolution). While the spreading of the evidence from a single pixel over a region reduces the effect of errors, oversampling of the parameter space may lead to the spreading of the evidence from multiple feature points belonging to the same shape. The result in this case would be the broadening of the peak that localizes the shape in the image.

Convolution filtering in parameter space can also be used to enhance the peaks identifying the shapes. Let us consider the task of detecting thin lines in images that also contain thick lines. Using the normal line parameterization, we assume a high angular resolution (as an approximation to the line segment cell) with the abscissas representing the $\rho$ parameter with a resolution of $\Delta\rho$. The convolution with the filter $[-1, 2, -1]_\rho$

$$\mathcal{H}(\rho, \theta) * [-1, 2, -1]_\rho \tag{7.46}$$

corresponds to an image space template with a positive central bar surrounded by two negatively weighted side bars. As the value of a cell in parameter space is the weighted sum of all feature points inside the template, we see that the parameter space contribution of a line whose thickness is greater than $3\Delta\rho$ is lower than that of a $\Delta\rho$ thick line.

We pointed out that the Hough transform presents computational advantages over the Radon transform and straightforward template matching approaches. Let us now consider the issue in some detail for the case of line detection with the normal parameterization of Equation 7.22. Let us employ a uniform quantization for the two parameters $\theta$ and $\rho$ using $n_\theta$ and $n_\rho$ cells for the ranges $[0, \pi)$ and $[-R, R]$, $R$ being the maximum possible value of $\rho$ due to the finite extent of images. For each edge point in the image we compute $n_\theta$ values for $\rho$ to draw the sinusoid in accumulator space (neglecting the overhead of drawing lines connecting the points lest some of the sinusoids be drawn unconnected). Given $k$ edge points in the image, the computations required are $kn_\theta$ after which the $n_\theta n_\rho$ cells must be inspected to locate points of high support. The complexity, therefore, increases linearly with the number of edge points instead of quadratically as in the exhaustive procedure of considering all $k(k-1)/2$ pairs of points to compute all possible lines. Note, however, that this computational advantage is closely related to the fact that a discrete parameter space is employed. In fact, the Hough transform considers for each edge point only $n_\theta$ distinct lines instead of the complete set of $k-1$ that would be considered in the exhaustive approach. Further computational savings derive from knowledge of the direction of the gradient. Perfect knowledge of such information would reduce $kn_\theta$ to 1: a single cell in accumulator space would need updating.

A way to improve the efficiency of the transform is to employ an adaptive coarse-to-fine discretization of parameter space. The formal definition of the transform shows that the effect of a coarse discretization of parameter space corresponds to the sum of the corresponding cells at a finer resolution. As the storage needed for the accumulator scales exponentially with the resolution of discretization of the parameter axes, significant storage savings can be achieved by performing the transform on a coarse grid and progressively refining the discretization of parameter space only at those cells whose counts exceed an appropriate threshold. Let the linear number of cells per parameter axis at the highest resolution be $n_f$ and that at the coarse resolution $n_c$. If we assume that a single peak must be located, the savings in required storage are of the order of $O((n_f/n_c)^q)$ where $q$ is the dimension of parameter space and only $O(\log n_f \log n_c)$ refinement iterations are needed to get to the maximum prescribed resolution. The savings are extended to the step of locating the maxima in accumulator space, as only the cells used need to be scanned. However, the incrementing procedure itself does not necessarily improve. As an example, if gradient direction information is enough to restrict the number of cells to be updated to a finite number of points when the continuous parameter space is considered, no improvements are to be expected.

Let us conclude the treatment of the Hough transform by establishing a link to the approach presented in Chapter 3, considering the problem of line detection in an image corrupted by additive Gaussian noise of known variance $\sigma_n^2$. At each point $(\rho_i, \theta_j)$ in accumulator space we face a two-hypothesis problem: $H_0$, the value of the accumulator is due to noise; and $H_1$, the value is due to the presence of a line. The optimal choice can be made by considering the likelihood ratio. Due to the fact that the adopted discretization of parameter space may result in image space templates that are significantly larger than the line we want to detect, we are actually facing a detection problem even if we restrict consideration to a single cell in parameter space. Let us denote by $\pi_{ij}$ the cell in parameter space whose center is $(\rho_i, \theta_j)$, and $\mathrm{k}((\rho, \theta); \boldsymbol{x})$ the subset of $\mathrm{K}(\pi_{\rho_i \theta_j}; \boldsymbol{x})$ representative of a single line. Let us further assume that for $(\rho, \theta) \in \pi_{ij}$ the probability $p(\rho, \theta) = p(\rho_i, \theta_j) = p_{ij}$. The likelihood ratio is then

$$p(I \mid H_0) = \exp\left[-\frac{1}{2\sigma_n^2} \sum_{i,j \in \mathrm{K}} I_{ij}^2\right] \tag{7.47}$$

$$p(I \mid H_1(\rho, \theta)) = \exp\left[-\frac{1}{2\sigma_n^2}\left(\sum_{i,j \in \mathrm{k}} (I_{ij} - \mathrm{S}_{ij}(\rho, \theta))^2 + \sum_{i,j \in \mathrm{K}-\mathrm{k}} I_{ij}^2\right)\right] \tag{7.48}$$

$$\mathrm{L}(\pi_{ij}) = \left[\iint_{\pi_{ij}} p(I \mid H_1(\rho, \theta)) p_{ij} \, d\theta \, d\rho\right] \Big/ p(I \mid H_0) \tag{7.49}$$

where $\mathrm{S}_{ij}(\rho, \theta)$ represents a line in image space whose pixels have value 1. The summation in $p(I \mid H_1(\rho, \theta))$ is split into two terms: the first one over the pixels belonging to the line,

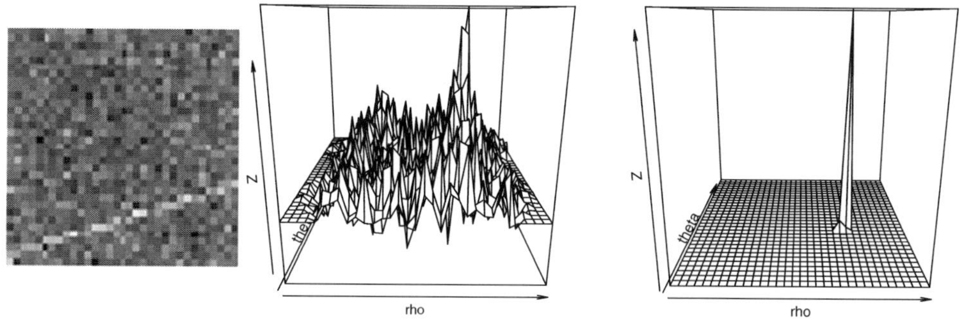


**Figure 7.7.** The Hough transform in noisy images benefits from statistical signal processing insights. The likelihood ratio map (right) based on the Hough transform (middle) provides a better signal to noise ratio for the detection of lines in noisy images.

and the second one over the complement, as the line we want to detect may be thinner than the image space template associated with parameter space. This means that

$$\sum_{i,j\in k} (I_{ij} - S_{ij}(\rho, \theta))^2 = \sum_{i,j\in k} I_{ij}^2 + \sum_{i,j\in k} S_{ij}^2 - 2 \sum_{i,j\in k} I_{ij} S_{ij} \tag{7.50}$$

$$\approx \sum_{i,j\in k} I_{ij}^2 + \sum_{i,j\in k} S_{ij}^2 - 2 \sum_{i,j\in K} I_{ij} \tag{7.51}$$

where we have exploited the fact that noise is zero on average to extend the summation from k to K, the complete image space cell template. Exploiting the normalization of $p(\rho, \theta)$ and the fact that the length of the lines falling within the image space template have approximately the same length $\|S\|$, we obtain the following approximation for the likelihood ratio:

$$L(\pi_{ij}) \approx \exp\left[-\frac{1}{\sigma_n^2}\left(\sum_{i,j\in K} I_{ij} - \frac{1}{2}\|S\|\right)\right]. \tag{7.52}$$

The result shows that the likelihood ratio can be derived simply from the value of the Hough accumulator ($\sum_{i,j\in K} I_{ij}$), by subtracting the expected line length ($\|S\|$) and performing a nonlinear, exponential mapping exploiting knowledge of the noise distribution. As shown in Figure 7.7, the probabilistic approach greatly enhances the signal to noise ratio of the transform.

Let us now apply the machinery of the Hough transform to two different detection tasks related to face processing: face detection and iris localization. While not a perfect ellipse, the outline of a face may often be approximated with good accuracy by an elliptical contour. Deviations from a perfect ellipse can be considered as a form of geometrical noise or estimation errors and can be accommodated by the Hough transform as such using the approach outlined in Equation 7.41. We also assume the eccentricity of the ellipse is fixed so that the only real parameter is the location of the ellipse. In order to apply the transform in an efficient way, we need to exploit the information provided by the direction of the gradient to focus the update in parameter space. Let us consider the parametric definition of an ellipse whose axes are parallel to the coordinate ones (an upright face) and located at the origin:

$$x = a \cos\theta \tag{7.53}$$

$$y = b \sin\theta. \tag{7.54}$$

The tangent at the ellipse is characterized by the angle $\varphi$:

$$\tan\varphi = \frac{dy}{dx} = \frac{dy}{d\theta}\left(\frac{dx}{d\theta}\right)^{-1} \tag{7.55}$$

$$= -\frac{b}{a\tan\theta} \tag{7.56}$$

so that

$$\theta_\varphi = \arctan\left(-\frac{b}{a\tan\varphi}\right) \tag{7.57}$$

which allows us to map back a feature point $(x_0, y_0)$ in image space to the centers of the two compatible shapes that themselves lie on an ellipse with the same parameters but whose center is located at the feature point. The coordinates of the two compatible centers are given by

$$(x_1, y_1) = (x_0 - a\cos\theta_\varphi, y_0 - b\sin\theta_\varphi) \tag{7.58}$$

$$(x_2, y_2) = (x_0 + a\cos\theta_\varphi, y_0 + b\sin\theta_\varphi). \tag{7.59}$$

As previously noted, we need to consider a range of possible ellipse sizes and the uncertainty in the estimation of the gradient direction $\varphi$ at the feature point. These uncertainties also serve the purpose of accommodating the deviations of a face outline from the hypothesized elliptical shape and from the assumed upright pose. We have seen that filterings in parameter and image space are closely related and can be chosen to improve the response of the transform to the specific shape considered. In this case we can choose a weighted incrementation step: each image feature point contributes to accumulator space an amount $w_\theta$ that is inversely proportional to its angular deviation from the computed $\theta_\varphi$ scaled by an appropriate $\Delta\theta$:

$$w_\theta = 1 - \frac{2\arccos(\hat{\boldsymbol{\theta}} \cdot \hat{\boldsymbol{\theta}}_\varphi)}{\Delta\theta}. \tag{7.60}$$

If we further assume that the ellipse axes $a$, $b$ are such that $a \in [a_1, a_2]$ and $b \in [b_1, b_2]$ we may restrict the update in accumulator space to the shaded area of Figure 7.8. Let us now consider the problem of detecting the center of the eyes, i.e. the center of the pupil. While the parameter space of a translating shape is two dimensional, this space can sometimes be decomposed into two one-dimensional spaces, one for each translating dimension, by exploiting information on gradient direction. The detection of circles of known radius $r$

$$(x - x_0)^2 + (y - y_0)^2 = r^2 \tag{7.61}$$

is such a case. Differentiation with respect to $x$ leads to

$$\frac{dy}{dx} = \frac{(x - x_0)}{(y - y_0)} \tag{7.62}$$

and

$$x_0 = x \pm \frac{r}{\sqrt{1 + (dx/dy)^2}} \tag{7.63}$$

$$y_0 = y \pm \frac{r}{\sqrt{1 + (dy/dx)^2}}. \tag{7.64}$$

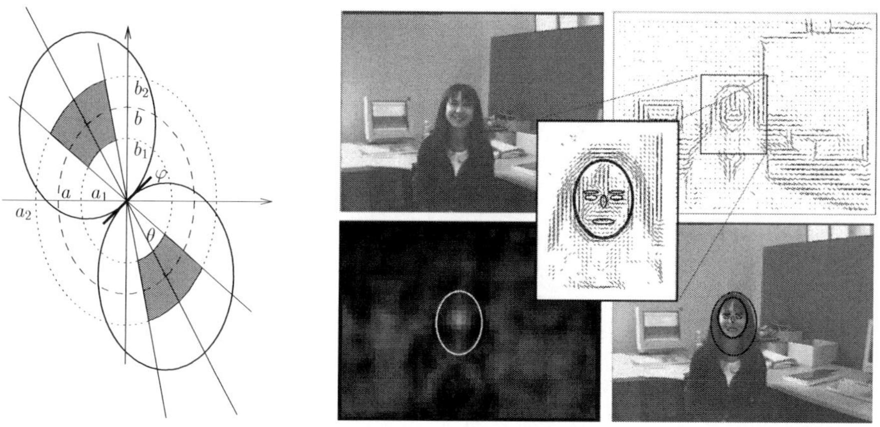


**Figure 7.8.** The outline of a frontal face can be approximated in many situations by an ellipse. A modification of the circle Hough transform can be used to spot ellipses. Non-constant ellipse size, deviation from exact elliptical contour, and edge direction uncertainty require area updating in parameter space. Source: adapted from Maio and Maltoni (2000).

We can employ two one-dimensional accumulator spaces: $\mathcal{H}_x^r$ for estimating $x_0$ and $\mathcal{H}_y^r$ for $y_0$. A major feature of the iris is its being enclosed by two concentric boundaries: the outer sclera/iris circle and the inner pupil limit. We may then accumulate evidence for all circles with $r \in [r_m, r_M]$

$$\mathcal{H}_x = \sum_{r=r_m}^{r_M} \mathcal{H}_x^r \tag{7.65}$$

$$\mathcal{H}_y = \sum_{r=r_m}^{r_M} \mathcal{H}_y^r \tag{7.66}$$

so that the location of the center of the eye is finally obtained as $(x_0, y_0)$ where

$$x_0 = \underset{x}{\operatorname{argmax}} (\mathcal{H}_x) \tag{7.67}$$

$$y_0 = \underset{y}{\operatorname{argmax}} (\mathcal{H}_y). \tag{7.68}$$

An example is shown in Figure 7.9.

## 7.4. The Generalized Hough Transform

The shape examples considered so far, i.e. lines and circles, are characterized by the availability of an explicit, analytic expression that is exploited to determine which cells in parameter space should be updated. We have also seen that knowledge of directional information is key to the efficiency of the Hough transform. Can the procedure be extended to arbitrary shapes, for which no analytic expression is available, preserving the advantages of the approach? The answer is affirmative and the solution is easily obtained starting from the detection of circles with given radius $\boldsymbol{r}_0$. The shape parameters are then limited to the two-dimensional shape location coordinates. Let us consider filled circles, so that for each point

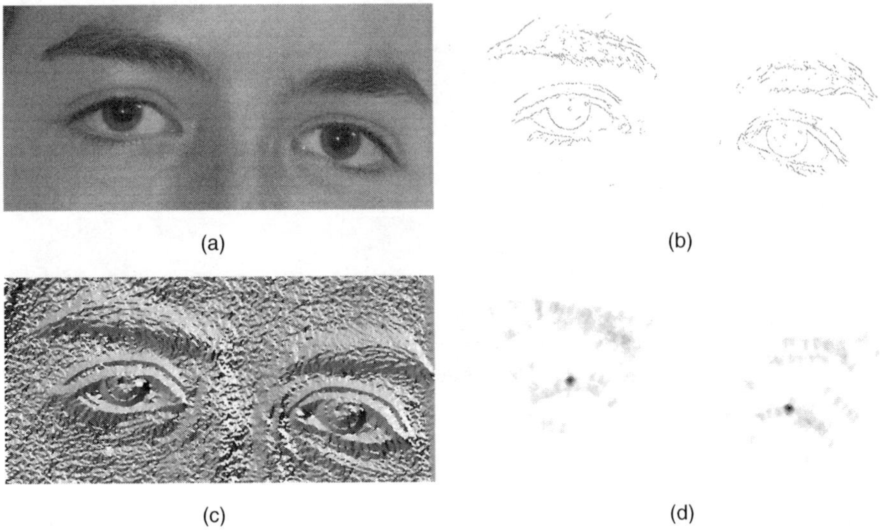


**Figure 7.9.** Irises can be detected using an Hough transform whose target shape is a dark circle over a light background (a). After detecting the most significant edges (b), we can exploit the associated directional information (c) to perform efficiently a circle Hough transform (d) accumulating evidence from multiple radii.

**Intermezzo 7.3.** Robustness properties of the Hough transform

The Hough transform is an example of a fixed-band estimator. These estimators maximize the number of points within a given band from the value of the parameter to be estimated. In the case of the Hough transform, the band corresponds to a cell $\pi_i$ in parameter space, whose associated shape parameter $\boldsymbol{q}$ usually corresponds to its center. The band identifies the so-called inlier fraction, i.e. the points belonging to the cell, and the operating way of the Hough transform (see Equation 7.33) corresponds to the maximization of such a number.

Any perturbation affecting a number of points smaller than the number of inlier points has no effect on the correct localization of the shape (or even shapes, if multiple maxima are considered). This type of robustness makes it possible to detect shapes represented by less than half of the data.

It is possible to view the Hough transform as an M-estimator with a discontinuous $\rho(\cdot)$ function

$$\rho(\boldsymbol{q} - \boldsymbol{q}_0) = \begin{cases} 0 & \boldsymbol{q} \in \pi(\boldsymbol{q}_0) \\ 1 & \boldsymbol{q} \notin \pi(\boldsymbol{q}_0). \end{cases} \tag{7.69}$$

The Hough transform can also be used in a (multiple) regression context and a rigorous statistical analysis of its properties shows that it has a cube root convergence rate and a 50% breakdown point (Goldenshluger and Zeevi 2004; Stewart 1997, 1999).

on the boundary we know the direction of the local normal (see Figure 7.10). Neglecting errors, we only need to update a single point in accumulator space for each image feature point:

$$\boldsymbol{q} = \boldsymbol{x} + r_0\hat{\boldsymbol{n}}(\boldsymbol{x}) \tag{7.70}$$

and the normal versor $\hat{\boldsymbol{n}}(\boldsymbol{x})$ points from the boundary to the center. In the case of a general shape, of fixed scale and orientation, such as the one represented in Figure 7.10, the vector $\boldsymbol{r} = \boldsymbol{q} - \boldsymbol{x}$ pointing from the boundary point to the shape center will not be necessarily aligned to $\hat{\boldsymbol{n}}(\boldsymbol{x})$ and its size will not be constant. The variability of $\boldsymbol{r}$ is best captured by means of a

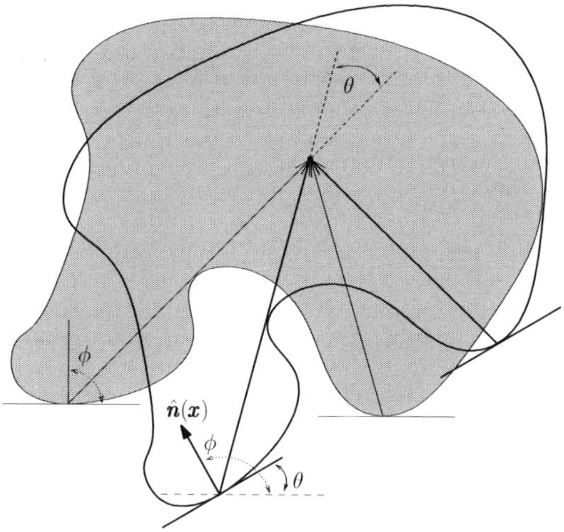


**Figure 7.10.** The generalized Hough transform represents a generic shape by means of an $R$-table which permits efficient update of accumulator cells using gradient information.

**Table 7.1.** The $R$-table format, where $\partial$ represents the boundary of the shape. The set of $\boldsymbol{r}$ for each angular value may contain multiple values.

| $i$ | $\phi_i$ | $R_{\phi_i}$ |
|---|---|---|
| 0 | 0 | $\{\boldsymbol{r}_{0j}\} = \{\boldsymbol{r} \mid \boldsymbol{q} - \boldsymbol{r} = \boldsymbol{x},\, \boldsymbol{x} \in \partial,\, \phi(\boldsymbol{x}) = 0\}$ |
| 1 | $\Delta\phi$ | $\{\boldsymbol{r}_{1j}\} = \{\boldsymbol{r} \mid \boldsymbol{q} - \boldsymbol{r} = \boldsymbol{x},\, \boldsymbol{x} \in \partial,\, \phi(\boldsymbol{x}) = \Delta\phi\}$ |
| 2 | $2\Delta\phi$ | $\{\boldsymbol{r}_{2j}\} = \{\boldsymbol{r} \mid \boldsymbol{q} - \boldsymbol{r} = \boldsymbol{x},\, \boldsymbol{x} \in \partial,\, \phi(\boldsymbol{x}) = 2\Delta\phi\}$ |
| $\vdots$ | $\vdots$ | $\vdots$ |

table, called the $R$-table, explicitly giving $\boldsymbol{r}$ for each possible orientation $\hat{\boldsymbol{n}}$. The construction of the $R$-table is based on the following steps:

1. Choose a reference point $\boldsymbol{O}$ for the shape.
2. Compute the gradient direction $\phi(\boldsymbol{x})$ for each point $\boldsymbol{x}$ on the shape boundary and compute $\boldsymbol{r} = \boldsymbol{O} - \boldsymbol{x}$.
3. Store $\boldsymbol{r}$ as a function of $\phi$ in the $R$-table.

A few remarks are in order here:

- each row of the $R$-table is characterized by a single value of $\phi$ but may have multiple $\boldsymbol{r}$ values;
- the choice of the point representing the shape origin may affect the precision of the localization (see Figure 7.11).

The approach can be generalized to shapes of varying scale $s$ and orientation $\theta$. Rotation can be easily accommodated by changing the index in the table from $\phi$ to $[(\phi + \theta) \mod 2\pi]$

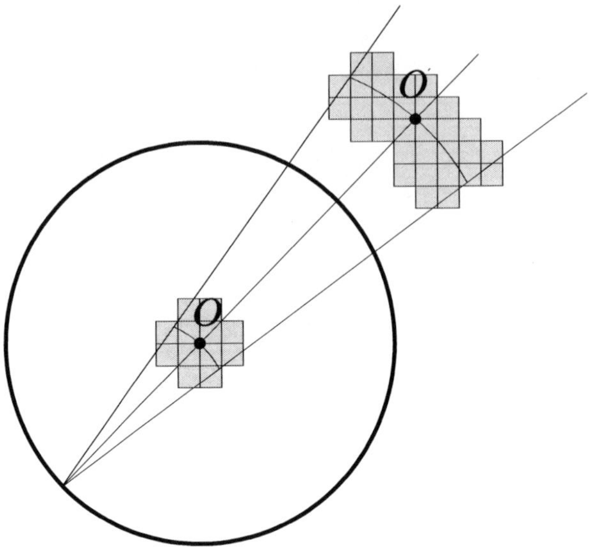


**Figure 7.11.** The choice of the reference point used in the generalized Hough transform may have a significant impact on the accuracy of localization. The reason is that errors in position deriving from errors in the estimation of gradient direction depend linearly on the distance from the reference point. The figure shows the effect of angular uncertainty for two different shape origins.

and then rotating all vectors $\boldsymbol{r}$ by $\theta$ (see Figure 7.10). Shape contrast reversal can also be easily accommodated by transforming $\phi$ to $[(\phi + \pi) \mod 2\pi]$ without any modification of the $\boldsymbol{r}$ values, while scaling can be taken care of by scaling the $\boldsymbol{r}$ values. So far we have applied the Hough transform, both standard and generalized, to the detection of a single, possibly parameterized, shape. It is natural to ask whether a population of related shapes can be detected by the same means. We saw in Chapter 6 that it is sometimes possible to pack multiple templates into a single one. The description of shapes by means of the $R$-table supports a similar strategy. Given a set of shapes and a discretization $\Delta\phi$, we may construct a cumulative $R$-table by summing all the $R$-tables of the single shapes (Table 7.1). It may happen that the sets of points for a given direction have non-null intersections for some of the shapes:

$$\exists\, j, k : \boldsymbol{r}_{\alpha ij} = \boldsymbol{r}_{\beta ik} \quad \alpha \neq \beta \tag{7.71}$$

where the Greek index identifies the shapes. Such entries should not be collapsed to a single element in the final $R$-table: as the corresponding contribution is more likely to happen, the point should contribute with the larger weight derived from its multiple instances. We have already noted that the computation of the shape parameters providing the coordinates in parameter space is affected by errors and that a solution to the problem is to increment all cells within an appropriate neighborhood of the estimated parameter. This naturally extends to use of the $R$-table where the uncertainty region may take care of both estimation errors and shape variability at the same time. The vectors of each set $R_{\phi_i} = \bigcup_{\alpha k} \{\boldsymbol{r}_{\alpha ik}\}$ can be considered as a sample of a random distribution and the values to be used when updating the accumulator can be estimated using adaptive kernel methods. The resulting weighting kernels can be stored in the $R$-table and used in place of the original vector sets.

## 7.5. Bibliographical Remarks

The literature on edge detection is extremely vast. Among the many proposed approaches we chose the one by Elder and Zucker (1998), further elaborated in Elder (1999). These papers contain the detailed computations whose results are presented in Section 7.1, and also include an extensive set of references that the reader interested in edge detection will find useful.

The Radon transform was originally introduced by Radon (1917) and the Hough transform by Hough (1962). The connection between the two was recognized by several authors starting with Deans (1981). The approach followed in the chapter to connect the two transforms is based on the papers by Luengo Hendriks *et al.* (2005). The notion of sampling invariant computations presented by Verbeek (1985) is also of interest. Besides the reviews on the Hough transform (Illingworth and Kittler 1988; Leavers 1993), the early paper of Duda and Hart (1972) remains required reading. The connection of the Hough transform to template matching was remarked upon by Stockman and Agrawala (1977) and Sklansky (1978). The details of the transform in the presence of noise and its relation to signal processing have been investigated by Sklansky (1978) and Hunt *et al.* (1988) on which our treatment is based (see also Hunt *et al.* (1990)). A detailed analysis of the effects of cell shapes on localization is presented by Princen *et al.* (1992) while quantization effects are considered by Lam *et al.* (1994) and discretization errors, including parameter space oversampling, are described by van Veen and Groen (1981). The generalized Hough transform was introduced by Ballard (1981) who also addresses the equivalence of parameter space smoothing and the multicell incrementation strategy to cope with errors. The generalization of the $R$-table approach to multiple shapes is considered by Garrido and Perez De La Blanca (1998) and Ecabert and Thiran (2004).

An interesting statistical study of the Hough transform presenting results on its robustness, convergence rates, and consistency can be found in Goldenshluger and Zeevi (2004).

This chapter focused on the Hough transform as a 1-to-$n$ mapping, whereby a single feature point votes for $n$ points (a curve) in parameter space. While we have seen that 1-to-1 or 1-to-$m$ mappings with $m < n$ can be achieved by exploiting structural edge information, a different perspective is offered by the $n$-to-1 methods which use $n$ image feature points to solve the shape defining constraints and update a single parameter space point providing infinite range and high resolution. The randomized Hough transform, first proposed by Xu *et al.* (1990), is an efficient way of performing these computations. Extensions of the approach are reviewed in the paper by Kalviainen *et al.* (1995) while the application of $n$-to-1 methods to shapes under affine transformations are investigated in the paper by Aguado *et al.* (2002).

The Hough transform has been used in face processing tasks mainly to detect eyes and as a first screening step for face detection. The description in the chapter is based on the papers by Benn *et al.* (1997) and Maio and Maltoni (2000).

## References

Aguado A, Montiel E and Nixon M 2002 Invariant characterisation of the Hough transform for pose estimation of arbitrary shapes. *Pattern Recognition* **35**, 1083–1097.

Ballard D 1981 Generalizing the Hough transform to detect arbitrary shapes. *Pattern Recognition* **13**, 111–122.

Benn D, Nixon M and Carter J 1997 Robust eye centre extraction using the Hough transform. *Proceedings of the 1st International Conference on Audio- and Video-Based Biometric Person Authentication*, pp. 3–9.

Deans S 1981 Hough transform from the Radon transform. *IEEE Transactions on Pattern Analysis and Machine Intelligence* **3**, 185–188.

Duda R and Hart P 1972 Use of the Hough transform to detect lines and curves in pictures. *Communications of the ACM* **15**, 11–15.

Ecabert O and Thiran J 2004 Adaptive Hough transform for the detection of natural shapes under weak affine transformations. *Pattern Recognition Letters* **25**, 1411–1419.

Elder J 1999 Are edges incomplete? *International Journal of Computer Vision* **34**, 97–122.

Elder J and Zucker S 1998 Local scale control for edge detection and blur estimation. *IEEE Transactions on Pattern Analysis and Machine Intelligence* **20**, 699–716.

Garrido A and Perez De La Blanca N 1998 Physically-based active shape models initialization and optimization. *Pattern Recognition* **31**, 1003–1017.

Goldenshluger A and Zeevi A 2004 The Hough transform estimator. *Annals of Statistics* **32**, 1908–1932.

Hough P 1962 Method and means for recognizing complex patterns. US Patent No. 3069654.

Hunt D, Nolte L and Reibman A 1990 Hough transform and signal detection theory performance for images with additive noise. *Computer Vision, Graphics and Image Processing* **52**, 386–401.

Hunt D, Nolte L and Ruedger W 1988 Performance of the Hough transform and its relationship to statistical signal detection theory. *Computer Vision, Graphics and Image Processing* **43**, 221–238.

Illingworth J and Kittler J 1988 A survey of the Hough transform. *Computer Vision, Graphics and Image Processing* **44**, 87–116.

Kalviainen H, Hirvonen P, Xu L and Oja E 1995 Probabilistic and non-probabilistic Hough transforms: overview and comparisons. *Image and Vision Computing* **13**, 239–252.

Lam W, Lam L, Yuen K and Leung D 1994 An analysis on quantizing the Hough space. *Pattern Recognition Letters* **15**, 1127–1135.

Leavers V 1993 Which Hough transform? *CVGIP: Image Understanding* **58**, 250–264.

Luengo Hendriks C, van Ginkel M, Verbeek P and van Vliet L 2005 The generalized Radon transform: sampling, accuracy and memory considerations. *Pattern Recognition* **38**, 2494–2505.

Maio D and Maltoni D 2000 Real-time face location on gray-scale static images. *Pattern Recognition* **33**, 1525–1539.

Princen J, Illingworth J and Kittler J 1992 A formal definition of the Hough transform: properties and relationships. *Journal of Mathematical Imaging and Vision* **1**, 153–168.

Radon J 1917 Über die Bestimmung von Funktionen durch ihre Integralwerte längs gewisser Mannigfaltigkeiten. *Berichte über die Verhandlungen der Sächsischen Akademie* **69**, 262–277.

Sklansky J 1978 On the Hough technique for curve detection. *IEEE Transactions on Computers* **C-27**, 923–926.

Stewart C 1997 Bias in robust estimation caused by discontinuities and multiple structures. *IEEE Transactions on Pattern Analysis and Machine Intelligence* **19**, 818–833.

Stewart C 1999 Robust parameter estimation in computer vision. *SIAM Review* **41**, 513–537.

Stockman G and Agrawala A 1977 Equivalence of Hough curve detection to template matching. *Communications of the ACM* **20**, 820–822.

van Veen T and Groen F 1981 Discretization errors in the Hough transform. *Pattern Recognition* **14**, 137–145.

Verbeek P 1985 A class of sampling-error free measures in oversampled band-limited images. *Pattern Recognition Letters* **3**, 287–292.

Xu L, Oja E and Kultanen P 1990 A new curve detection method: randomized Hough transform (RHT). *Pattern Recognition Letters* **11**, 331–338.

# 8 LOW-DIMENSIONALITY REPRESENTATIONS AND MATCHING

Will this description satisfy him?

*Antony and Cleopatra*
WILLIAM SHAKESPEARE

The representation of images of even moderate resolution requires a significant amount of numeric data, usually one value for monochromatic images, three values for color images, per pixel if the usual array-based method is used. This chapter investigates the possibility of alternative ways to represent iconic data so that a large variety of images can be faithfully described using vectors of reduced dimensionality. Besides significant storage savings, these approaches provide significant benefits to template detection and recognition algorithms, improving their efficiency and effectiveness.

## 8.1. Principal Components

In the previous chapters we appreciated the fact that the computational complexity of template matching is related to the resolution at which images are represented. Finding an alternative representation of images providing (nearly) the same accuracy of representation while at the same time requiring less data is then a natural goal. A convenient and effective solution exists if we do not require that the increased representational efficiency be general, but limit ourselves to images whose contents are visually similar. There are several important side effects of representing data more parsimoniously:

1. A classifier working on them often has fewer parameters to estimate.
2. Fewer samples are needed to represent the associated probability distribution to a good approximation.
3. Pattern matching can be performed with higher speed as less data must be worked on.

All these advantages are related to the fact that the volume of a data space increases exponentially with the number of dimensions: if we fix the resolution with which we want to describe our data distribution, the number of samples also increases exponentially, the so-called curse of dimensionality. An important example from Chapter 3 is the Bayes classifier for normally distributed data that needs to estimate the mean vector and the covariance matrix

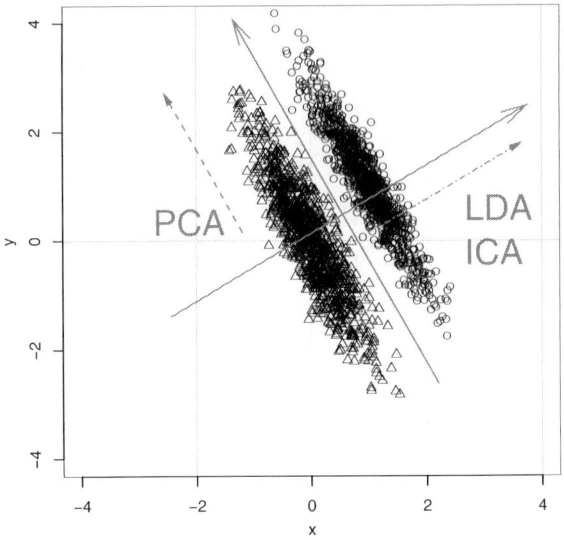


**Figure 8.1.** Two different ways are shown of choosing a new representation basis: by maximizing the projected variance (principal component analysis or PCA) or by maximizing the discriminability of the different classes (linear discriminant analysis or LDA). The direction chosen in the second case also represents the direction of maximal non-Gaussianity and would therefore be chosen by independent component analysis (ICA), another criterion discussed in this chapter.

from the available sample: the lower the dimensionality of the pattern space, the smaller the number of parameters to be estimated.

Let us consider face images: even if they (usually) require thousands of pixels for their representation, they do not fill the associated high-dimensional space which contains all possible images. It is not unreasonable to suppose that they all reside in a limited region of the complete multidimensional space. In order to devise an algorithm to find such a subspace, we need a quantitative criterion by which the relative merits of the candidate subspaces can be gauged. A convenient choice is to search for a linear subspace which provides the projection minimizing the squared residual error. The solution to the problem is given by principal component analysis (Figure 8.1), also known as Karhunen–Loeve transform. As we will see, the solution is very convenient in that it provides a sequence of linear spaces providing decreasing residual error, each of them including those with higher residual error.

We are now ready to formalize the problem and to find the solution by exploiting a few linear algebra theorems. Let $X \in \mathbb{R}^{M\times N}$ be a matrix whose columns represents the $N$ samples $\boldsymbol{x}_i \in \mathbb{R}^M$ from the class of patterns, e.g. face images, whose representation we want to optimize. Let us now compute the sample average of our patterns

$$\bar{\boldsymbol{x}} = \frac{1}{N}\sum_{i=1}^{N} \boldsymbol{x}_i, \tag{8.1}$$

and let $\boldsymbol{x}'_i = \boldsymbol{x}_i - \bar{\boldsymbol{x}}$. The unbiased sample covariance matrix $\Sigma$ can be expressed as

$$\Sigma = \frac{1}{N-1}\sum_{i=1}^{N} \boldsymbol{x}'_i \boldsymbol{x}'^T_i = \frac{1}{N-1} X' X'^T \tag{8.2}$$

where the primed matrices are built from the centered vectors. A fundamental result, known as (reduced) singular value decomposition, states that any real matrix belonging to $\mathbb{R}^{M\times N}$, with $M > N$, can be decomposed into the product of three matrices:

$$X' = UDV^T \tag{8.3}$$

where $U \in \mathbb{R}^{M\times N}$, $D \in \mathbb{R}^{N\times N}$ is a diagonal matrix, and $V \in \mathbb{R}^{N\times N}$. Both $U$ and $V$ have orthonormal columns. From

$$X'X'^T = (UDV^T)(UDV^T)^T = UD^2U^T \tag{8.4}$$

we have, exploiting the orthonormality of $U$,

$$(X'X'^T)U = (N-1)\Sigma U = UD^2 \tag{8.5}$$

so that the columns of $U$ are eigenvectors of $\Sigma$. The space spanned by the eigenvectors is called the eigenspace. As the columns of $U$ are orthonormal, the transformation

$$X'' = U^T X' \tag{8.6}$$

is actually a rotation and the covariance matrix of the rotated vectors is diagonal

$$\Sigma' = (U^T X')(U^T X')^T = D^2. \tag{8.7}$$

By Equation 8.6 each column of $X''$ contains the projections of the centered samples $X'$ on the eigenvectors of $\Sigma$, i.e. the columns of $U$. The values of the projections are called the principal components of the (centered) vectors, while the columns of $U$ are also dubbed principal directions. As the transformation described by Equation 8.6 preserves the norm of the vectors, we may consider our representational criterion in the rotated space (see Figure 8.2). Let us sort the eigenvectors by decreasing eigenvalues so that $\lambda_1^2 \geq \lambda_2^2 \geq \cdots$ and let us sort the columns of $U$ accordingly. The diagonal covariance matrix gives, for each of the coordinates, the corresponding average squared value

$$(N-1)\Sigma_{ii} = \sum_{q=1}^{N} (x''_{q,i})^2 = \lambda_i^2 \tag{8.8}$$

so that the average approximation error that arises if we use only the first $k$ coordinates is

$$\Delta_k = \sum_{i=k+1}^{N} \lambda_i^2 \tag{8.9}$$

and is actually the minimal possible value given $k$, due to the fact that $\lambda_1^2 \geq \lambda_2^2 \geq \cdots$ (see also Figure 8.3). If we choose $k$ so that $\sum_{i=1}^{k} \lambda_i^2 \gg \Delta_k$, the linear space spanned by the principal directions allows us to describe our data well and we may neglect the complementary space of dimension $N-k$. Under the assumption that $M > N$ the matrix $X'^T X' \in \mathbb{R}^{N\times N}$ may be significantly smaller than $X'X'^T \in \mathbb{R}^{M\times M}$. Exploiting the singular value decomposition we obtain

$$(X'^T X')V = D^2 V \tag{8.10}$$

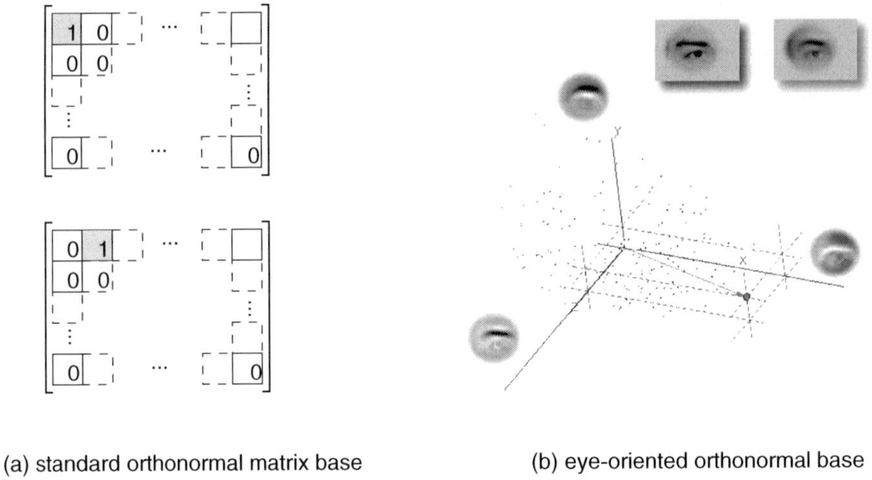


**Figure 8.2.** The commonly employed matrix orthonormal base (a) and a pattern-specific orthonormal base optimized for the description of eyes (b). The direction vectors of the latter have a more complex (visual) structure tuned to the pattern class they describe.

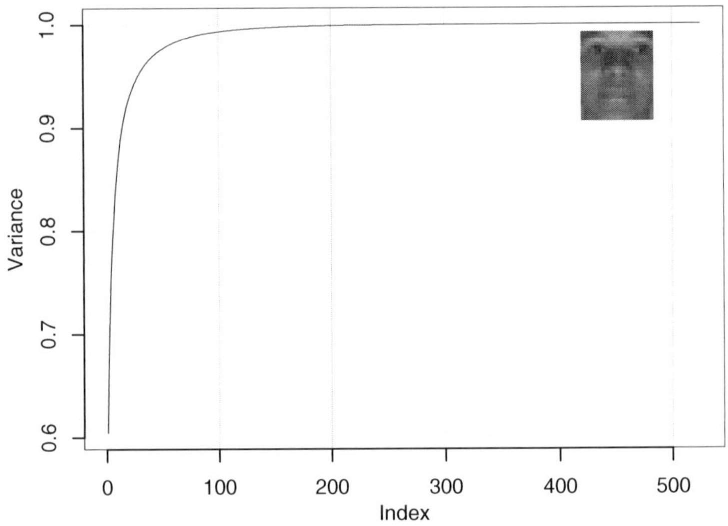


**Figure 8.3.** A plot of the fraction $1 - \Delta_k/\Delta_N$ captured by an increasing number of principal components computed from a set of 800 low-resolution face images.

and

$$(X'^T X')VX'^T = D^2VX'^T \tag{8.11}$$

so that we may compute the eigenvectors of $X'^T X'$, which is smaller, and create the eigenvectors in the original space with the transformation $VX'^T$. When the number of samples is greater than the dimension of their space we may work directly on the $M \times M$

covariance matrix exploiting the spectral theorem decomposition

$$\Sigma = E \Lambda E^T \tag{8.12}$$

where the columns of $E$ are the eigenvectors of $\Sigma$ and $\Lambda$ is a diagonal matrix with $\Lambda_{ii} = \lambda_i^2$. The coordinates of the transformed vectors

$$X'' = E^T X' \tag{8.13}$$

are their principal components in a way completely analogous to Equation 8.6. If we introduce the notation $\boldsymbol{e}_i$ for the $i$th eigenvector, the principal components explicitly given by

$$x_i'' = (\boldsymbol{e}_i \cdot \boldsymbol{x}') \tag{8.14}$$

represent the projections onto the new coordinate axes that are identified by the eigenvectors $\boldsymbol{e}_i$. The reconstruction of a vector $\boldsymbol{x}$ in terms of its principal components is then

$$\boldsymbol{x} = \bar{\boldsymbol{x}} + \sum_{i=1}^{n_d} (\boldsymbol{e}_i \cdot \boldsymbol{x})\boldsymbol{e}_i \tag{8.15}$$

$$\approx \bar{\boldsymbol{x}} + \sum_{i=1}^{k} (\boldsymbol{e}_i \cdot \boldsymbol{x})\boldsymbol{e}_i \tag{8.16}$$

where the last expression is the optimal approximation using only $k$ components, and $n_d = \min(M, N - 1)$ represents the dimension of the space spanned by the samples. The quality of the approximation can be gauged by the reconstruction error:

$$\delta_k^2(\boldsymbol{x}) = \left\| \boldsymbol{x} - \left( \bar{\boldsymbol{x}} + \sum_{i=1}^{k} (\boldsymbol{e}_i \cdot \boldsymbol{x})\boldsymbol{e}_i \right) \right\|^2. \tag{8.17}$$

The covariance matrix $\Sigma$ is the key element in PCA and the best possible estimate for it should be obtained. Two important sources of estimation inaccuracy are:

1. The limited number of samples available. This can be partially cured by resorting to shrinkage estimators (see Section 3.6.3 and Figure 8.4). The final remark of Intermezzo 3.6 is particularly important: while the sample estimate of the covariance matrix is usually sufficient to get a good estimate of the principal subspace, it is often not accurate enough in providing the correct eigenvalues, a fact that has a negative impact on the estimation of pattern probability.

2. The presence of outliers, due to noise, acquisition errors, optical phenomena, or other causes. This can be cured using a robust covariance estimator (see Section 4.4).

An important characteristic of PCA is its non-parametric nature: it can be applied (nearly) blindly to any kind of data. Whenever available, a priori knowledge may be exploited in parametric algorithms that can provide better results. Application of a nonlinear transformation to data before attempting PCA is a simple way to incorporate a priori information. The strategy is followed by kernel PCA that can then be considered a parametric variation of PCA (see Section 8.2). Let us summarize the main features of the whole PCA process:

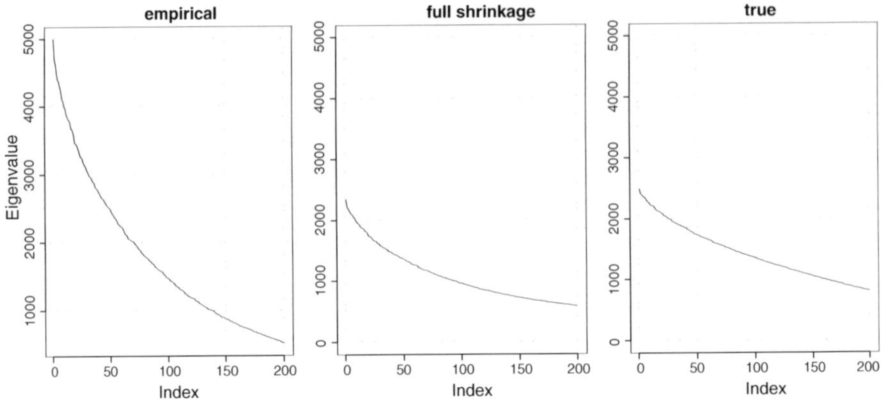


**Figure 8.4.** Plots of the eigenvalues estimated using the empirical and shrinkage estimators and the true ones from a random covariance matrix for $n_d = 625$ and $N = 400$, a typical situation in many low-resolution face processing tasks. The advantage of the shrinkage estimator over the empirical one is significant (and would be even more dramatic for lower values of $N$).

1. It is a linear process and it reduces to a change of basis. A simple and often very effective way to introduce nonlinearity is provided by kernel PCA, which will be described in Section 8.2.

2. The principal directions are orthogonal; this simplifies their computation and their usage.

3. In many situations of interest, large variance is associated with significant data variation and not to noise: principal components with larger associated variances represent interesting data variability while those with lower variances are associated with noise.

4. The whole process is based only on the mean and the covariance of data. The underlying assumption is that these statistics represent all there is to know about the data to represent them optimally. In other words, these statistics are assumed to be sufficient, fully characterizing the probabilistic process underlying data generation. However only distributions of the exponential family (e.g. Gaussian, exponential, etc.) are fully described given their first two moments (Pitman–Koopman–Darmois theorem).

From the above description it is clear that PCA is mainly a descriptive tool. The fact that it minimizes the $L_2$ approximation error (for a given number of directions) suggests that it is the correct approach for normally distributed data. In many cases, the assumption of normality does not hold. In the case of faces, the assumption holds only approximately for frontal faces and does not hold if we consider faces from different orientations. Another case where the assumption does not hold is in the case of log-polar imaging: even for frontal faces, spatio-variant imaging limits the effectiveness of a direct PCA. A solution is described in Intermezzo 8.1 and is commonly known as modular eigenspaces: the pattern space is subdivided into regions (e.g. poses or field view positions) and PCA is computed separately for each region, as if it were a different pattern class.

The characteristics of PCA can be exploited also in pattern detection and recognition tasks. Let us consider the problem of comparing a single, known face to a large set of faces in order to ascertain its identity. PCA can be used in the following way:

1. Compute the principal directions from the large face set (see Figure 8.8) and decide the number $k$ of components to be used (see Section 8.1.2 and Section C.2).
2. Compute the principal components $\boldsymbol{c}_i \in \mathbb{R}^k$ of each face in the set after having removed the average face.
3. Compute the principal components $\boldsymbol{c}$ for the unknown face.
4. Compare in turn the unknown face to all faces in the database by means of the quantity $\|\boldsymbol{c} - \boldsymbol{c}_i\|_{L_2}$. As the principal directions are obtained using a rotation in the original space, Euclidean pattern distances are preserved and PCA gives the best possible approximation to the complete distance with a linear projection onto a $k$-dimensional space.
5. The database face with minimal distance provides the identification.

PCA is then mainly used as a dimensionality reduction devised to speed up the comparison step. A more effective application, based on a probabilistic approach, is investigated in the next section.

**Intermezzo 8.1.** PCA in space-variant images

In most cases, the PCA representation of a face does not depend on the position of the face within the image: it is translationally invariant (at least approximately, neglecting perspective effects). However, if a log-polar image representation is used (see Section 2.4.3) the visual appearance of an object within the field of view can no longer be considered translationally invariant. A possibility is to use a single PCA and warp it, using the log-polar projection, to each location within the field of view, and then proceed in the usual way. A more effective strategy, however, is that of performing a different PCA at each different position, taking advantage of the decreasing resolution from the center to the periphery: the lower the resolution, the more similar the patterns and the fewer the components needed for an accurate representation of the patterns (Traver *et al.* 2004).

### 8.1.1. PROBABILISTIC PCA

The derivation and usage of PCA explored in the previous section does not rely on a probabilistic model of the data. While this is not a problem in tasks such as face recognition, where minimizing the distance between patterns can solve the task, it can be a problem in template detection. As PCA identifies a whole linear subspace, we cannot rely on a small reconstruction error to detect instances of the template class: the linear subspace identified by PCA may be rich enough to describe well other pattern classes (see Figure 8.5).

This limitation can, however, be easily removed, enhancing the power of PCA by permitting the application of Bayesian methods. Let us first consider the detection of patterns belonging to a given class. Given a large enough set we may perform PCA and project any pattern onto the first $k$ principal directions. We then compute the residual reconstruction error and compare it to the expected one: if it is much larger, we may be confident that the pattern does not belong to the class that we are considering. If the residual error is small enough,

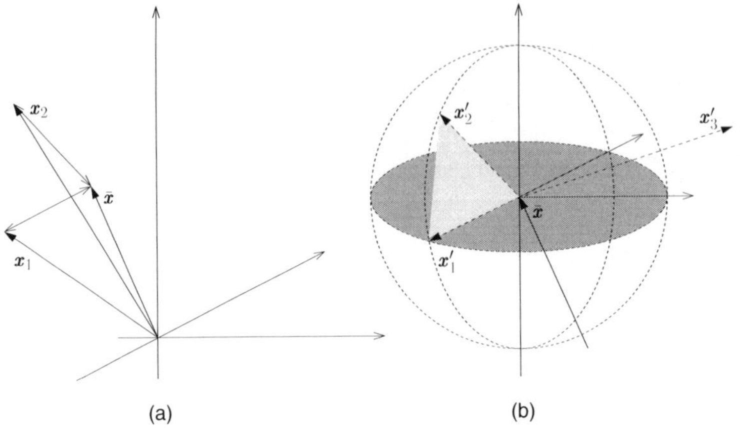


**Figure 8.5.** This shows the difference between computing the distance from a single prototype pattern (a) and computing the distance from a linear space, such as the one spanned by the principal components (b). In the first case points $\boldsymbol{x}_1$ and $\boldsymbol{x}_2$ are considered to be equidistant from $\bar{\boldsymbol{x}}$, while in the second case $\boldsymbol{x}'_1$ is considered to be very close to $\bar{\boldsymbol{x}}$ as it lies in the space spanned by the patterns represented by $\bar{\boldsymbol{x}}$.

and the vector of its principal components has a high enough probability, we accept it as an instance of the class, detecting it. Formalizing the above reasoning requires a probabilistic formulation of PCA and shows how to obtain it.

The likelihood of a pattern $\boldsymbol{x} \in \Omega$, where $\Omega$ represents the set of all patterns of the given class, is given by

$$P(\boldsymbol{x}|\Omega) = \frac{\exp[-\frac{1}{2}(\boldsymbol{x} - \bar{\boldsymbol{x}})^T \Sigma^{-1} (\boldsymbol{x} - \bar{\boldsymbol{x}})]}{(2\pi)^{n_d/2} |\Sigma|^{1/2}} \tag{8.18}$$

and the sufficient statistics to characterize the corresponding likelihood are given by the Mahalanobis distance

$$d_\Sigma(\boldsymbol{x}) = (\boldsymbol{x} - \bar{\boldsymbol{x}})^T \Sigma^{-1} (\boldsymbol{x} - \bar{\boldsymbol{x}}) = \sum_{i=1}^{n_d} \frac{x''^2_i}{\lambda_i^2} \tag{8.19}$$

with $\boldsymbol{x}''$ representing the pattern in the diagonalized PCA space where the covariance matrix is diagonal. We now assume that our class $\Omega$ is effectively described by $k$ principal components and we want to estimate the above sufficient statistics using only the projections onto the relevant space. We can decompose the full pattern space into two complementary subspaces: the features space $\mathbb{F}^k$, corresponding to the space spanned by the first $k$ principal directions, and its complement $\mathbb{F}_\perp^{n_d-k}$. We then consider the following estimator:

$$\hat{d}(\boldsymbol{x}) = \sum_{i=1}^{k} \frac{x''^2_i}{\lambda_i^2} + \frac{1}{\rho} \sum_{i=k+1}^{n_d} x''^2_i \tag{8.20}$$

$$= \sum_{i=1}^{k} \frac{x''^2_i}{\lambda_i^2} + \frac{1}{\rho} \epsilon^2(\boldsymbol{x}) \tag{8.21}$$

where $\rho$ is a free parameter and $\epsilon^2(\boldsymbol{x})$ is the corresponding $k$-component reconstruction error, representing the distance from our feature space. We may then write the probability of $\boldsymbol{x}$ as

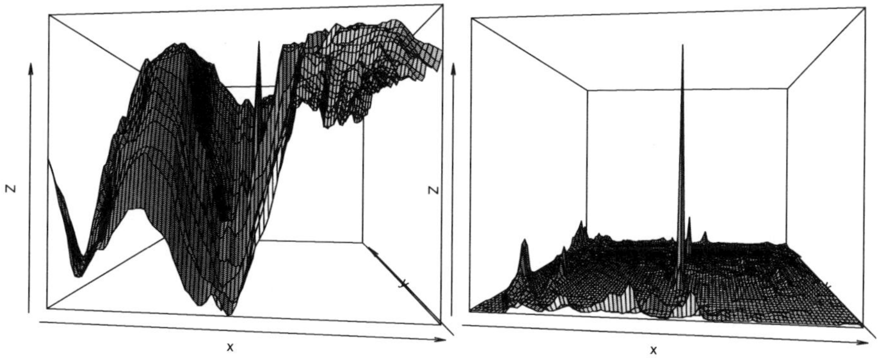


(a) (b)

**Figure 8.6.** This shows the difference between the map of the reconstruction residual (a) (the higher the value, the lower the residual) and the probability map (b) when detecting an eye using the PCA approach based on Equation 8.18 (the higher the value, the more probable the presence of the template). The latter has a significantly higher signal to noise ratio than the former.

the product of two independent Gaussian densities:

$$\hat{P}(\boldsymbol{x}|\Omega) = \left[\frac{\exp(-\frac{1}{2}\sum_{i=1}^{k} x_i''^2/\lambda_i^2)}{(2\pi)^{k/2}\prod_{i=1}^{k}\lambda_i}\right]\left[\frac{\exp(-\epsilon^2(\boldsymbol{x})/2\rho)}{(2\pi\rho)^{(n_d-k)/2}}\right] \tag{8.22}$$

$$= P_F(\boldsymbol{x}|\Omega)P_{F^\perp}(\boldsymbol{x}|\Omega). \tag{8.23}$$

In order to select an optimal $\rho$ we must introduce an optimality criterion. From an information-theoretic perspective we should choose $\rho$ in such a way that the difference between the true density $P(\boldsymbol{x}|\Omega)$ and its estimate $\hat{P}(\boldsymbol{x}|\Omega)$ is minimized. An appropriate estimator of the difference of two distributions is the Kullback–Leibler divergence:

$$d_{\mathrm{KL}}(\rho) = E\left[\log\frac{P(\boldsymbol{x}|\Omega)}{\hat{P}(\boldsymbol{x}|\Omega)}\right] = \frac{1}{2}\sum_{i=k+1}^{n_d}\left(\frac{\lambda_i^2}{\rho} - 1 + \log\frac{\rho}{\lambda_i^2}\right). \tag{8.24}$$

The optimal weight $\hat{\rho}$ is found by minimizing the criterion function $d_{\mathrm{KL}}(\rho)$ with respect to $\rho$. By imposing $\partial_\rho d_{\mathrm{KL}}(\rho) = 0$, we get

$$\hat{\rho} = \frac{1}{n_d - k}\sum_{i=k+1}^{n_d}\lambda_i^2 \tag{8.25}$$

corresponding to the average of the eigenvalues in $F^\perp$. While apparently only a minor gain, the above derivation provides a probabilistic connection between the dimension of PCA space and the probabilities of patterns (Figure 8.6).

### 8.1.2. HOW MANY COMPONENTS?

PCA is a descriptive tool and does not directly aim at pattern recognition. The choice of the optimal number $\hat{k}$ of components depends on the task for which they are used. If the task is a description of the patterns, it is natural to choose $\hat{k}$ by constraining the expected reconstruction error $\Delta_k$. A first estimate of the expected reconstruction error, also known as the predicted residual error sum of squares or PRESS (see also Intermezzo 8.2), is given, from Equation 8.9, by

$$\Delta_k^0 = \sum_{i=k+1}^{n_d} \lambda_i^2. \tag{8.26}$$

As we used the data to compute our components, this does not provide a reliable estimate of the expected error on data outside the training set. In order to get a more realistic estimate we must resort to a technique such as cross-validation (see Section C.2). In the case of leave-one-out cross-validation, we must in turn exclude one of the data points and compute PCA and the reconstruction error for the left-out point as a function of the number of components. The expected reconstruction error is then obtained by averaging the resulting $N$ estimates.

**Intermezzo 8.2.** PRESS estimation in the presence of outliers

When a robust procedure is employed to compute PCA, the usual PRESS is not itself robust as outliers fully contribute to its value, in spite of a robust computation of PCA. In this case we should substitute the PRESS with a weighted sum, whose sample-specific weights are computed before the cross-validation procedure is started and after the computation of a robust PCA, whose knowledge is necessary to spot the outliers. Naive leave-one-out cross-validation is, however, extremely computationally expensive, particularly when robust estimates must be computed. For some robust PCA algorithms, an approximated leave-one-out estimate can be obtained efficiently without full recomputation of PCA for each leave-one-out estimation (Hubert and Engelen 2007).

If the task is not purely descriptive, it is sometimes possible to determine an appropriate number of components by looking at the distribution of the eigenvalues. If we consider Gaussian white noise with given $\sigma^2$, the eigenvalues found by PCA will all be equal to $\sigma^2$ and a plot of the sorted eigenvalues as a function of their rank will be a straight line. If we add white noise to a population of patterns of dimension $k$ embedded in a space of dimension $n_d > k$ and look at the eigenvalues' spectrum we will see that a region of negative slope (the variance of the templates) is followed by a nearly flat region (the noise). The boundary of the two regions is approximately located at $k$, starting from which noise is the only contribution. In many cases the plot of a generic indicator function against the number of components may exhibit regions characterized by different slopes. A particularly interesting example is described in Intermezzo 8.3.

In order to automate the choice of a proper $k$, several indicator functions have been proposed. The principle underlying the use of indicator functions to select the dimension of feature space is that samples are corrupted by some kind of noise. Therefore, there will be a point, i.e. a value $k$ of the dimension of feature space, where the improvement in the fidelity of the pattern description obtained is smaller than the amount of noise, which, from that point, becomes the most significant effect described by PCA. It is expected that the scaling laws of the eigenvalues corresponding to the two different phenomena, i.e. signal and noise, will differ. The optimal value $\hat{k}$ is then found by identifying the boundary point between the regions corresponding to signal/noise dominance respectively. A simple (empirical) function is the factor indicator function:

$$\mathrm{IND_F}(k) = \frac{1}{(n_d - k)^2} \sqrt{\frac{\sum_{j=k+1}^{n_d} \lambda_j^2}{N(n_d - k)}} \tag{8.27}$$

where $N$ is the number of samples. The $\mathrm{IND_F}$ function is a minimum when $k$ equals the true number of components (see Figure 8.7). The above indicator function can be subjected to slope analysis by introducing the ratio of derivatives function:

$$\mathrm{ROD}(k) = \frac{\mathrm{IND_F}(k) - \mathrm{IND_F}(k+1)}{\mathrm{IND_F}(k+1) - \mathrm{IND_F}(k+2)}. \tag{8.28}$$

The rationale behind the ROD indicator is that of finding a break-even point in the plot by inspecting the derivative of the indicator function. The number of significant components can then be determined as the value corresponding to the first maximum of ROD, or to the first minimum of the indicator function $\mathrm{IND_F}$. It is also possible to use multiple criteria, deriving a single value with a simple aggregating rule, such as taking the average or median.

**Intermezzo 8.3.** The dimensionality of face space

Given a set of $N$ independent templates described by vectors in $\mathbb{R}^{n_d}$, $n_d > N$, we need $N$ components to represent them perfectly. This is true also for face images, but, as in many other cases, using $N$ components is not really necessary to maximize detection and recognition performance. A set of interesting psychophysical experiments has been performed to get a bound on the number of components necessary to achieve human performance. PCA was performed on a set of (approximately) 2000 frontal images of Caucasian faces, each represented by $n_d = 3496$ pixels. A set of face images, half of them belonging to the set used for PCA (in-population), half of them not (out-population), were presented to a group of human observers that had to rate their degree of familiarity on a six-degree scale, from 1, certainly unfamiliar, to 6, certainly familiar. The returned familiarity can be used to construct an ROC curve: each degree corresponds to a score binarization threshold, and the human performance can then be computed and used to construct the curve. In order to test the number of components required, the quality of each face reconstruction was quantified by a signal to noise ratio

$$\mathrm{SNR} = \log_2\left(\frac{\|\boldsymbol{x}_i\|}{\delta_k^2}\right) \tag{8.29}$$

where $\delta_k^2$ is the reconstruction error, defined in Equation 8.17. The higher the SNR, the higher the number $k$ of components necessary to attain it. Sets of faces with the same SNR were used to test human recognition. Different faces required slightly different values of $k$ to attain the same quality: testing faces not used for the PCA computation consistently required more components. The best human subjects achieved a 75% accuracy with an SNR of 7.24, corresponding to 107 components for in-population faces and 124 for out-population faces. A closely matching value of 120 is reported in the literature using the Bayesian estimator of Equation 8.31 (Meytlis and Sirovich 2007).

A more sophisticated approach is based on Bayesian reasoning. From a Bayesian perspective we may score each model of the data by associating to it the probability it assigns to observed data. Different models will assign different probabilities and the higher the probability, the better the model. This is the same procedure followed in Bayesian classification as described in Chapter 3. We now consider the number $k$ of directions of feature space as a parameter of the model and we select the value maximizing data probability. The a posteriori probability of data $P(D|M)$, the so-called evidence for model $M$, is given by

$$p(D|M) = \int_{\boldsymbol{\theta}} p(D|\boldsymbol{\theta})p(\boldsymbol{\theta}|M)\, d\boldsymbol{\theta} \tag{8.30}$$

where $D$ represents the data, $M$ the model, and $\boldsymbol{\theta}$ the unknown model parameters. In the case of probabilistic PCA, from Equation 8.23, we have $\boldsymbol{\theta} = (\boldsymbol{\mu}, \Sigma_k, \rho_k)$, where we indexed with $k$ the parameters depending on the dimension of feature space.

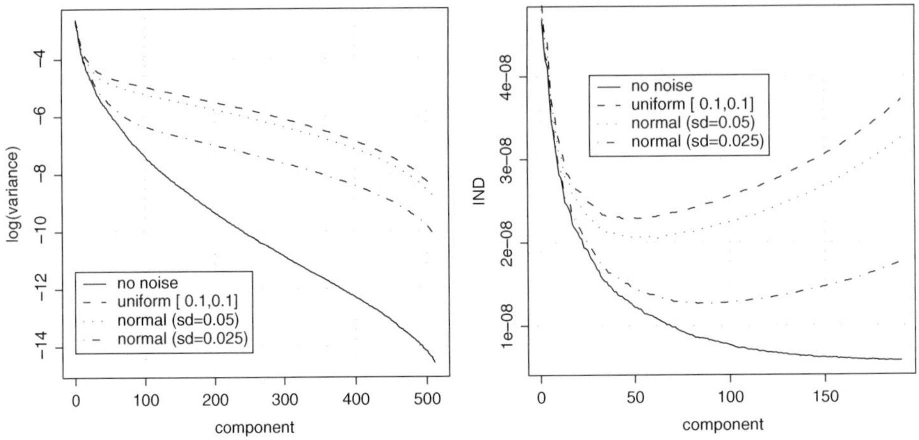


**Figure 8.7.** The PCA spectrum (left) for a set of 800 face images with different amounts of noise. The use of the indicator function $\mathrm{IND_F}(\cdot)$ provides a clear indication of the number of components that are not dominated by noise (right).

To proceed with the estimation it is necessary to assign non-informative prior probabilities $p(\boldsymbol{\theta}|M)$ in order to carry out the integration. Making explicit the dependence on $k$, which indexes our models $M$, we can obtain $p(D|M) = p(D|k)$ which we need to maximize over $k$. The computation is somewhat complex (see Section 8.6 for references) and the final result is

$$
\begin{aligned}
p(D|k) &\approx p(E)\bigg(\prod_{j=1}^{k} \lambda_j^2\bigg)^{-N/2} \hat{\rho}_k^{-N(n_d-k)/2}(2\pi)^{(m+k)/2} N^{-k/2}|A|^{-1/2} \\
p(E) &= 2^{-k}\prod_{i=1}^{k}\Gamma((n_d-i+1)/2)\pi^{-(n_d-i+1)/2} \\
m &= n_d k - k(k+1)/2 \\
|A| &= \prod_{i=1}^{k}\prod_{j=i+1}^{n_d} N(\hat{\lambda}_j^{-2}-\hat{\lambda}_i^{-2})(\lambda_i^2-\lambda_j^2) \\
\hat{\lambda}_i^2 &= \begin{cases} \lambda_i^2 & i \le k \\ \hat{\rho} & i > k \end{cases}
\end{aligned}
\tag{8.31}
$$

where the notation corresponds to that of Section 8.1.1. If we drop the terms that do not increase with $N$, we obtain the Bayesian information criterion (or BIC) approximation

$$
p(D|k) \approx \bigg(\prod_{j=1}^{k} \lambda_j^2\bigg)^{-N/2} \hat{\rho}_k^{-N(n_d-k)/2} N^{-(m+k)/2} \tag{8.32}
$$

whose performance is reported to be markedly inferior.

## 8.2. A Nonlinear Approach: Kernel PCA

One of the limitations of PCA is that it is limited to the space of orthogonal linear transformations. Relaxing the orthogonality requirement leads to the field of independent components analysis, which will be addressed in the next section. Relaxing the linearity assumption leads to the so-called nonlinear PCA. As an example, let us consider the case where two-dimensional data points are distributed on a circle, that represents the one-dimensional manifold, embedded in $\mathbb{R}^2$, where our data reside. The principal curve, the nonlinear counterpart of a principal direction, is the circle, and the principal component is obtained by a nonlinear mapping $f(\cdot) : \mathbb{R}^2 \to \mathbb{R}$. The computation of the approximation in the original space given the principal component requires yet another nonlinear mapping $g(\cdot) : \mathbb{R} \to \mathbb{R}^2$:

$$y = f(\boldsymbol{x}), \quad \boldsymbol{x} \approx \boldsymbol{g}(y) \tag{8.33}$$

where $y$ represents the coordinate along the circle and $\boldsymbol{x}$ the point in the original space. A few pointers to the literature on the argument are provided in Section 8.6 and a simple idea based on the use function approximation is described in Intermezzo 8.4.

An important alternative is based on the idea of applying a linear PCA on a nonlinear transformation of the samples, using the so-called kernel trick. As a simple example, let us consider the following nonlinear transformation of the plane:

$$\phi(\boldsymbol{x}) : (x_1, x_2) \to (x_1^2, \sqrt{2}x_1x_2, x_2^2) \equiv (z_1, z_2, z_3). \tag{8.34}$$

The transformation has the property that points lying on a given ellipse are mapped onto a plane

$$\left(\frac{x_1}{a}\right)^2 + \left(\frac{x_2}{b}\right)^2 \to \frac{z_1}{a^2} + \frac{z_3}{b^2} = 1 \tag{8.35}$$

so that we can discriminate points lying inside the ellipse from those lying outside by means of a planar surface in the transformed space. An interesting property of $\phi$ is that the dot product in the transformed space can be computed easily in the original space

$$K(\boldsymbol{x}, \boldsymbol{y}) = \langle \phi(\boldsymbol{x}), \phi(\boldsymbol{y}) \rangle = (\langle \boldsymbol{x}, \boldsymbol{y} \rangle)^2 \tag{8.36}$$

where $K(\cdot, \cdot)$ is the so-called kernel function and $\langle \cdot, \cdot \rangle$ is the (extended) notation for the dot product operation. Testing whether one point lies on one side or the other of the separating plane in transformed space requires only dot products that can then be computed in the original space by means of Equation 8.36. It turns out that PCA can be formulated completely in terms of dot products. We can therefore project our data in an appropriate space, even infinite dimensional, and, provided that a convenient formula for the computation of the dot product exists based on the data in the original space, derive the nonlinear principal components as the coefficients of the linear one in the transformed space. Before proceeding to the derivation, let us state the conditions under which we can associate a projection to a kernel function.

**Definition 8.2.1.** *Given a set $\mathfrak{X}$, a continuous, symmetric, and positive definite function $K : \mathfrak{X} \times \mathfrak{X} \to \mathbb{R}$ is said to be a Mercer kernel.*

**Intermezzo 8.4.** Nonlinear PCA as an optimization problem

The problem of nonlinear PCA can be formalized as follows:

Given two parametric function classes $\Omega_f = \{f(\boldsymbol{\theta}_f; \cdot) : \mathbb{R}^{n_d} \to \mathbb{R}^k\}$, $\Omega_g = \{g(\boldsymbol{\theta}_g; \cdot) : \mathbb{R}^k \to \mathbb{R}^{n_d}\}$, a set of patterns $\{\boldsymbol{x}\}_{i=1}^N$, $\boldsymbol{x}_i \in \mathbb{R}^{n_d}$, and the number $k$ of required components, the nonlinear principal components of a pattern $\boldsymbol{x}$ can be defined as $f(\hat{\boldsymbol{\theta}}; \boldsymbol{x}) \in \mathbb{R}^k$ where

$$(\hat{\boldsymbol{\theta}}_f, \hat{\boldsymbol{\theta}}_g) = \operatorname*{argmin}_{\boldsymbol{\theta}_f, \boldsymbol{\theta}_g} \sum_{i=1}^N \|\boldsymbol{x}_i - g(f(\boldsymbol{x}_i))\|. \tag{8.37}$$

A commonly employed class of functions is that of three-layer, feedforward, sigmoidal neural networks. The optimization problem of Equation 8.37 corresponds to a network composed of two cascaded three-layer networks, with the dimensionality of the output of the first (and the matching one of the input of the second) being equal to the required number of nonlinear components (Bishop 1995).

**Theorem 8.2.2.** *If $K(x, y)$ is a Mercer kernel, then there exists a Hilbert space $\mathfrak{H}_K$ of real-valued functions defined on $\mathfrak{X}$ and a feature map $\phi : \mathfrak{X} \to \mathfrak{H}_K$ such that*

$$\langle \phi(x), \phi(y) \rangle_K = K(x, y) \tag{8.38}$$

*where $\langle \cdot, \cdot \rangle_K$ is the dot product on $\mathfrak{H}_K$.*

Among Mercer kernels, the following are often used:

$$K(x, y) = (\langle x, y \rangle)^q, \quad \text{polynomial} \tag{8.39}$$

$$K(x, y) = \exp\left(-\frac{\|x - y\|^2}{2\sigma^2}\right), \quad \text{radial basis function (RBF)} \tag{8.40}$$

$$K(x, y) = f(d(x, y)), \quad d \text{ a distance function, } f \text{ a function on } \mathbb{R}_0^+. \tag{8.41}$$

We now need to reformulate PCA using exclusively a dot product notation. A preliminary step is centering in the high-dimensional feature space:

$$\phi'(\boldsymbol{x}_i) = \phi(\boldsymbol{x}_i) - \frac{1}{N} \sum_{j=1}^N \phi(\boldsymbol{x}_j) \tag{8.42}$$

from which we derive the covariance matrix

$$\Sigma_\phi = \frac{1}{N} \sum_{j=1}^N \phi'(\boldsymbol{x}_j) \phi'(\boldsymbol{x}_j)^T \tag{8.43}$$

and we need to find the associated eigenvectors corresponding to non-zero eigenvalues

$$\Sigma_\phi \boldsymbol{v}_\phi = \lambda \boldsymbol{v}_\phi. \tag{8.44}$$

The eigenvectors with non-null eigenvalue must reside in the space spanned by the transformed samples as

$$\Sigma_\phi \boldsymbol{v}_\phi = \frac{1}{N} \sum_{j=1}^N (\phi'(\boldsymbol{x}_j) \cdot \boldsymbol{v}_\phi) \phi'(\boldsymbol{x}_j) = \lambda \boldsymbol{v}_\phi. \tag{8.45}$$

This simple fact has two important consequences: the first one is that $\boldsymbol{v}_\phi$ can be expressed as a linear combination of the transformed data vectors

$$\boldsymbol{v}_\phi = \sum_{j=1}^{N} \alpha_j \phi'(\boldsymbol{x}_j) \tag{8.46}$$

and the second one is that we can transform Equation 8.44 into a set of $N$ equations

$$\phi'(\boldsymbol{x}_k) \cdot \Sigma_\phi \boldsymbol{v}_\phi = \lambda \phi'(\boldsymbol{x}_k) \cdot \boldsymbol{v}_\phi \quad k = 1, \ldots, N. \tag{8.47}$$

Substitution of Equation 8.46 into Equation 8.45, followed by the dot product with $\phi'(\boldsymbol{x}_i)$, gives

$$\lambda \sum_{k=1}^{N} \alpha_k \phi'(\boldsymbol{x}_k) \cdot \phi'(\boldsymbol{x}_i) = \left[ \frac{1}{N} \sum_{k=1}^{N} \alpha_k \left( \phi'(\boldsymbol{x}_k) \cdot \sum_{j=1}^{N} \phi'(\boldsymbol{x}_j) \right) \phi'(\boldsymbol{x}_j) \right] \cdot \phi'(\boldsymbol{x}_i). \tag{8.48}$$

Introducing matrix $K'$ by

$$K'_{ij} - \phi'(\boldsymbol{x}_i) \cdot \phi'(\boldsymbol{x}_j) \tag{8.49}$$

we finally obtain

$$K'^2 \boldsymbol{\alpha} = N\lambda K' \boldsymbol{\alpha} \tag{8.50}$$

where $\boldsymbol{\alpha} = (\alpha_1, \ldots, \alpha_N)^T$. Instead of solving Equation 8.50 we may solve

$$K' \boldsymbol{\alpha} = N\lambda \boldsymbol{\alpha} \tag{8.51}$$

for non-zero eigenvalues. Simple substitution shows that the solutions of Equation 8.51 are solutions of Equation 8.50 but they do not exhaust them. In fact, we may add a zero-value eigenvector and obtain a different solution of Equation 8.50 with the same eigenvalue. However, a zero-eigenvalue solution of Equation 8.51 when used in Equation 8.46 provides a vector that is orthogonal to all $\phi'(\boldsymbol{x}_j)$ and, being a combination of them, must be $\boldsymbol{0}$. $K'$ has $N$ eigenvalues $\lambda_1 \geq \lambda_2 \geq \cdots \geq \lambda_N$ (corresponding to $N\lambda$ of Equation 8.51) and a corresponding complete set of eigenvectors $\boldsymbol{\alpha}_i$, $i = 1, \ldots, N$. Let $p$ be the index of the last non-null eigenvalue. The normalization

$$\boldsymbol{v}_{\phi k} \cdot \boldsymbol{v}_{\phi k} = 1 \quad k \leq p \tag{8.52}$$

imposes the following normalization on $\boldsymbol{\alpha}_k$, $k \leq p$:

$$1 = \sum_{i,j=1}^{N} \alpha_{ki}\alpha_{kj}(\phi'(\boldsymbol{x}_i) \cdot \phi'(\boldsymbol{x}_j)) = \sum_{i,j=1}^{N} \alpha_{ki}\alpha_{kj} K'_{ij} = \boldsymbol{\alpha}_k \cdot K' \boldsymbol{\alpha}_k = \lambda_k \boldsymbol{\alpha}_k \cdot \boldsymbol{\alpha}_k. \tag{8.53}$$

Given a point $\boldsymbol{x}$, its $k$th principal components in the transformed space are then

$$(\boldsymbol{v}_{\phi,k} \cdot \phi'(\boldsymbol{x})) = \sum_{i=1}^{N} \alpha_{ki}(\phi'(\boldsymbol{x}_i) \cdot \phi'(\boldsymbol{x})). \tag{8.54}$$

We recall that in the above treatment the transformed vectors were assumed to be centered. As we do not want to work directly on the transformed vectors, we must express the equations

**Algorithm 8.1**: Kernel PCA

**Data**: A dataset $\boldsymbol{x}_1, \ldots, \boldsymbol{x}_N$, a Mercer kernel $K$.

1 Compute the centered kernel matrix $K'$

$$K'_{ij} = \left( K(\boldsymbol{x}_i, \boldsymbol{x}_j) - \frac{1}{N} \sum_{k=1}^{N} K(\boldsymbol{x}_k, \boldsymbol{x}_j) - \frac{1}{N} \sum_{k=1}^{N} K(\boldsymbol{x}_k, \boldsymbol{x}_i) + \frac{1}{N^2} \sum_{k=1}^{N} \sum_{l=1}^{N} K(\boldsymbol{x}_k, \boldsymbol{x}_l) \right). \tag{8.55}$$

2 Compute $\{\boldsymbol{\alpha}_k\}_{k=1}^N$, the set of the unit-norm eigenvectors corresponding to the $k$th largest eigenvalues $\lambda_k$ of $K'$.

3 Compute the set $\{c_k\}_{k=1}^N$ of the kernel PCA principal components of $\boldsymbol{x}$

$$c_k(\boldsymbol{x}) = \sum_{i=1}^{N} \frac{\alpha_{ki}}{\sqrt{\lambda_k}} \left( K(\boldsymbol{x}_i, \boldsymbol{x}) - \frac{1}{N} \sum_{l=1}^{N} K(\boldsymbol{x}_l, \boldsymbol{x}) \right). \tag{8.56}$$

involving centered quantities in terms of the non-centered ones. This is straightforward and leads us to Algorithm 8.1.

A few concluding remarks are in order. There are no general recipes for the choice of the parameter of kernel PCA, namely the kernel used. Often, several kernels can be compared using cross-validation techniques (see Appendix C) and the best performing one is used. The very useful characteristic of standard PCA of Equation 8.16 does not carry over to kernel PCA. Given an approximate reconstruction in the transformed space it may be impossible to find an exact preimage of it, i.e. a pattern in the original space whose projection matches the given reconstruction.

The complexity of kernel PCA is higher than that of the standard version as can be appreciated by inspection of Equation 8.54: for each kernel component we must compute $N$ dot products instead of one. Often this additional complexity is rewarded by the possibility of using classifiers of reduced complexity. While in standard PCA the maximum number of components is limited by the dimensionality of the pattern space, in kernel PCA it is limited by the number of data points: the resulting richer description may be beneficial in recognition tasks. The interpretation of kernel PCA in the original pattern space is more complex but, for some kernels such as the polynomial ones, it can be interpreted in terms of higher order features, reversing the kernel trick of Equation 8.34. Once the kernel principal components $c_k(\boldsymbol{x})$ have been computed, we can use them to detect patterns (using the probabilistic approach described in Section 8.1.1) and to recognize patterns (following the procedure described at the end of Section 8.1).

## 8.3. Independent Components

From a probabilistic point of view, we would like to describe patterns by means of independent components as this would ease many estimation tasks required for successful pattern detection. Unfortunately, PCA provides uncorrelated components: uncorrelatedness is a less stringent requirement than independence, the latter implying the former. In fact, statistical independence of two random variables $y_1$ and $y_2$ implies that the expected value of the product $h_1(y_1)h_2(y_2)$ is the product of the expected values

$$E\{h_1(y_1)h_2(y_2)\} = E\{h_1(y_1)\}E\{h_2(y_2)\} \tag{8.57}$$

for any function $h_1$ and $h_2$. When $h_1 = h_2 =$ identity, it implies

$$E\{y_1 y_2\} - E\{y_1\}E\{y_2\} = 0 \tag{8.58}$$

corresponding to zero covariance and null correlation. It is then natural to ask ourselves whether it is possible to generate independent components and, if it is possible, when we can expect approaches based on independent components to beat PCA.

The computation of independent components is indeed possible, and independent component analysis (ICA) is by itself a very important research field. ICA defines a generative model for the observed multivariate data: the data variables are assumed to be linear mixtures of some unknown (latent) variables, and the mixing matrix is itself unknown. Let us consider a set of $N$ samples $\boldsymbol{x}_i \in \mathbb{R}^{n_d}$ as column vectors, and assume that they can be expressed as

$$\boldsymbol{x}_i = A\boldsymbol{s}_i, \quad i = 1, \ldots, N \tag{8.59}$$

where $A \in \mathbb{R}^{n_d \times n_d}$ is the so-called mixing matrix and the vectors $\boldsymbol{s}_i \in \mathbb{R}^{n_d}$ are chosen in such a way that their $n_d$ components are statistically independent. We may also rewrite the previous equation in a different way, highlighting the fact that $\boldsymbol{x}_i$ is written as a linear combination of (column) vectors $\boldsymbol{a}_j$ representing the columns of $A$:

$$\boldsymbol{x}_i = \sum_{j=1}^{n_d} \boldsymbol{a}_j s_{i,j}. \tag{8.60}$$

The goal of ICA is to find an unmixing matrix $U$ such that

$$U\boldsymbol{x}_i = \boldsymbol{s}_i \tag{8.61}$$

and use $\boldsymbol{s}_i$ instead of $\boldsymbol{x}_i$ for classification. The proposed model can be written in a more compact form as

$$X = AS \tag{8.62}$$

where the columns of matrix $X$ represent the samples $\boldsymbol{x}_i$, the columns of $S$ represent $\boldsymbol{s}_i$, and the rows of $S$ are termed sources. Both $A$ and $S$ are unknown so that we cannot determine $S$ uniquely. In fact, we can perform an arbitrary scaling of the rows of $S$ and compensate for it by redefining $A$:

$$X = (AD^{-1})(DS) \tag{8.63}$$

where $D$ is an arbitrary diagonal matrix $D = \text{diag}(c_1, \ldots, c_{n_d})$. Another source of ambiguity is related to the fact that we can change the sign of any $c_i$ without affecting the independence of the sources. Finally, we can insert an arbitrary permutation matrix $P$ in the model also without affecting the independence of the sources:

$$X = (AP^{-1})(PS). \tag{8.64}$$

Before proceeding with ICA, it is convenient to sphere the original data. This can be achieved most easily by means of the sphering (or whitening) transformation

$$\boldsymbol{x}'_i = E\Lambda^{-1/2}E^T(\boldsymbol{x}_i - \bar{\boldsymbol{x}}) \tag{8.65}$$

where $E^T$ is the matrix whose rows are the eigenvectors of the covariance matrix of the data and $\Lambda$ the diagonal matrix whose elements are the corresponding eigenvalues (see Equation 8.12).

The independent components are the $p$ rows of $UX'$ which we assume to be normalized to unit standard deviation (exploiting the transformation freedoms stated above). As the components are independent, they are also uncorrelated so that the covariance matrix of $UX'$ must be the identity matrix. As $E\{UX'\} = UE\{X'\} = 0$ (Equation 8.65), this covariance matrix can be written as

$$I = E\{(UX')(UX')^T\} = UE\{X'(X')^T\}U^T = UU^T \tag{8.66}$$

because $X'$ is whitened (Equation 8.65). The unmixing matrix is then an orthogonal matrix and, without loss of generality, we can assume it to be a rotation matrix (exploiting the degrees of freedom in choosing the independent components). ICA algorithms are based on the central limit theorem, which states that a sum of independent random variables tends toward a Gaussian distribution under appropriate conditions. Let us consider a candidate independent direction $\boldsymbol{w}$ and $\boldsymbol{z} = A^T\boldsymbol{w}$. We have that $y = \boldsymbol{w}^T\boldsymbol{x} = \boldsymbol{w}^T(A\boldsymbol{s}) = \boldsymbol{z}^T\boldsymbol{s}$ is a linear combination of $s_i$ with weights $z_i$. As the sum of independent variables is more Gaussian than any of the original variables, $y$ is more Gaussian than any $s_i$, being minimally Gaussian when it equals one of the $s_i$ and all but one of the weights are 0. This means that the vector $\boldsymbol{w}$ maximizing non-Gaussianity identifies one of the independent components. The possibility of flipping the sign of an independent direction increases the number of local maxima in the non-Gaussianity landscape. Finding all independent components is made easier by the fact that they must be uncorrelated: each new direction can be searched for in the space orthogonal to that identified by the directions previously found. If the ICA model holds, optimizing the ICA non-Gaussianity measures produces independent components; if the model does not hold, what we get are the projection pursuit directions that are interesting because of their non-Gaussianity. ICA algorithms are particularly sensitive to outliers as the latter appear to them as promising directions where non-Gaussianity can be maximized. Robust techniques, similar to those used for the computation of robust covariance matrices, can be used (see Intermezzo 8.5).

We now need to find a convenient measure of non-Gaussianity and an efficient algorithm to maximize it in order to find all the directions associated to the independent components.

There exist many measures of non-Gaussianity but there is one based on the notion of differential entropy $H(\boldsymbol{y})$ of a density distribution $p(\boldsymbol{y})$

$$H(\boldsymbol{y}) = -\int p(\boldsymbol{y}) \log p(\boldsymbol{y})\, d\boldsymbol{y} \tag{8.67}$$

that is particularly effective. A basic result of information theory is that a Gaussian variable has the largest entropy among all random variables of equal variance. This implies that the negentropy $J_H(\boldsymbol{y})$ satisfies

$$J_H(\boldsymbol{y}) = H(\boldsymbol{y}_g) - H(\boldsymbol{y}) \geq 0 \tag{8.68}$$

where $\boldsymbol{y}_g$ is a Gaussian random variable of the same covariance matrix as $\boldsymbol{y}$. A problem with Equation 8.68 is that it can be computed only if we know the distribution of $\boldsymbol{y}$, which is not the usual case, requiring us to perform a non-parametric estimation of it. A convenient solution relies on the possibility of approximating $J(y)$ for a zero-mean and unit-variance variable $y$ as

$$J(y) \approx \sum_{i=1}^{k} c_i [E\{g_i(y)\} - E\{g_i(\eta)\}]^2 \tag{8.69}$$

where $\eta$ is a zero-mean, unit-variance Gaussian random variable and the functions $g_l$ are non-quadratic (a quadratic function would not be useful due to the fact that we already whitened our data). The approximation need not be very accurate and good results can be obtained with a single term expansion based on one of the two contrast functions

$$g_1(u) = \frac{1}{a_1} \log(\cosh(a_1 u)) \tag{8.70}$$

$$g_2(u) = -e^{-u^2/2} \tag{8.71}$$

where $1 \leq a_1 \leq 2$. One of the most widely adopted ICA algorithms is FastICA. It assumes sphered data and solves the constrained optimization problem

$$\hat{\boldsymbol{w}} = \operatorname*{argmin}_{\boldsymbol{w}} \; E\{g(\boldsymbol{w}^t \boldsymbol{x})\}, \quad \|\boldsymbol{w}\|^2 = 1 \tag{8.72}$$

using a fixed point iteration scheme (see Algorithm 8.2).

This algorithm finds a single independent direction; if we want to find all the independent components we need to find all the corresponding directions. In order to prevent the algorithm from converging repeatedly towards the same solution, we must decorrelate any new direction from those already found. This can be achieved by modifying the second step by preceding the normalization of Equation 8.79 with

$$\boldsymbol{w}_{t,k+1} \rightarrow \boldsymbol{w}_{t,k+1} - \sum_{j=1}^{k} (\boldsymbol{w}_{t,k+1}^T \boldsymbol{w}_j) \boldsymbol{w}_j \tag{8.73}$$

where $k$ represents the number of computed directions. The expected values appearing in Equation 8.72 can be computed over a subset of the data to improve computational efficiency, resorting to larger data subsets only when convergence is not satisfactory. Contrary to PCA there is no intrinsic way to sort the resulting directions, but, from a classification perspective, we may sort them based on their classification effectiveness or, as in the case of PCA, by the amount of described variance.

Given a set of $N$ images with $n_d$ pixels we may formulate ICA in two different ways:

1. The rows of $X$ are vectorized images and $X$ is an $N \times n_d$ matrix (ICA I, see Figure 8.8). Data centering corresponds to zero average images and not to zero average coordinates.

2. The columns of $X$ are vectorized images and $X$ is a $n_d \times N$ matrix (ICA II, see Figure 8.8). Data centering corresponds to that of standard PCA: the average vector is the average image and the average of each coordinate (over the data set) is zero.

The difference is not only formal and the results of ICA differ significantly:

**Intermezzo 8.5.** Robustification of ICA

ICA is even more susceptible to outliers than PCA: the quest for non-Gaussianity which is the driving force of the representation also makes it eager to align its directions to outliers. An apparently simple idea to robustify ICA is that of removing outliers prior to the computation. Outlier removal by means of traditional approaches, such as those based on robust estimation of location and scale, may cause problems for the convergence of algorithms such as FastICA: valid points are often flagged as outliers due to the sensitivity of the procedure for outlier detection to the non-Gaussianity of the bulk data distribution. The effect is then to remove what we want to detect: non-Gaussianity.

It is, however, possible to limit the impact of non-Gaussianity on the procedure for outlier detection by using rejection criteria that are asymmetric with respect to the robustly estimated location parameter. This is possible by using a robust estimator of the skewness of a distribution, the so-called medcouple (MC) that has a breakdown point of 25%:

$$\mathrm{MC}(x_1, \ldots, x_n) = \operatorname*{med}_{ij} \frac{(z_j - \mathrm{med}_k z_k) - (z_i - \mathrm{med}_k z_k)}{z_j - z_i} \tag{8.74}$$

where $i$ and $j$ are restricted to observations lying on different sides of the median value $z_i \leq \mathrm{med}_k(z_k) \leq z_j$ and $z_i \neq z_j$. When the distribution is symmetric, $\mathrm{MC} = 0$; when it is right skewed (left skewed) we have $\mathrm{MC} > 0$ ($\mathrm{MC} < 0$). We can use this information to establish the range $[c_1, c_2]$ outside of which a value is considered to be an outlier:

$$[c_1, c_2] = [z_{(n/4)} - 1.5e^{-3.5\mathrm{MC}}\mathrm{IQR}, z_{(3n/4)} + 1.5e^{4\mathrm{MC}}\mathrm{IQR}] \quad \text{if } \mathrm{MC} \geq 0 \tag{8.75}$$

and

$$[c_1, c_2] = [z_{(n/4)} - 1.5e^{-4\mathrm{MC}}\mathrm{IQR}, z_{(3n/4)} + 1.5e^{3.5\mathrm{MC}}\mathrm{IQR}] \quad \text{if } \mathrm{MC} < 0 \tag{8.76}$$

where $\mathrm{IQR} = (z_{(3n/4)} - z_{(n/4)})$ is the interquartile range. We can use $c_1$ and $c_2$ to compute an outlyingness value for each point $\boldsymbol{x}_i$, projecting it over a random set of directions $\{\boldsymbol{v}_h\}$:

$$\mathrm{AO}_i = \max_{\boldsymbol{v}_h} \frac{|\Delta_{\boldsymbol{v}}|}{(c_2(\boldsymbol{v}) - \mathrm{med}_{\boldsymbol{v}})\theta(\Delta_{\boldsymbol{v}}) + (\mathrm{med}_{\boldsymbol{v}} - c_1(\boldsymbol{v}))\theta(-\Delta_{\boldsymbol{v}})} \tag{8.77}$$

where $\mathrm{med}_{\boldsymbol{v}} = \mathrm{med}_j(\boldsymbol{v}^T\boldsymbol{x}_j)$, $\Delta_{\boldsymbol{v}} = \boldsymbol{v}^T\boldsymbol{x}_i - \mathrm{med}_{\boldsymbol{v}}$, and $c_1$ and $c_2$ are computed over the set $\{\boldsymbol{v}^T\boldsymbol{x}_i\}_{i=1}^N$. If the outlyingness of a point exceeds the cutoff value $\theta_{\mathrm{AO}} = \mathrm{AO}_{(3n/4)} + 1.5e^{4\mathrm{MC}}\mathrm{IQR}$, that point should not be used for ICA (Brys *et al.* 2005).

**ICA I** results in independent images contained in matrix $S$ while the mixing coefficients are contained in matrix $A$. It produces spatially localized image features that are influenced only by a small region of the image. The fact that ICA I produces localized image features suggests that it is better suited to tasks where templates exhibit localized changes, such as the recognition of facial expressions, representing localized modifications of faces. This advantage has been experimentally verified.

**ICA II** results in independent coefficients while the ICA image basis is contained in $A$. It corresponds to a factorial code representation, i.e. a vector coding of data with code components statistically independent, producing global features influenced by the whole pattern.

The different structure of ICA I and ICA II should be considered also when comparing them to other, similar, projection approaches such as PCA. In particular, it is unfair to compare ICA (both versions) to plain PCA, without data sphering: the comparison should be performed with a similarly normalized PCA.

The description of the FastICA algorithm suggested data sphering as an appropriate ICA preprocessing. The specific ICA transformation is then given by an orthogonal matrix that we may further restrict to belong to rotations. It is then apparent that, if we limit ourselves to a

**Algorithm 8.2**: The FastICA algorithm

**Data**: A dataset $\{\boldsymbol{x}_i\}_{i=1}^{N}$, a contrast function $g(\cdot)$.
**Initialization**: Generate an initial random vector $\boldsymbol{w}_0$, $t = 0$.

1 Update the candidate independent direction:

$$\boldsymbol{w}'_{t+1} = E\{\boldsymbol{x}g'(\boldsymbol{w}_t^T\boldsymbol{x})\} - E\{g''(\boldsymbol{w}_t^T\boldsymbol{x})\}\boldsymbol{w}_t. \tag{8.78}$$

2 Enforce the normalization constraint

$$\boldsymbol{w}_{t+1} = \frac{\boldsymbol{w}'_t}{\|\boldsymbol{w}_t\|}. \tag{8.79}$$

3 **if** $\boldsymbol{w}_{t+1}^T\boldsymbol{w}_t < 1 - \epsilon$ **then**
  | let $t \rightarrow t + 1$ and go to Line 1
**else**
  | **return** *The independent direction* $\boldsymbol{w}_{t+1}$
**end**

rotation-invariant classifier and we keep all of the components, there cannot be any significant advantage in using ICA over PCA, provided that the two approaches are compared on an equal footing: for each ICA architecture the corresponding PCA base whitening computation must be performed. The situation becomes different when we select a subset of the directions computed by ICA or PCA for classification. Even if data are whitened, the significance of the directions for classification varies in the two approaches. In a classification task a practical feature selection strategy is that of progressively incorporating the directions based on their classification effectiveness. However, the resulting sequential feature selection process is not necessarily optimal. The performances of ICA and PCA are no longer necessarily similar and experimental results confirm this. However, even in this case, there is no clear winner. Let us close the section by remarking that ICA, as PCA, is valuable as a dimensionality reduction tool: classification in the reduced space is then performed by an appropriate classifier that may be a simple nearest neighbor classifier based on the $L_1$ or $L_2$ metric, the Mahalanobis distance, the cosine similarity, or a more complex classifier, e.g. a Bayes classifier (described in Chapter 3) or a support vector machine (described in Chapter 12).

## 8.4. Linear Discriminant Analysis

The criteria optimized by the two approaches considered so far did not explicitly address classification performance. The subset of directions chosen was not selected to optimize classification but rather to minimize reconstruction error (PCA) or maximize component independence (ICA). Both PCA and ICA can provide a complete basis for the representation of data and they both offer a way to sort the basis vectors to perform dimensionality reduction, preserving the most descriptive directions. From the perspective of pattern recognition, a natural goal is to transform the patterns from the original high-dimensional space into a lower-dimensional space hopefully preserving all available discriminant information. If we restrict consideration to linear transformations of the original data, the approach is known as

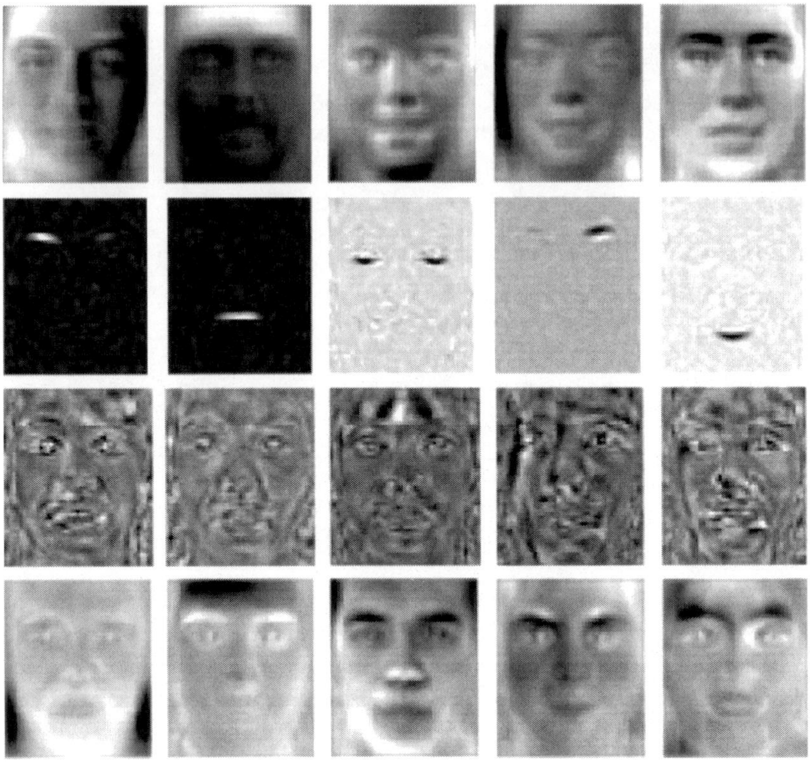

**Figure 8.8.** Respectively from top to bottom, the images corresponding to the directions computed by PCA, ICA I, ICA II, LDA. Source: Delac *et al.* (2005).

linear discriminant analysis and was originally proposed by Fisher for the two-class problem and generalized by Rao to the multiclass problem. The first step is the introduction of an appropriate criterion whose maximization will provide the desired transformation. As we want to maximize the discriminability of the different classes, we can maximally spread the different classes while at the same time collapsing each of them. This can be quantified with the introduction of the (unbiased) between- and within-class covariance matrices, $S_b$ and $S_w$, from which the (unbiased) total scatter matrix $S_t$ can be computed:

$$S_b = \frac{1}{n_c - 1} \sum_{c=1}^{n_c} N_c(\boldsymbol{\mu}_c - \boldsymbol{\mu})(\boldsymbol{\mu}_c - \boldsymbol{\mu})^T \tag{8.80}$$

$$S_w = \frac{1}{N - n_c} \sum_{c=1}^{n_c} \sum_{\boldsymbol{x}_i \in C_c} (\boldsymbol{x}_i - \boldsymbol{\mu}_c)(\boldsymbol{x}_i - \boldsymbol{\mu}_c)^T \tag{8.81}$$

$$S_t = \frac{1}{N - 1}[(n_c - 1)S_b + (N - n_c)S_w] \tag{8.82}$$

where $n_c$ is the number of classes, $N$ is the number of samples, and $N_c$ is the number of samples of class $C_c$. The definitions reported above show that $\text{rank}(S_b) \leq n_c - 1$, and that $\text{rank}(S_w) \leq N$ as it is the sum of $N$ rank 1 matrices. We want to find a set of directions $\boldsymbol{w}_i$

maximizing the following separability criterion:

$$J(\boldsymbol{w}) = \frac{\boldsymbol{w}^T S_b \boldsymbol{w}}{\boldsymbol{w}^T S_w \boldsymbol{w}}. \tag{8.83}$$

The number of different directions is limited by $\text{rank}(S_b) = q$, so that we cannot find more than $n_c - 1$ directions. The optimization problem can then be formalized as

$$\hat{\boldsymbol{w}}_i = \underset{\boldsymbol{w}}{\text{argmax}} \ J(\boldsymbol{w}), \quad i = 1, \ldots, n_c - 1. \tag{8.84}$$

We temporarily assume that $S_w$ is non-singular. The problem can be solved by using the method of Lagrange multipliers and introducing the constraint $\boldsymbol{w}^T S_w \boldsymbol{w} = 1$ that fixes the arbitrary scale factor of the problem. This leads to

$$(S_b - \lambda S_w)\boldsymbol{w} = 0 \tag{8.85}$$

and, for the assumed invertibility of $S_w$,

$$(S_w^{-1} S_b - \lambda I)\boldsymbol{w} = 0. \tag{8.86}$$

The eigenvectors corresponding to the non-null eigenvalues provide the required basis from which we can construct a transformation matrix $W \in \mathbb{R}^{q \times n_d}$ so that $\boldsymbol{y} = W\boldsymbol{x}$ represents the projection of a pattern from the original space $\mathbb{R}^{n_d}$ into the lower-dimensional one $\mathbb{R}^q$, $q < n_c$. The between and within scatter matrices in the transformed space are

$$S'_b = W^T S_b W \tag{8.87}$$

$$S'_w = W^T S_w W. \tag{8.88}$$

The solution found by Equation 8.86 also solves the following optimization problems:

$$\hat{W}_1 = \underset{W}{\text{argmax}} \frac{\det(W^T S_b W)}{\det(W^T S_w W)} \tag{8.89}$$

$$\hat{W}_2 = \underset{W}{\text{argmax}} \ \text{tr}\left(\frac{W^T S_b W}{W^T S_w W}\right) \tag{8.90}$$

which are often found as alternative optimality criteria for linear discriminant analysis. In the two-dimensional case, where we can compute a single direction $\boldsymbol{w}_1$, we have from Equation 8.80 that

$$(n_c - 1)S_b = S_b = \frac{N_1 N_2}{N}(\boldsymbol{\mu}_1 - \boldsymbol{\mu}_2)(\boldsymbol{\mu}_1 - \boldsymbol{\mu}_2)^T \tag{8.91}$$

and from Equation 8.86 that

$$S_w^{-1}(\boldsymbol{\mu}_1 - \boldsymbol{\mu}_2)(\boldsymbol{\mu}_1 - \boldsymbol{\mu}_2)^T \boldsymbol{w}_1 = \lambda \boldsymbol{w}_1 \tag{8.92}$$

leading to

$$\boldsymbol{w}_1 = S_w^{-1}(\boldsymbol{\mu}_1 - \boldsymbol{\mu}_2), \tag{8.93}$$

which corresponds to the Bayes solution reported by Equation 3.50 for the case of two signals distributed according to $N(\boldsymbol{\mu}_1; S_w)$ and $N(\boldsymbol{\mu}_2; S_w)$. The equivalence to the Bayes solution does not hold in general. It is possible to show that the between-class scatter matrix $S_b$ can be written as

$$\frac{(n_c - 1)}{N} S_b = \sum_{i=1}^{n_c-1} \sum_{j=i+1}^{n_c} \frac{N_i}{N} \frac{N_j}{N} (\boldsymbol{\mu}_i - \boldsymbol{\mu}_j)(\boldsymbol{\mu}_i - \boldsymbol{\mu}_j)^T \tag{8.94}$$

and that the Fisher criterion can be decomposed into $n_c(n_c - 1)/2$ two-class Fisher criteria

$$J_F(W) = \sum_{i=1}^{n_c-1} \sum_{j=i+1}^{n_c} p_i p_j \text{tr}[(W S_w W^T)^{-1} (W S_{ij} W^T)] \tag{8.95}$$

where $S_{ij} = (\boldsymbol{\mu}_i - \boldsymbol{\mu}_j)(\boldsymbol{\mu}_i - \boldsymbol{\mu}_j)^T$ and $p_i = N_i/N$. If we further assume that the within-class covariance is the identity matrix $I$ and that the transformation is orthonormal

$$J_F(W) = \sum_{i=1}^{n_c-1} \sum_{j=i=1}^{n_c} p_i p_j \text{tr}(W S_{ij} W^T) \tag{8.96}$$

$$= \sum_{i=1}^{n_c-1} \sum_{j=i=1}^{n_c} p_i p_j (W\boldsymbol{\mu}_i - W\boldsymbol{\mu}_j)(W\boldsymbol{\mu}_i - W\boldsymbol{\mu}_j)^T. \tag{8.97}$$

The Fisher criterion is then reduced to the spread of the class means, clearly not the same thing as the classification error. A way to adapt the Fisher approach to Bayesian classification is considered in Intermezzo 8.6.

Let us now turn to the eigenvalue problem whose solution provides us with the required transformation. The between scatter matrix, by definition, is the sum of $n_c$ matrices of rank 1 but only $n_c - 1$ of them are independent (since they are computed with respect to the mean of the means): the dimensionality of the arrival space is at most $n_c - 1$. This means that a lot of information might get discarded in the process. In many cases, things are actually worse. When $N < n_d$, i.e. the number of samples is less than the dimension of the original space, the covariance matrix $S_w$ is singular. The first consequence of this is that the generalized eigenproblem of Equation 8.85 cannot be transformed into a standard one. Furthermore, it is not uncommon that $n_d \gg N$, making even the storage of the matrices a challenge. And, of course, the problems highlighted in Chapter 3 regarding the estimation of the covariance matrices remain. The problems can be solved by proper interpretation of the original Fisher criterion. By construction, the matrices $S_b$, $S_w$, and $S_t$ are positive semidefinite so that the null space $\ker(S_t) = \ker(S_b) \cap \ker(S_w)$. If we represent a vector as the sum of two components lying respectively in the kernel of $S_t$ and in its complement, we can see that only the latter contributes to the value of the criterion: the one in the kernel is also in the kernel of $S_b$, therefore nullifying the value of the criterion. We can then perform a singular value decomposition of the $S_t$ total scatter matrix and project onto its eigenvectors with non-null eigenvalues without losing any discriminatory information (with respect to the criterion employed). This step takes care of the situation $n_d > N$ as $\text{rank}(S_t) = m < N$. Let us now consider $S_w$ and let $l = \text{rank}(\ker(S_w) \cap \text{range}(S_t))$. Since we have removed the null space of $S_t$, i.e. the intersection of the null spaces of $S_b$ and $S_w$, for each vector $\boldsymbol{c} \in \ker(S_w)$ we have that $\boldsymbol{c}^T S_w \boldsymbol{c} = 0$ and $\boldsymbol{c}^T S_b \boldsymbol{c} > 0$ so that the Fisher criterion can be replaced by a simpler one

**Intermezzo 8.6.** A classification-tuned LDA

In an attempt to improve the correspondence of the Fisher criterion to that of optimal classification we can modify Equation 8.95 by weighting the contribution of the between-class matrices

$$J_F(A) = \sum_{i=1}^{n_c-1} \sum_{j=i+1}^{n_c} p_i p_j \omega(\Delta_{ij}) \mathrm{tr}[(A S_w A^T)^{-1}(A S_{ij} A^T)] \tag{8.98}$$

where $\omega(\cdot)$ is a positive weighting function whose argument $\Delta_{ij}$ is the Mahalanobis distance of the means of class $i$ and $j$:

$$\Delta_{ij} = \sqrt{(\boldsymbol{\mu}_i - \boldsymbol{\mu}_j)^T S_w^{-1} (\boldsymbol{\mu}_i - \boldsymbol{\mu}_j)}. \tag{8.99}$$

We may choose $\omega(\cdot)$ in such a way that, in the case of two Gaussian classes whose within-class scatter matrix is the identity, we obtain a quantity proportional to the accuracy of the corresponding Bayes classifier. This is achieved if we set

$$\omega(\Delta_{ij}) = \frac{1}{2\Delta_{ij}^2} \mathrm{erf}\left(\frac{\Delta_{ij}}{2\sqrt{2}}\right). \tag{8.100}$$

In fact, if we set $\boldsymbol{\mu}_{ij} = \boldsymbol{\mu}_i - \boldsymbol{\mu}_j$ and select direction $\boldsymbol{v} = \boldsymbol{\mu}_{ij}/\|\boldsymbol{\mu}_{ij}\|$, we obtain

$$\omega(\Delta_{ij})\, \mathrm{tr}(\boldsymbol{v} S_{ij} \boldsymbol{v}^T) = \frac{1}{2\Delta_{ij}^2} \mathrm{erf}\left(\frac{\Delta_{ij}}{2\sqrt{2}}\right) \frac{\mu_{ij}^T}{\|\mu_{ij}\|} \boldsymbol{\mu}_{ij} \boldsymbol{\mu}_{ij}^T \frac{\boldsymbol{\mu}_{ij}}{\|\boldsymbol{\mu}_{ij}\|} = \frac{1}{2} \mathrm{erf}\left(\frac{\Delta_{ij}}{2\sqrt{2}}\right) \tag{8.101}$$

corresponding (up to an additive constant 1/2) to the accuracy of the Bayes classifier in the two-class problem. The reasoning can be easily extended to the case where the within-class scatter matrix is not the identity matrix but is the same for all classes. In this case we simply whiten the original vectors by applying the linear transform $S_w^{-1/2}$ to recover the case discussed. Denoting by $L'$ the dimensionality reduction operator for the whitened data, we obtain the projection operator for the original vectors as $L = L' S_w^{-1/2}$ (Loog *et al.* 2001).

in terms of $S_b$. When $\boldsymbol{c} \in \ker^{\perp}(S_w)$ the original Fisher criterion is well defined and can be maximized following the original recipe. The computations are summarized in Algorithm 8.3 and a visual representation of the directions found in a face classification task is reported in Figure 8.8.

**Algorithm 8.3**: Linear discriminant analysis

1. Compute the projection matrix $P_t \in \mathbb{R}^{m \times n_d}$ from the original space onto range($S_t$).
2. Compute the projection matrix $P_w \in \mathbb{R}^{r \times m}$ from the original space onto range($S_w$).
3. Compute the projection matrix $P_{w0} \in \mathbb{R}^{(m-r) \times m}$ from range($S_t$) onto ker($S_w$).
4. Compute the $l$ optimal eigenvectors from $P_{w0}^T P_t S_b P_t P_{w0}$; due to the projection onto the null space of $S_w$, we have that $l \leq (n_c - 1)$.
5. If $l < (n_c - 1)$, extract the remaining eigenvectors from

$$\frac{P_w^T P_t^T S_b P_t P_w}{P_w^T P_t^T S_w P_t P_w}. \tag{8.102}$$

Efficient implementations of the algorithm exist (see the references in Section 8.6) and improved classification performances have been reported for several tasks including face recognition.

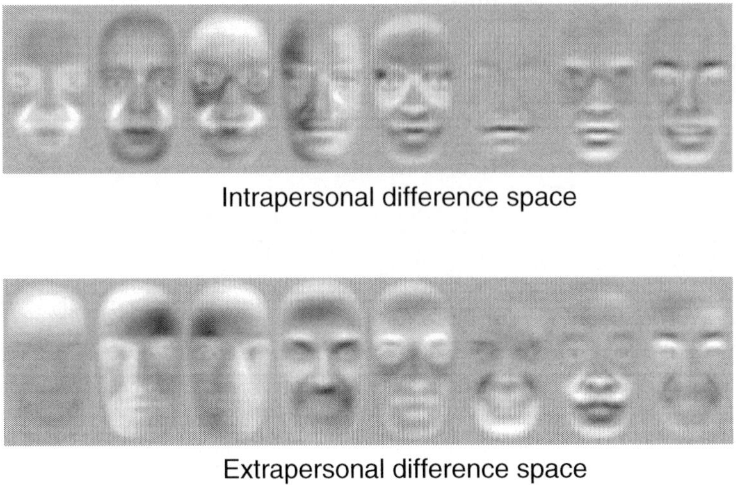


**Figure 8.9.** The principal components of the intrapersonal and extrapersonal difference space. Source: Moghaddam *et al.* (2000).

### 8.4.1. BAYESIAN DUAL SPACES

Fisher LDA focuses on the distinction of between-class scatter and within-class scatter. This type of LDA has two major limitations:

1. It shares with standard PCA the lack of a probabilistic foundation.

2. It is optimal, from a classification perspective, only when discriminating between two Gaussian distributions of equal variance.

However, the idea of separately modeling a space of within-class and between-class differences can be pushed further. Let us consider our favorite example, face recognition. It is plausible that the differences between images of the same person can be distributed in a way that differs significantly from the differences between images of different persons. If we limit ourselves to frontal images, we may expect intrapersonal differences to be more subtle and due to changes in expression and illumination, whereas extrapersonal differences are richer in structure spanning a larger subspace. A natural approach is then to model separately the spaces of interpersonal $\Omega_I$ and extrapersonal variations $\Omega_E$ using a PCA approach (see Figure 8.9).

Let $\{I^j_k\}$ be a set of images where $j$ identifies the individual and $k$ one of the corresponding images. The space of interpersonal variations $\Omega_I$ models the distribution of the set $\{\Delta^{jj}_{kl} = (I^j_k - I^j_l)\}_{k\neq l}$, $j = 1, \ldots, m$, while $\Omega_E$ models the distributions of $\{\Delta^{ji}_{kl} = (I^j_{k_j} - I^i_{l_i})\}_{j\neq i}$. If we assume that the two probability densities can be approximated by a multivariate Gaussian distribution, we obtain

$$\begin{aligned} P(\boldsymbol{\Delta}|\Omega_I) &= \frac{e^{-\frac{1}{2}\boldsymbol{\Delta}^T \Sigma_I^{-1} \boldsymbol{\Delta}}}{(2\pi)^{n_d/2}|\Sigma_I|^{1/2}} \\ P(\boldsymbol{\Delta}|\Omega_E) &= \frac{e^{-\frac{1}{2}\boldsymbol{\Delta}^T \Sigma_E^{-1} \boldsymbol{\Delta}}}{(2\pi)^{n_d/2}|\Sigma_E|^{1/2}} \end{aligned} \tag{8.103}$$

characterized by the two covariance matrices $\Sigma_I$ and $\Sigma_E$. An important characteristic of the two distributions is that they are both zero mean due to the fact that for each $\boldsymbol{\Delta}_{kl}^{ji}$ there is $\boldsymbol{\Delta}_{lk}^{ji} = -\boldsymbol{\Delta}_{kl}^{ji}$. An immediate consequence of this is that linear discriminant techniques cannot be expected to work (the between-class scatter matrix $S_b$ is singular) while MAP decision rules prove effective. The number of components $N_I$ required to model $\Omega_I$ accurately is lower than $N_E$, the number of components required for $\Omega_E$, in accordance with intuition. The $m$-ary classification problem for an unknown image $I$ reduces to the computation of $I - I_k^j$ for each image in the database, and to the application of an MAP rule to determine its identity:

$$\hat{k} = \underset{k}{\operatorname{argmax}} \frac{P(\boldsymbol{\Delta}_{l_k}^k|\Omega_I)P(\Omega_I)}{P(\boldsymbol{\Delta}_{l_k}|\Omega_I)P(\Omega_I) + P(\boldsymbol{\Delta}_{l_k}|\Omega_E)P(\Omega_E)} \tag{8.104}$$

where $l_k$ runs over the image set of person $k$ and $\hat{k}$ is the assigned identity. The computations required by Equation 8.104 can be carried out efficiently by precomputing some of the terms. As the computation of the conditional probabilities (the Mahalanobis distance) can be carried out efficiently in the diagonalized PCA representation, all images stored in the database can be preprojected onto the corresponding spaces and the resulting projections whitened by multiplication by $\lambda_i^{-1}$, $\lambda_i^2$ being an eigenvalue of the appropriate covariance matrix. If we denote by $\boldsymbol{y}_\Omega$ the resulting sphered vectors we have that

$$\begin{aligned} P(\boldsymbol{\Delta}_{l_k}^k|\Omega_I) &= \frac{e^{-\frac{1}{2}\|\boldsymbol{y}_\Omega - \boldsymbol{y}_{l_k,\Omega_I}^k\|}}{(2\pi)^{N_I/2}|\Sigma_I|^{1/2}} \\ P(\boldsymbol{\Delta}_{l_k}^k|\Omega_E) &= \frac{e^{-\frac{1}{2}\|\boldsymbol{y}_\Omega - \boldsymbol{y}_{l_k,\Omega_E}^k\|}}{(2\pi)^{N_E/2}|\Sigma_E|^{1/2}} \end{aligned} \tag{8.105}$$

where only the norms in the two subspaces need to be computed for the new image.

## 8.5. A Sample Application: Photographic-quality Facial Composites

The treatment of low-dimensional representations for patterns has been focused on detecting and recognizing them. In the case of PCA, while the goal of the representation is actually that of minimizing the reconstruction error of the patterns with a given number of components, the reconstruction itself has been mainly instrumental in the detection process, e.g. to compute the reconstruction error as an effective hypothesis testing tool. There is, however, an application of PCA where the synthesis of patterns from their coefficients, according to Equation 8.16, assumes a more central role while remaining linked to recognition. The task we are referring to is the generation of photographic-quality composites, like those compiled in police departments based on eyewitnesses reports. Before going into the details of how PCA can be exploited, let us briefly analyze the task.

Eyewitnesses play an important role in the investigative aspects of police work. When a suspect is unknown to the witness and not in custody, there are two main methods to obtain a description:

1. Showing to the eyewitness a set of photographs from the available police archives.

2. Creating a pictorial reconstruction of the facial features of the suspect based on the visual memory of the eyewitness.

The first approach has two major drawbacks:

1. If the suspect is not within the archives it is useless.

2. Successful inspection of a large photographic archive requires an uncommon ability to focus the attention for a long period.

The second approach, the pictorial reconstruction of the suspect, also has two significant drawbacks: the inherent difficulty of recalling a good mental image, and making this description explicit, textualizing it so that a skilled artist can translate it back to visual data. Traditional systems for the creation of photographic-quality composites assume faces to be perceived as a set of independent features. This simplifying assumption conflicts with reported psychophysical evidence suggesting the existence of a global comparison process coupled to a sequential one. Furthermore, some systems do not even allow the position of the features to be changed within the face. Another major drawback of traditional systems is the limited number of alternatives available for each facial feature. Using photo-retouching tools to modify the results obtained with these systems destroys the possibility of using the identifying codes associated with the specific features as a comparison key with other mug-shots built from the same set. In many cases, alternative features cannot be browsed in their original context, a full face, but as independent items, e.g. a photo-album of noses, a fact that may hinder their correct selection. The expansion of patterns into a restricted set of principal components allows all of these limitations to be overcome:

1. Changing the expansion coefficients and reconstructing the coded pattern provide an unlimited number of alternatives;

2. PCA provides a very compact representation that can be used to compare images at high speed, supporting real-time feedback from the database to the user.

Let us consider how such a system can be built. A set of $N$ images, from police archives, is geometrically normalized (fixing the length and orientation of the interocular segment) and their overall intensity adjusted (e.g. using histogram equalization or simple scaling so that reference image patches have predefined average intensity values). Following the idea of modular eigenspaces, each face can be decomposed into a set of $n_F$ features: the hair, eyes, nose, mouth, and chin. To each feature $F$ we associate a region $r_F$ whose size corresponds to the average size of the feature in the database used and we define an easily detectable point to be used as the feature center, e.g. the center of the pupil or of the bridge of the nose. For each feature, we extract from the $N$ database images a set of subimages $\{F_i\}_{i=1}^{N}$ that we describe in terms of the first $K$ principal components:

$$F_i = \sum_{j=1}^{K} c_{F_i,j} \boldsymbol{e}_j + \boldsymbol{e}_0, \quad i = 1, \ldots, N. \tag{8.106}$$

Each face image can then be described by a set of vectors $\{\boldsymbol{c}_F\}_{F=1,\ldots,n_F}$. The effectiveness of the description increases if the database is partitioned according to race, sex, and age before applying PCA as the resulting pattern distributions are more compact and can be described with fewer principal components. If the database originally used for the computation of the eigenvectors is representative, there is no need to recompute the eigenvectors when new images are added. The compactness of the PCA description and the speed at which a pattern can be reconstructed from its coefficients can be exploited by a graphical interface building a photographic-quality composite by adding all available features to the average database face.

Each feature is associated to a set of sliders, whose position is represented by a number $x_{Fi} \in [0, 1]$ that corresponds to an expansion coefficient $c_{Fi}$ such that $P(c_{Fi}) = x_{Fi}$ where $P(\cdot)$ is the cumulative probability distribution of the given coordinate. This association maximizes the selectivity of the slider as intervals (of $x_{Fl}$) of the same size cover the same number of database images. The feature is reconstructed and redisplayed whenever the slider is moved. The synthesis of a photographic-quality image by means of the composition of a set of features requires that the region representing each of them be blended with a background face image to avoid annoying artifacts. This can be achieved by using regions $R_F \supset r_F$ when synthesizing the patterns according to Equation 8.106 while restricting to regions $r_F$ the computation of the eigenfeatures $\boldsymbol{e}_i$. The frame between $r_F$ and $R_F$ is used to blend the features. Each of them is characterized by a map of blending weights $W_F$ whose size corresponds to that of the whole face and by a layer (or priority) value $L_F$ ($L_{\text{eyes}} > L_{\text{mouth}} > L_{\text{nose}} > L_{\text{hair}} > L_{\text{chin}} > L_{\text{face}}$). Each map has three different regions: a region external to $R_F$ filled with zeros, a region corresponding to $r_F$ filled with ones, and a transition region $R_F$–$r_F$ whose values belong to the interval (0, 1). Let us consider, to simplify notation, a one-dimensional feature: $r_F = [x_1, x_2]$, $R_F = [x_0, x_3]$, with $x_0 < x_1 < x_2 < x_3$. The two-dimensional case is handled by taking the product of two one-dimensional weight functions $W(x)$:

$$W(x) = \begin{cases} 0 & x < x_0, x > x_3 \\ 1 & x \in [x_1, x_2] \\ \frac{1}{2}\left[1 - \cos\left(\pi \frac{x - x_0}{x_1 - x_0}\right)\right] & x \in [x_0, x_1] \\ \frac{1}{2}\left[1 - \cos\left(\pi \frac{x - x_3}{x_2 - x_3}\right)\right] & x \in [x_2, x_3]. \end{cases} \tag{8.107}$$

If we denote by $W_1, W_2, \ldots$ the maps sorted by increasing layer (or priority) value, when the user modifies one of the sliders, the corresponding feature is synthesized and the composite image is built by sequential integration of features with increasing priority:

$$I = (((((1 - W_1)F_0 + W_1F_1)(1 - W_2) + W_2F_2)(1 - W_3) + W_3F_3) \cdots). \tag{8.108}$$

Alternative features are therefore built in their natural context, a face, overcoming one of the drawbacks of traditional systems. The low-dimensional vectors $\boldsymbol{c}_F$ can be used to compare

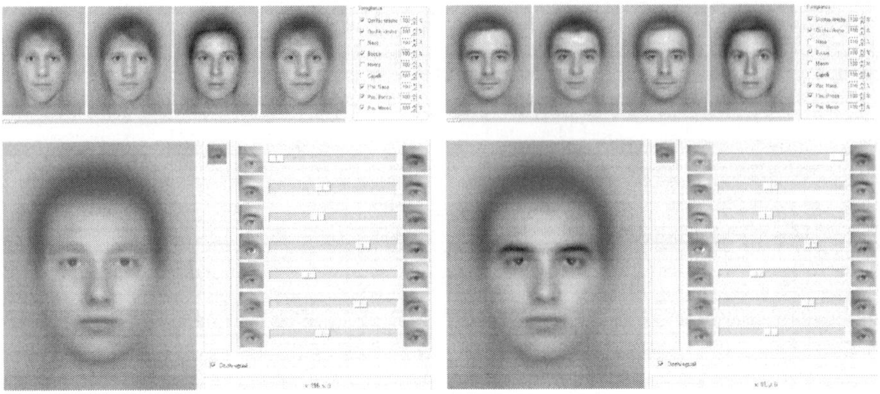

**Figure 8.10.** Variations of a single face, obtained by changing the first principal component of the eyes, and the corresponding most eye-similar images from a sample database.

the composite image being built to a database of images that can be presented to the user by decreasing similarity (see Figure 8.10). The composite image can then be used as an access key to the database, which in turn can be exploited to improve the building process: the coefficients of a facial feature from the database can be copied into the current description of the composite. Such a system can also accommodate a photo-retouching step without losing effectiveness as shown in Figure 8.11. The photo re-touched image can be itself described in terms of principal components and enter the loop.

## 8.6. Bibliographical Remarks

A detailed presentation of PCA and LDA can be found in the book by Fukunaga (1990) and also in Bishop (1995). The application of PCA to face representation is considered in the paper by Sirovich and Kirby (1987) and extended to cover modeling with partial data in Everson and Sirovich (1995). While the idea underlying PCA most naturally fits the $L_2$ norm, extensions to other metrics have been considered, most notably the $L_1$ case for its associated robustness (Gao 2008; Ke and Kanade 2005). A general approach to robustifying PCA from a representation perspective (as opposed to the computation of robust covariance matrices) is addressed in De la Torre and Black (2001, 2003). Nonlinear PCA by means of neural networks is described in Bishop (1995) and nonlinear kernel PCA is presented by Scholkopf *et al.* (1998). One of the first applications of PCA to face recognition is reported in Turk and Pentland (1991). The use of PCA in a Bayesian framework for face recognition is considered in Moghaddam and Pentland (1995) and Moghaddam *et al.* (2000) while the framework of probabilistic PCA is introduced by Tipping and Bishop (1999). The application of PCA to the creation of photographic-quality composites for police work is discussed in Brunelli and Mich (1996). The compact representation provided by PCA enables the use of techniques commonly used in image retrieval systems to exploit histogram-based image descriptions such as those analyzed in Brunelli and Mich (2000, 2001).

The optimal number of principal components is often chosen using cross-validation techniques which are computationally expensive. Bayesian estimation based on Laplace

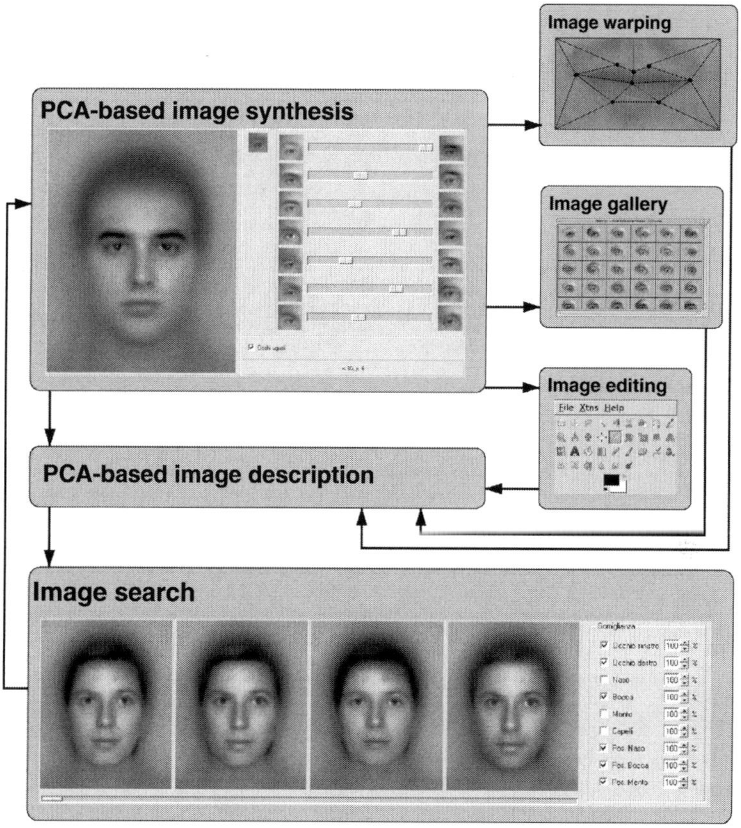


**Figure 8.11.** The architecture of SpotIt!, a system based on PCA for the generation of photographic-quality composites and for mug-shot database browsing. Besides modifying a facial feature by changing its principal component expansion, an image can be modified by moving a set of control points. The image is then automatically projected onto the principal components and used to improve the photo-composite or to browse a photo-album. The similarity of each facial feature to the stored ones can be separately weighted, e.g. to match eyewitness confidence.

approximation is described by Minka (2001) while the use of indicator functions is reviewed by Wasim and Brereton (2004).

The generalization of LDA considered in the chapter is due to Yang and Yang (2003) while the associated computational issues are discussed extensively by Tebbens and Schlesinger (2007). A comparison of this approach to competing solutions can be found in Park and Park (2008) where a generalization of LDA using kernel functions is also discussed. The reformulation of the multiclass Fisher criterion as a sum of pairwise criteria is presented by Loog *et al.* (2001) from which the classification-tuned LDA approach is also taken.

ICA, especially FastICA, is clearly described by Hyvarinen and Oja (2000). An extensive analysis of ICA applied to face recognition appears in Bartlett *et al.* (2002) and a comparative performance analysis of PCA, ICA, and LDA in face analysis tasks is reported by Delac *et al.* (2005). The equivalence of PCA and ICA approaches in appearance-based object recognition tasks is considered in Vicente *et al.* (2007), and Yang *et al.* (2007).

## References

Bartlett M, Movellan J and Sejnowski T 2002 Face recognition by independent component analysis. *IEEE Transaction on Neural Networks* **13**, 1450–1464.

Bishop C 1995 *Neural Networks for Pattern Recognition*. Oxford University Press.

Brunelli R and Mich O 1996 SpotIt! An Interactive Identikit System. *Graphical Models and Image Processing* **58**, 399–404.

Brunelli R and Mich O 2000 Image retrieval by examples. *IEEE Transactions on Multimedia* **2**, 164–171.

Brunelli R and Mich O 2001 Histogram analysis for image retrieval. *Pattern Recognition* **34**, 1625–1637.

Brys G, Hubert M and Rousseeuw P 2005 A robustification of independent component analysis. *Journal of Chemometrics* **19**, 364–375.

De la Torre F and Black M 2001 Robust principal component analysis for computer vision. *Proceedings of the 8th International Conference on Computer Vision and Pattern Recognition (ICCV'01)*, vol. 1, pp. 362–369.

De la Torre F and Black M 2003 A framework for robust subspace learning. *International Journal of Computer Vision* **54**, 117–142.

Delac K, Grgic M and Grgic S 2005 Independent comparative study of PCA, ICA, and LDA on the FERET data set. *International Journal of Imaging Systems and Technology* **15**, 252–260.

Everson R and Sirovich L 1995 The Karhunen–Loeve transform for incomplete data. *Journal of the Optical Society of America A* **12**, 1657–1664.

Fukunaga K 1990 *Statistical Pattern Recognition* 2nd edn. Academic Press.

Gao J 2008 Robust $L_1$ principal component analysis and its Bayesian variational inference. *Neural Computation* **20**, 555–572.

Hubert M and Engelen S 2007 Fast cross-validation of high-breakdown resampling methods for PCA. *Computational Statistics and Data Analysis* **51**, 5013–5024.

Hyvarinen A and Oja E 2000 Independent component analysis: algorithms and applications. *Neural Networks* **13**, 411–430.

Ke Q and Kanade T 2005 Robust $L_1$ norm factorization in the presence of outliers and missing data by alternative convex programming. *Proceedings of the IEEE Conference on Computer Vision and Pattern Recognition (CVPR'05)*, vol. 1, pp. 739–746.

Loog M, Duin R and Haeb-Umbach R 2001 Multiclass linear dimension reduction by weighted pairwise Fisher criteria. *IEEE Transactions on Pattern Analysis and Machine Intelligence* **23**, 762–766.

Meytlis M and Sirovich L 2007 On the dimensionality of face space. *IEEE Transactions on Pattern Analysis and Machine Intelligence* **29**, 1262–1267.

Minka T 2001 Automatic choice of dimensionality for PCA. *Proceedings of Advances in Neural Information Processing Systems*, vol. 13, pp. 598–604.

Moghaddam B and Pentland A 1995 Probabilistic visual learning for object recognition. *Proceedings of the 5th International Conference on Computer Vision and Pattern Recognition (ICCV'95)*, pp. 786–793.

Moghaddam B, Jebara T and Pentland A 2000 Bayesian face recognition. *Pattern Recognition* **33**, 1771–1782.

Park C and Park H 2008 A comparison of generalized linear discriminant analysis algorithms. *Pattern Recognition* **41**, 1083–1097.

Scholkopf B, Smola A and Muller K 1998 Nonlinear component analysis as a kernel eigenvalue problem. *Neural Computation* **10**, 1299–1319.

Sirovich L and Kirby M 1987 Low-dimensional procedure for the characterization of human faces. *Journal of the Optical Society of America A* **4**, 519–524.

Tebbens J and Schlesinger P 2007 Improving implementation of linear discriminant analysis for the high dimension/small sample size problem. *Computational Statistics and Data Analysis* **52**, 423–437.

Tipping M and Bishop C 1999 Probabilistic principal component analysis. *Journal of the Royal Statistical Society: Series B (Statistical Methodology)* **61**, 611–622.

Traver V, Bernardino A, Moreno P and Santos-Victor J 2004 Appearance-based object detection in space-variant images: a multi-model approach. *Proceedings of the International Conference on Image Analysis and Recognition*, Lecture Notes in Computer Science, vol. 3212, pp. 538–546. Springer.

Turk M and Pentland A 1991 Eigenfaces for recognition. *Journal of Cognitive Neuroscience* **3**, 71–86.

Vicente M, Hoyer P and Hyvärinen A 2007 Equivalence of some common linear feature extraction techniques for appearance-based object recognition tasks. *IEEE Transactions on Pattern Analysis and Machine Intelligence* **29**, 896–900.

Wasim M and Brereton R 2004 Determination of the number of significant components in liquid chromatography nuclear magnetic resonance spectroscopy. *Chemometrics and Intelligent Laboratory Systems* **72**, 133–151.

Yang J and Yang J 2003 Why can LDA be performed in PCA transformed space? *Pattern Recognition* **36**, 563–566.

Yang J, Zhang D and Yang J 2007 Constructing PCA baseline algorithms to reevaluate ICA-based face-recognition performances. *IEEE Transactions on Systems, Man, and Cybernetics, Part B* **37**, 1015–1021.

# 9 DEFORMABLE TEMPLATES

Play'd foul play with our oaths: your beauty, ladies,
Hath much deform'd us, fashioning our humours
Even to the opposed end of our intents:

*Loves Labours Lost*
WILLIAM SHAKESPEARE

There are cases that are not easily reduced to pattern detection and classification. One such a case is the detailed estimation of the parameters of a parametric curve: while Hough/Radon techniques may be sufficient, accurate estimation may benefit from specific approaches. Another important case is the comparison of anatomical structures, such as brain sections, across different individuals or even across time for the same person. Instead of modeling the variability of the patterns within a class as a static multidimensional manifold, we may focus on the constrained deformation of a parameterized model and measure similarity by the deformation stress.

## 9.1. A Dynamic Perspective on the Hough Transform

While the Hough transform is a flexible tool for the location of shapes, even when they are affected by some variability, accurate segmentation of the shapes it finds may be difficult. This difficulty is not specific to the Hough transform but affects all methods that rely on global information to locate shapes as they do not accurately model shape variations at small scales. Accurate segmentation can be achieved after the shape has been reliably located by using information at a pixel level. The Hough transform can then be seen as the first stage of a two-step process:

1. Locate a shape.
2. Segment the shape by locally driven refinement.

An important class of local shape refinement methods is based on the physical analogy of energy minimization. Let us consider again the Radon/Hough transform reported in Equation 7.23:

$$\mathcal{R}_{s(\boldsymbol{q})}(\mathrm{I}; \boldsymbol{q}) = \int_{\mathbb{R}^d} \mathrm{I}(\boldsymbol{x})\delta(\mathcal{K}(\boldsymbol{x}; \boldsymbol{q}))\, d\boldsymbol{x} \tag{9.1}$$

recalling that $\mathcal{K}(\boldsymbol{x}; \boldsymbol{q})$ is the set of constraints characterizing a given shape. Determining the optimal value $\hat{\boldsymbol{q}}$

$$\hat{\boldsymbol{q}} = \underset{\boldsymbol{q}}{\operatorname{argmax}}\ \mathcal{R}_{s(\boldsymbol{q})}(\mathrm{I}; \boldsymbol{q}) \tag{9.2}$$

*Template Matching Techniques in Computer Vision: Theory and Practice* Roberto Brunelli

**Intermezzo 9.1.** Random function minimization

Let $f(\boldsymbol{x})$ be the function to be minimized. Vector $\boldsymbol{x} \in \mathbb{R}^d$ represents the set of minimization parameters and we want to find a vector $\boldsymbol{x}_0$ belonging to the function domain for which the value of the function is minimal. In order to find such a value the domain must be explored and the function evaluated at several points. Let $\boldsymbol{x}$ be the point at which the function has just been computed. We try to get a lower value by perturbing in turn each of the coordinates $x_i$ by a quantity $r_i$. If the perturbation results in a lower value, $r_i$ will be doubled in the next iteration. Otherwise the reverse perturbation $-r_i$ is tried. If the function decreases, $r_i$ is halved if it does not, it is divided by 4. After modification of all of the coordinates a further perturbation $\boldsymbol{\Delta}(t)$ is applied

$$\boldsymbol{\Delta}(t) = \frac{\alpha}{W} \sum_k e^{-(t-t_k)^2/\mu^2} \boldsymbol{\delta}_k \tag{9.3}$$

where $\boldsymbol{\delta}_k$ is the jump of the probe point from the position at the beginning of the $k$th perturbation cycle to the final one, time $t = d^{-1} n_{\text{fe}}$ with $n_{\text{fe}}$ the number of function evaluations, and

$$W = \sum_k e^{-(t-t_k)^2/\mu^2}. \tag{9.4}$$

If the displacement of the probe point by $\boldsymbol{\Delta}(t)$ results in a lower function value, the algorithm memory defined by $(\alpha, \mu)$ is expanded by increasing both values. Otherwise it is decreased. The algorithm is similar to the Nealder–Mead simplex method but performs better in a variety of tasks (Brunelli 1994; Brunelli and Tecchiolli 1995).

with high accuracy requires a high-density sampling of parameter space.

The Radon/Hough transform receives a (positive) contribution only from image points that satisfy the constraints defining the shape: when they do not satisfy the constraints, even if they are very close to a valid shape, they do not contribute at all. This on/off response reduces the efficiency with which we can explore parameter space and it is one of the reasons why we need high-density sampling for accurate localization. Let us focus on a two-dimensional spatial domain and slightly modify the above equation into

$$\mathcal{R}_{s(\boldsymbol{q})}(\text{I}; \boldsymbol{q}) = \int_{\mathbb{R}^d} \bigg( \int \text{I}(\boldsymbol{x}')\text{V}_\epsilon(\|\boldsymbol{x} - \boldsymbol{x}'\|)\, d\boldsymbol{x}' \bigg) \delta(\mathcal{K}(\boldsymbol{x}; \boldsymbol{q}))\, d\boldsymbol{x} \tag{9.5}$$

where $\text{V}_\epsilon(\|\boldsymbol{x} - \boldsymbol{x}'\|) \xrightarrow{\epsilon \to 0} \delta(\boldsymbol{x} - \boldsymbol{x}')$ and $\text{V}_\epsilon(0) = 1$. The function

$$U(\boldsymbol{x}) = -\int \text{I}(\boldsymbol{x}')\text{V}_\epsilon(\|\boldsymbol{x} - \boldsymbol{x}'\|)\, d\boldsymbol{x}' \tag{9.6}$$

can be considered as a potential (energy) field and we recover the original formulation in the limit $\epsilon \to 0$. What we have gained with the new formulation is that the Radon/Hough transform gathers contributions from points near the correct shape and the closer they are to the shape, the higher their contribution. We may establish a physical analogy mapping (binary) image values to positive electric charges and points on the shape to negative electric charges: the shape senses the electric field generated by image values and the closer it is to the image charge distribution, the lower is its potential energy. The analogy also suggests how we can make use of the new formulation: we modify the parameters of the shape so that its potential energy decreases, and the shape itself gets closer to its supporting image positions:

$$\hat{\boldsymbol{q}} = \operatorname*{argmin}_{\boldsymbol{q}} \bigg( -\int_{\mathbb{R}^d} \text{U}(\boldsymbol{x})\delta(\mathcal{K}(\boldsymbol{x}; \boldsymbol{q}))\, d\boldsymbol{x} \bigg) \tag{9.7}$$

where the minimization can be performed using gradient descent techniques starting from an initial estimate or using the stochastic minimization technique presented in Intermezzo 9.1.

If we do not insist on $\mathrm{V}_\epsilon$ being a nascent delta function to recover the original formulation, we may conveniently assume the following functional dependence:

$$V(x;\, t) = \frac{w^2(t)}{x^2 + w^2(t)} = \left(1 + \frac{x^2}{w^2(t)}\right)^{-1} \tag{9.8}$$

where we have introduced a temporal dependence and

$$w(t) = b + (a - b)e^{-t/c} \tag{9.9}$$

introduces a scale factor that effectively determines the range of the potential (see Figure 9.1). The form of $V(x;\, t)$ allows us to carefully control the potential field in order to optimize the chances that energy minimization by gradient descent leads to a good shape parameter choice. Note that $w(0) = a$, $w(\infty) = b$, and $V(x;\, t) \to 1$ for $w \to \infty$ independent of $\boldsymbol{x}$ and $\boldsymbol{q}$. We can then devise a minimization strategy not dissimilar from simulated annealing, a technique that simulates the stochastic evolution of a system towards the state of minimum energy at the limit of low temperature. If $\Delta x$ represents the linear dimension of a pixel, we can choose $b \approx \Delta x$ and $a \gg b$. At the beginning of the minimization procedure all local minima will be washed out and only the major structures of the potential field will be detected. The time scale is fixed by $c$, the relaxation time, that should be larger than the mean convergence time of the energy minimization algorithm in order to let the system avoid the nuisance of local minima emerging when $w$ is small. The strategy can be easily modified to support the detection of multiple tracks. The problem in this case is that multiple tracks can be attracted towards the same image structure. In order to prevent this from happening, we need to add a repulsive field for each track found:

$$U_j(\boldsymbol{x}) = +\int \delta(\mathcal{K}(\boldsymbol{x}';\, \boldsymbol{q}))\mathrm{V}_\epsilon(\|\boldsymbol{x} - \boldsymbol{x}'\|)\, d\boldsymbol{x}', \tag{9.10}$$

leading to the following minimization problem:

$$\begin{aligned}\{\hat{\boldsymbol{q}}_1, \ldots, \hat{\boldsymbol{q}}_n\} = \operatorname*{argmin}_{\{\boldsymbol{q}_1,\ldots,\boldsymbol{q}_n\}} \Bigg( &- \sum_i \int_{\mathbb{R}^d} \mathrm{U}(\boldsymbol{x})\delta(\mathcal{K}(\boldsymbol{x};\, \boldsymbol{q}_i))\, d\boldsymbol{x} \\ &+ \sum_{i \neq j} \int_{\mathbb{R}^d} \mathrm{U_j}(\boldsymbol{x})\delta(\mathcal{K}(\boldsymbol{x};\, \boldsymbol{q}_i))\, d\boldsymbol{x}\Bigg).\end{aligned} \tag{9.11}$$

The minimization problem of Equation 9.11 can be solved in at least two different ways:

1. All shape estimates $\boldsymbol{q}_i$ move simultaneously, in parallel, resembling the trajectories of a multibody dynamical problem.

2. Trajectories are considered sequentially: each shape is considered in turn.

The first case often presents additional local minima that make the minimization process more complex. The second case is usually easier and it results in a sequence of minimization problems where each located shape adds a repulsive contribution to the original image

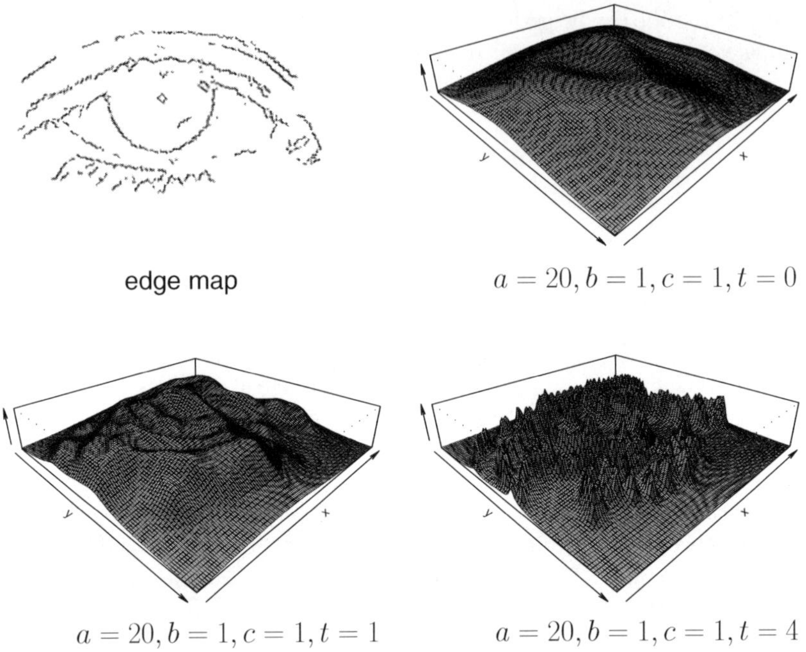


**Figure 9.1.** The potential field from the reported edge map at different time values: each instant corresponds to a different range $w$ of the potential defined according to Equation 9.9. When $t$ is small, the potential field is less structured but it is non-null over the complete image. When $t$ increases, a more detailed structure appears but the field is null or negligible over some regions.

potential field:

$$\hat{\boldsymbol{q}}_j = \underset{\boldsymbol{q}}{\operatorname{argmin}} \left( - \sum_i \int_{\mathbb{R}^d} \mathrm{U}(\boldsymbol{x}) \delta(\mathcal{K}(\boldsymbol{x}; \boldsymbol{q}_i)) \, d\boldsymbol{x} + \sum_{i<j} \int_{\mathbb{R}^d} \mathrm{U_j}(\boldsymbol{x}) \delta(\mathcal{K}(\boldsymbol{x}; \boldsymbol{q})) \, d\boldsymbol{x} \right). \tag{9.12}$$

The modification of the shape parameters, and the consequent deformation of the shape, in order to minimize the potential energy associated with the image, justifies the term *deformable template*. While this is not so engaging for a line template, where deformable reduces to rotation and translation, the eye template considered in the next section will better champion the idea. The main advantage of deformable templates over traditional template matching is that sampling of the multidimensional parameter volume is substituted with a more computationally convenient path-following strategy.

## 9.2. Deformable Templates

The Radon/Hough transform is usually applied to binary images and the definition of shape is purely geometrical: no textural information is used. Consequently, the potential energy driving shape deformation is defined only in terms of the presence/absence of edge information. However, template matching applications often require the detection of templates that are better characterized by the joint use of geometrical and textural information. The energy minimization strategy outlined above for purely geometrical shapes can be generalized to gray-level images and richer shape templates. The key to the generalization is the introduction of multiple scalar potential fields, each one associated with a different low-level image feature: edges, peaks, valleys, corners, and so on. The deformable template will then be built from multiple substructures, each of them interacting in a specific way with one or more of the image potentials. The computation of feature potentials does not require that the corresponding feature be located with accuracy: only a value quantifying the confidence in the presence of the feature is required. Direct implementation of Equation 9.6 is feasible only when a limited number of image points act as potential sources since each evaluation would require a full scan of the complete feature maps. When the distribution of the sources is dense, as it would be in the case of the image potentials considered above, it is necessary to resort to an approximate but more efficient strategy. Let $I_i$ represent the $i$th feature image map. In order to transform $I_i$ into a useful scalar potential field $I_i^{\phi}$ we need to associate to each image point $(x_1, x_2)$ a value of the potential reflecting the static distribution of the feature sources so that we do not need to recompute $I_i^{\phi}(x_1, x_2)$ each time it is needed for the computation of the energy of the deformable template. Furthermore, as discussed in some detail in the previous paragraphs, controlling the potential during the minimization process is important in order to avoid being trapped in local minima. A solution is given by the computation of a sequence of maps

$$I_i^{\phi}(\boldsymbol{x}; t_k) = (e^{-t_k f(r)} * I_i)(\boldsymbol{x}) \tag{9.13}$$

where $r = \sqrt{x^2 + y^2}$, $f(\cdot)$ is an increasing function of its argument satisfying $f(0) = 0$, and $\{t_k\}$ is a sequence of increasing values. From a computational point of view, it would be desirable for $f(\cdot)$ to be separable in order to speed up the computation of the convolution that must be performed for each $t_k$. The choice $f(r) = \sqrt{r}$ results in long-range interactions of the template with feature potentials, but it is not separable. The simple alternative $f(r) = r^2$ leads to (separable) Gaussian smoothing, but it decays too fast. The fast decay may, however, be balanced by an appropriate choice of the time values $t_k$: very small values result in very large smoothing and long-range interactions. We have already tackled the problem of detecting image edges, but the loose requirements for feature potential fields suggest the use of a simpler, faster, and general way to compute them: we simply convolve the intensity image with small kernels resembling the feature we want to gauge. The kernels, representing very simplified feature templates for edges, peaks, and valleys, are given respectively by

$$k_e^x = \begin{pmatrix} -1 & 0 & 1 \\ -2 & 0 & 2 \\ -1 & 0 & 1 \end{pmatrix} \quad k_e^y = (k_e^x)^T \tag{9.14}$$

$$k_p = \begin{pmatrix} -1 & -1 & -1 \\ -1 & 8 & -1 \\ -1 & -1 & -1 \end{pmatrix} \tag{9.15}$$

$$k_v = \begin{pmatrix} 1 & 1 & 1 \\ 1 & -8 & 1 \\ 1 & 1 & 1 \end{pmatrix} \tag{9.16}$$

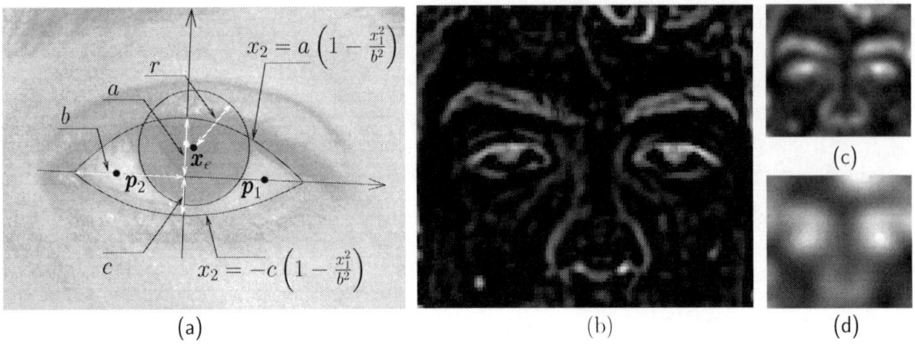


**Figure 9.2.** A deformable eye template (a) and the valley feature potentials (b) with different ranges obtained by smoothing the original map (c–d). The result is similar to the one presented in Figure 9.1 but it can be obtained much more efficiently by exploiting the separability of the smoothing Gaussian kernel.

and the corresponding feature maps are defined as

$$I_e = \sqrt{(k_e^x * I)^2 + (k_e^y * I)^2} \tag{9.17}$$

$$I_p = k_p * I \tag{9.18}$$

$$I_v = k_v * I. \tag{9.19}$$

As an illustrative example, let us consider the design of a deformable template for the detection of eyes. The main features of an eye are its bounding contour, the iris, and the sclera (see Figure 9.2). We can then split the eye model into five main parts:

1. The circle representing the iris, characterized by its radius $r$ and its center $\boldsymbol{x}_c$. The interior of the circle is attracted to the low-intensity values while its boundary is attracted to edges in image intensity.

2. Two points associated with the two visible regions of the sclera, separated by the iris, and further delimited by the upper and lower eyelids. They are attracted to peaks in the image intensity and they are identified by

$$\boldsymbol{x}_e + p_1(\cos\theta, \sin\theta)$$
$$\boldsymbol{x}_e + p_2(\cos\theta, \sin\theta)$$

where $p_1 \geq 0$ and $p_2 \leq 0$, the point $\boldsymbol{x}_e$ represents the center of the eye, and $\theta$ quantifies the orientation of the eye with respect to the horizontal axis.

3. The bounding contour of the eye, corresponding to the upper and lower eyelids, is characterized by two parabolic sections: it is centered at $\boldsymbol{x}_e$, it has maximum width $2b$, maximum height $a$, and minimum height $-c$, with $a, b, c > 0$; and it is attracted to edges. The equation of the upper boundary is

$$x_2 = a\left(1 - \frac{x_1^2}{b^2}\right) \tag{9.20}$$

while that of the lower one is

$$x_2 = -c\left(1 - \frac{x_1^2}{b^2}\right) \tag{9.21}$$

with $x_1 \in [-b, b]$.

The resulting template has 11 parameters that can be represented by the vector $\boldsymbol{q} = (\boldsymbol{x}_e, \boldsymbol{x}_c, p_1, p_2, r, a, b, c, \theta)$. In the previous section, where the potential-based version of the Radon/Hough transform was introduced, no constraints were (explicitly) imposed on the values of the parameters defining the shape: the path followed in parameter space was unconstrained. While for some simple shapes and image potential fields this poses no problems, it is usually necessary to limit the range of values that the parameters may assume in order to prevent the shape from collapsing onto itself and position itself erratically on the image. The constraints may be imposed explicitly or implicitly, with the introduction of internal energy terms that enforce a priori knowledge of the shape. The total energy of the deformable eye template can then be summarized as

$$E = E_v + E_e + E_i + E_p + E_{\text{prior}} \tag{9.22}$$

where:

1. The energy related to the valley potential is computed as the integral of $I_v^\phi$ over the interior of the iris circle divided by its area

$$E_v = -\frac{c_1}{|R_c|}\int_{R_c} I_v^\phi(\boldsymbol{x})\, d\boldsymbol{x}. \tag{9.23}$$

2. The edge potential is collected over the eyelid parabolas $\partial R_w$ (the external sclera boundary) and the iris boundary $\partial R_c$:

$$E_e = -\frac{c_2}{|\partial R_w|}\int_{\partial R_w} I_e^\phi(\boldsymbol{x})\, ds - \frac{c_3}{|\partial R_c|}\int_{\partial R_c} I_e^\phi(\boldsymbol{x})\, ds. \tag{9.24}$$

3. Image intensity must be maximized over the sclera $R_w$ and minimized over the iris $R_c$:

$$E_i = \frac{c_4}{|R_w|}\int_{R_w} I_i^\phi(\boldsymbol{x})\, d\boldsymbol{x} - \frac{c_5}{|R_c|}\int_{R_c} I_i^\phi(\boldsymbol{x})\, d\boldsymbol{x}. \tag{9.25}$$

4. The peak potential is evaluated at the two peak points

$$E_p = c_6[I_p^\phi(\boldsymbol{x}_e + p_1\boldsymbol{e}_1) + I_p^\phi(\boldsymbol{x}_e + p_2\boldsymbol{e}_2)]. \tag{9.26}$$

5. The internal, structural energy, preventing degeneration of the shape, is given by

$$\begin{aligned} E_{\text{prior}} = {} & k_1\|\boldsymbol{x}_e - \boldsymbol{x}_c\|^2 + k_2[p_1 - p_2 - (r + b)]^2 + k_3(b - 2r)^2 \\ & + k_4(b - 2a)^2 + (b - 2c)^2. \end{aligned} \tag{9.27}$$

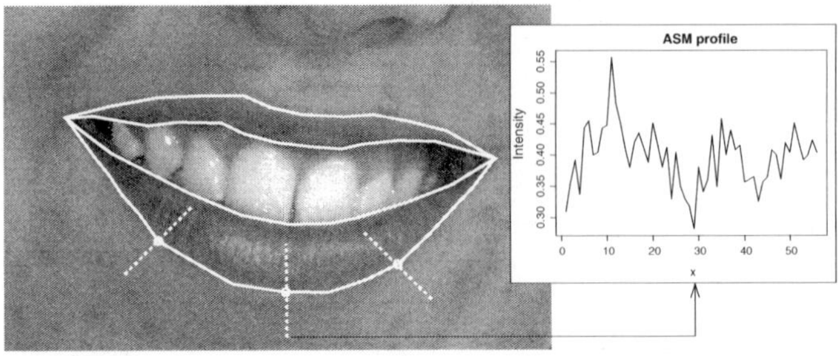


**Figure 9.3.** An active shape model for representing a mouth, with some of the corresponding control points and one intensity profile.

Given an initial parameter vector $\boldsymbol{q} = \boldsymbol{q}_0$, the shape is deformed by updating $\boldsymbol{q}$ using steepest descent on $E_c$

$$\left.\frac{dq_i}{dt}\right|_{t=t_k} = \left.\frac{\partial E_c(I_1^{\phi}, \ldots)}{\partial q_i}\right|_{t=t_k}. \tag{9.28}$$

The energy minimization process by steepest descent of Equation 9.28 takes advantage of potential warping described by Equation 9.13.

Successful design of a deformable template is often a complex task at the level of shape description, of image-derived potentials, and of the energy minimization process. The flexibility of deformable templates makes these systems inadequate for the task of localizing the shape within the image as a whole because they are often led astray by local image structure. Therefore, they are more appropriate as a refinement step than as fully fledged template detectors and the search in parameter space should be started close to the correct values. A task where the latter condition holds is that of feature tracking, after a successful initial feature localization step.

## 9.3. Active Shape Models

The complexity of design of the deformable templates considered in the previous section may be overcome using a radically different design known as active shape models (ASMs, see Figure 9.3 and Figure 9.4). They represent an object by:

- a set of points, each of them characterized by the corresponding position with respect to the object center;
- a visual signature, describing some visual characteristic of the point neighborhood.

Shape variability is modeled in a statistical way by applying PCA to the vector description of the shape obtained by concatenating all control point coordinates from a set of training images. The template is deformed by changing its principal components (also known as principal modes) and the optimal configuration is found by exploring the resulting principal deformation space and matching the visual signatures of the control points to the image.

Let us assume that we have a set of $N$ training images representing objects of the class we want to model, and that all images are geometrically normalized: overall scale and orientation

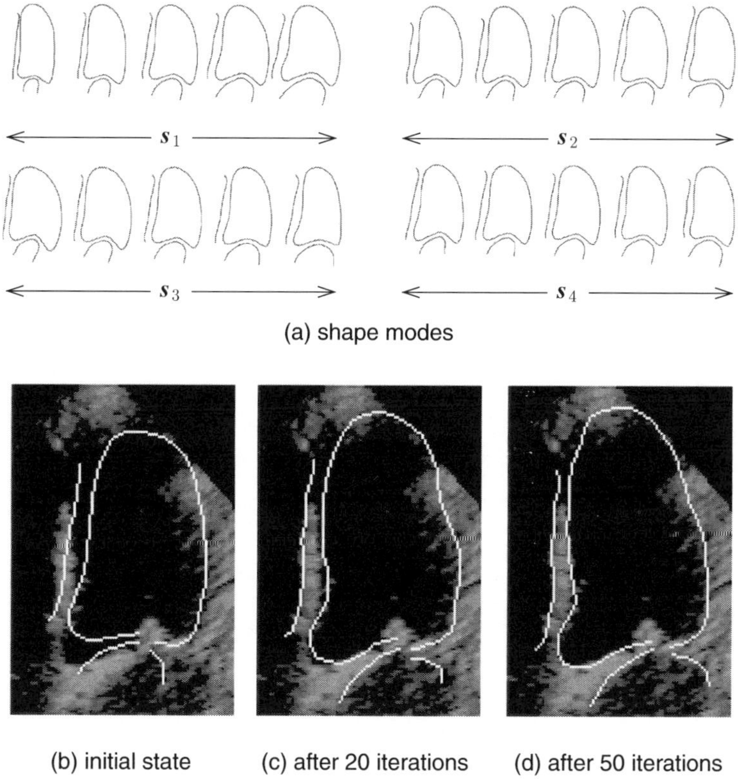


**Figure 9.4.** The principal directions of an ASM can be used for synthesis and analysis alike (see also Section 8.5). Constrained shape evolution allows the ASM to track images with low SNR. Source: Cootes *et al.* (1994).

are fixed. For each object instance we determine its boundary and select in a reproducible way a set of $n$ control points on it. A typical strategy is to first select a set of landmarks, and then to add additional points between them at regular intervals. Each shape in the training set is then represented by a $2n$-dimensional vector

$$z_i = (x_{i,0}, y_{i,0}, \ldots, x_{i,n-1}, y_{i,n-1})^T.$$

We impose the following normalization conditions on each shape:

$$\frac{1}{n}\sum_{j=1}^{n} x_{i,j} = 0 \quad \forall i = 1, \ldots, N$$

$$\frac{1}{n}\sum_{j=1}^{n} y_{i,j} = 0$$

$$\frac{1}{n}\sum_{j=1}^{n}(x_{i,j}^2 + y_{i,j}^2) = 0,$$

thereby fixing the origin and scale of the shape-centric reference coordinate system. Let us denote by $M(z; \alpha, \theta)$ the transformation that scales and rotates a shape $z_i$ acting on all constituent control points as

$$M(\boldsymbol{x}_{i,k}; \alpha, \theta, \boldsymbol{t}) = \begin{pmatrix} \alpha\cos(\theta)x_{i,k} - \alpha\sin(\theta)y_{i,k} + t_x \\ \alpha\sin(\theta)x_{i,k} + \alpha\cos(\theta)y_{i,k} + t_y \end{pmatrix} \tag{9.29}$$

where $\boldsymbol{t} = (t_x, t_y, \ldots, t_x, t_y)^T$ is the shape translation vector. From the set $\{\boldsymbol{x}_i\}_{i=1}^N$ we compute the average shape

$$\bar{z} = \frac{1}{N} z_i$$

the covariance matrix

$$\Sigma_s = \frac{1}{N-1}(z_i - \bar{z})(z_i - \bar{z})^T$$

and the matrix $S_{2n \times k_s}$ whose columns represent the first $k_s$ eigenvectors $\boldsymbol{s}_i$ of $\Sigma_s$, the so-called (shape) variation modes. Any shape in the class considered can then be approximated with a linear combination of the principal variation modes plus the average shape

$$z \approx \bar{z} + \sum_{i=1}^{k_s} c_i \boldsymbol{s}_i \tag{9.30}$$

where

$$\boldsymbol{c} = S^T(z - \bar{z}). \tag{9.31}$$

If we want to employ ASMs not only to describe shapes but also to spot them within an image, we need a matching function that tells us how good the image evidence is in supporting a given shape. ASMs can then be turned into local optimizers similar to the deformable templates considered in the previous section. We can associate to each control point a (pseudo-)intensity profile aligned along the boundary normal unit vector $\hat{\boldsymbol{n}}_i$ at point $i$

$$\boldsymbol{g}_i = (I^\phi(\boldsymbol{x}_i + (w_0 + 0)\hat{\boldsymbol{n}}_i), I^\phi(\boldsymbol{x}_i + (w_0 + 1)\hat{\boldsymbol{n}}_i), \ldots, I^\phi(\boldsymbol{x}_i + (w_0 + w - 1)\hat{\boldsymbol{n}}_i)) \tag{9.32}$$

where we assume that the normal is oriented outwards, $w_0$ represents the inner limit of the profile whose size is then $w$ pixels, and we have again introduced a feature map $I^\phi$. A common choice is $I^\phi = I_e^\phi$ representing image edges. In order to improve the local match of the ASM with image content we compute a new position for each control point by performing a linear search to optimize the match of its profile along its defining direction, i.e. the normal to the shape, obtaining a new candidate shape $z_1$.

As the shapes we are considering are geometrically aligned and normalized, translation, rotation, and scaling effects are not represented by the space of principal variations. In order to exploit the PCA representation of Equation 9.30 we need to associate to each shape the scale and rotation factor with respect to the original (normalized samples) so that we can go back to PCA space when needed.

Matching a shape $M(z_2; \alpha, \theta, \boldsymbol{t})$ to a shape $z_1$ requires the estimation of the four parameters $(\alpha, \theta, t_x, t_y)$ by minimization of the average squared error

$$\Delta_M = (z_1 - M(z_2; \alpha, \theta, \boldsymbol{t}))^T W (z_1 - M(z_2; \alpha, \theta, \boldsymbol{t})) \tag{9.33}$$

where $W$ is a diagonal matrix of weights for each point, possibly reflecting how good the profile match of Equation 9.32 is. The minimization of $\Delta_M$ leads to the following four linear equations:

$$\begin{pmatrix} X_2 & -Y_2 & W & 0 \\ Y_2 & X_2 & 0 & W \\ Z & 0 & X_2 & Y_2 \\ 0 & Z & -Y_2 & X_2 \end{pmatrix} \begin{pmatrix} a_x \\ a_y \\ t_x \\ t_y \end{pmatrix} = \begin{pmatrix} X_1 \\ Y_1 \\ C_1 \\ C_2 \end{pmatrix} \tag{9.34}$$

where

$$\begin{aligned} a_x &= \alpha \cos(\theta) & a_y &= \alpha \sin(\theta) \\ X_i &= \sum_{k=0}^{n-1} w_k x_{i,k} & Y_i &= \sum_{k=0}^{n-1} w_k y_{i,k} \\ Z &= \sum_{k=0}^{n-1} w_k (x_{2,k}^2 + y_{2,k}^2) & W &= \sum_{k=0}^{n-1} w_k \\ C_1 &= \sum_{k=0}^{n-1} w_k (x_{1,k} x_{2,k} + y_{1,k} y_{2,k}) & C_2 &= \sum_{k=0}^{n-1} w_k (y_{1,k} x_{2,k} - x_{1,k} y_{2,k}) \end{aligned}$$

which can be solved in a standard way. Having fixed the geometric transformation of the complete shape, we may further optimize shape matching by constrained deformation of the shape using its principal variation modes. As described at length in Chapter 8, this can be obtained by projecting the shape $z_1$ onto its principal component linear subspace. We need to undo the action of $M(z; \alpha, \theta, \boldsymbol{t})$ before computing the projection:

$$z' = M(z_1; \alpha^{-1}, -\theta, -\boldsymbol{t}) \tag{9.35}$$

$$\Delta_s = \|z' - \bar{z}\|^2 - \|S^T(z' - \bar{z})\|^2 = \|z' - \bar{z}\|^2 - \sum_{i=1}^{k_s} c_i^2 \tag{9.36}$$

and the quality of the match is gauged by $\Delta_s$. A variant of ASMs relies on a more pervasive use of subspace approximation. PCA can be applied to the profiles in the same way as it is applied to the coordinates: we simply concatenate the profile vectors, and compute the average (concatenated) profile $\bar{\boldsymbol{g}}$ and the matrix $P_{wn \times k_p}$ whose columns $\boldsymbol{p}_i$ represent the first $k_p$ principal directions of the profiles. If we denote by $\boldsymbol{g}$ the image shape profile, we have

$$\boldsymbol{g} \approx \bar{\boldsymbol{g}} + P^T(\boldsymbol{g} - \bar{\boldsymbol{g}}) \tag{9.37}$$

and we can use the following matching cost:

$$\Delta_g = \|\boldsymbol{g} - \bar{\boldsymbol{g}}\|^2 - \|P^T(\boldsymbol{g} - \bar{\boldsymbol{g}})\|^2. \tag{9.38}$$

The optimal shape can again be found attempting the minimization of an overall matching cost or the maximization of a probability closely resembling the probabilistic PCA. The actual probability maximization process requires the exploration of the shape parameter space $(\alpha, \theta, \boldsymbol{t}, \boldsymbol{c}, \boldsymbol{b})$ by means of techniques such as gradient descent or stochastic minimization (see Intermezzo 9.1).

## 9.4. Diffeomorphic Matching

In this section we want to shift our focus from the shapes being matched to the deformations that allow us to warp a shape into a similar one. If we consider the shapes resulting from linear combinations of the principal modes, we can see that they have a consistent quality as opposed, for instance, to the faces obtained by the linear combination of eigenfaces. This should come as no surprise, but the underlying reason is important: the vector descriptions of shapes are aligned in a true structural sense. Corresponding coordinates of shape descriptions identify structurally similar shape points. On the contrary, overlapping pixels of two face images do not necessarily share the same status due to the different shape, size, and location of the facial features. While this lack of alignment may not be so detrimental when templates must be located, it is a source of difficulty in many important tasks requiring the detailed comparison of complex shapes. Among these, the comparative analysis of medical images in general, and brain images in particular, stands out due to its importance. The complex structure of the brain, and its variability both across different individuals and throughout the life of a single human being, represent a challenge for the accurate image alignment required in medical studies. Image registration has many applications in medicine, among which we may cite deformable atlas registration and functional brain mapping. The former may be used to clarify the importance of dense, structural, matching of points between images representing the same structure. The template in this case is an anatomical atlas, e.g. an image representing a sample slice of the brain where important structures are identified. When a new image, the study, of the corresponding portion of the biological structure is considered, we want to map each point of the atlas into the study, so that all the information carried by the atlas is projected into the study. Additionally, we would like to know if the mapping is regular or if it presents some singularities, potential indicators of new structures such as tumors. In this case, the mapping between the atlas and the study is instrumental to the labeling of the study and it is the primary object of investigation for detecting anomalies. A different application of dense mapping between templates representing similar objects is that of graphical animation or image synthesis. In this case, knowledge of a dense, pixel-level correspondence enables, among other things, the animation of a face with the expressions of a different one (see Figure 9.5). Given two images $I_1$ and $I_2$, the computation of a dense mapping between them is not a trivial task and many techniques have been developed over the years. Among these we may cite:

1. Optical flow techniques, originally developed for matching moving images of the same scene and for stereo matching.

2. Elastic matching, finding the mapping field that minimizes a deformation cost computed as the elastic stress needed to deform one image, considered as a continuous medium, into another one.

3. Viscous fluid matching, modeling the mapping field as the motion of a fluid driven by local image mismatch.

4. Diffeomorphic matching, explicitly enforcing invertibility and smoothing by the minimization of an appropriate energy functional.

We briefly describe the idea behind a specific implementation of diffeomorphic matching and refer the reader to Section 9.5 for references to fluid matching and to Intermezzo 9.2 for optical flow.

**Intermezzo 9.2.** Optical flow

A task similar to the computation of the deformation flow between two different templates is that of optical flow. Time-varying images of a scene, be they due to changes in camera parameters or object motion, can be considered as multiple instances of similar templates and we may want to establish a dense correspondence between temporally subsequent images. The structure of the resulting deformation fields allows us to understand what is happening in the scene, e.g. how many different objects are moving and at what speed. A closely related, but more constrained, task is that of establishing the correspondence between two images of the same scene taken at the same time but from slightly displaced positions: stereo matching. In this case the local energy of the deformation field provides information on the distance of the corresponding point from the camera. The basic ingredient of an optical flow algorithm derives from the observation that

$$I(\boldsymbol{x}; t) = I(\boldsymbol{x} - \boldsymbol{v}t; 0), \tag{9.39}$$

expressing the fact that we are looking at the same world point, moving with velocity $\boldsymbol{v}$ in the image plane, at two different times. This results in the gradient constraint equation

$$\nabla I(\boldsymbol{x}; t) \cdot \boldsymbol{v} + \frac{\partial I(\boldsymbol{x}; t)}{\partial t} = 0 \tag{9.40}$$

with the usual notation $\nabla(I\boldsymbol{x}; t) = (\partial_x(\boldsymbol{x}; t), \partial_y(\boldsymbol{x}; t))^T$. Equation 9.40 constrains the component of motion normal to spatial contours of constant intensity. As we have two unknown components of $\boldsymbol{v}$ and a single equation the problem is under constrained. The solution proposed by Horn and Shunck is to impose a global smoothness constraint penalizing the magnitude of $\boldsymbol{v} = (v_1, v_2)$, leading to the following minimization problem:

$$\int_\Omega (\nabla I \cdot \boldsymbol{v} + \partial_t I)^2 + \lambda^2(\|\nabla v_1\|_2^2 + \|\nabla v_2\|_2^2)\, d\boldsymbol{x} \tag{9.41}$$

where $\Omega$ identifies the image domain and $\lambda$ controls the influence of the smoothness constraint. As in many other cases, a hierarchical approach is usually convenient: when working at lower resolution the distance between corresponding pixels in the two images is smaller, making the situation equivalent to that resulting from a smaller time interval. Equation 9.41 can be solved iteratively using the following update rules for iteration $k + 1$:

$$v_1^{k+1} = \bar{v}_1 - \frac{I_x[I_x\bar{v}_1^k + I_y\bar{v}_2^k + I_t]}{\alpha^2 + I_x^2 + I_y^2} \tag{9.42}$$

$$v_2^{k+1} = \bar{v}_2 - \frac{I_x[I_x\bar{v}_1^k + I_y\bar{v}_2^k + I_t]}{\alpha^2 + I_x^2 + I_y^2} \tag{9.43}$$

where $\bar{v}_1^k$ and $\bar{v}_2^k$ represent the average values of the corresponding velocity components over a suitable neighborhood, $v_1^0 = v_2^0 = 0$, and $I_x$ represents the partial derivative with respect to $x$ (or $y$ or $t$ with the corresponding change). The formulas used for the computation of the derivatives are important and Gaussian prefiltering in the spatial and temporal domain helps in obtaining accurate derivative values (with a typical value of $\sigma = 1.5$) and four-point central differences with a kernel

$$k_x = \frac{1}{12}(-1, 8, 0, -8, 1) \tag{9.44}$$

performing significantly better than first-order differences (Barron *et al.* 1994; Horn and Schunk 1981).

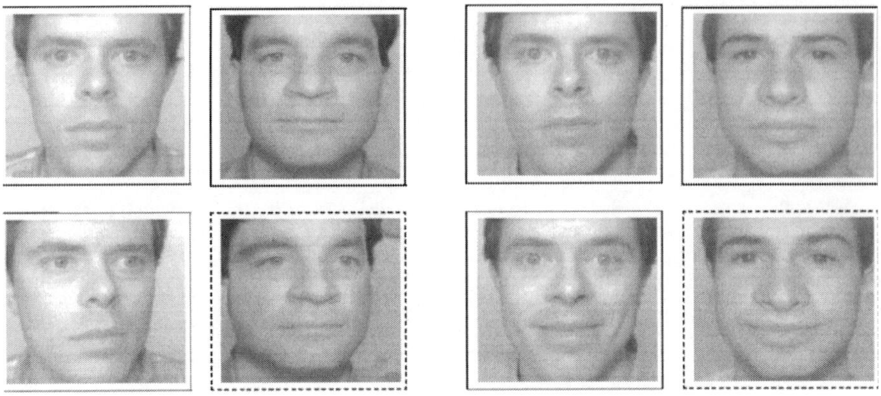

**Figure 9.5.** The exploitation of dense image mappings for graphical animation using Equations 9.62–9.63. The image in the dashed square is generated using the rotation/smiling deformation field of the prototype in the left column, after establishing a dense pixel correspondence that permits the transfer of the deformation to the image in the top right corner. Reproduced by permission of Tomaso Poggio (Beymer *et al.* 1993).

Let us consider images as positive real functions, defined not over a discrete set of points but over a subset $\Omega$ of the plane $\mathbb{R}^2$. Given two images $A$ and $B$, we search for a deformation field $\boldsymbol{u}(\boldsymbol{x}) = (u_1(\boldsymbol{x}), u_2(\boldsymbol{x}))$ such that

$$A \circ \boldsymbol{u}(\boldsymbol{x}) = A(\boldsymbol{u}(\boldsymbol{x})) \approx B(\boldsymbol{x}). \tag{9.45}$$

The deformation field should ideally be a diffeomorphism from the plane onto itself: a bijective map $\boldsymbol{u}(\boldsymbol{x})$ such that both it and its inverse $\boldsymbol{u}^{-1}$ are differentiable. This leads to requiring it to be the identity on the boundary of the image. We enforce smoothness on $\boldsymbol{u}$ by the introduction of a regularizing term

$$\Delta(\boldsymbol{u}) = \|\boldsymbol{u} - \boldsymbol{I_u}\|_{\Omega}^{H_1} \tag{9.46}$$

where $\boldsymbol{I_u}$ represents the identity, i.e. no displacement. The norm $H_1$ is given by

$$\|\boldsymbol{a}\|_{\Omega}^{H_1} = \int_{\boldsymbol{x}\in\Omega} \|\boldsymbol{a}(\boldsymbol{x})\|^2 + \|\partial(\boldsymbol{u})/\partial(\boldsymbol{x})\|_F^2 \, d\boldsymbol{x} \tag{9.47}$$

where $\partial(\boldsymbol{u})/\partial(\boldsymbol{x})$ denotes the Jacobian $J_{\boldsymbol{u}}$ of the transformation

$$J_{\boldsymbol{u}} = \frac{\partial(\boldsymbol{u})}{\partial(\boldsymbol{x})} = \left[\frac{\partial u_i}{\partial x_j}\right]_{1\le i,j\le 2} \tag{9.48}$$

and the associated norm is the Frobenius norm (see Equation 3.149). In order to fully define the computation of the mapping as a constrained minimization problem we need to choose a criterion that quantifies how good the match represented by Equation 9.45 is. The simplest possibility is given by the squared difference of the two images

$$\Delta(A, B) = \|A - B\|_{\Omega}^{L_2} = \int_{\Omega} (A(\boldsymbol{x}) - B(\boldsymbol{x}))^2 \, d\boldsymbol{x}.$$

This difference is not invariant to linear image rescaling and offset and a better choice is to use a local correlation. The latter can be computed between the two images at a given position $\boldsymbol{x}$ using a Gaussian window instead of giving all pixels the same weight as in the usual computation of correlation:

$$\Delta(A, B; \boldsymbol{x}) = \frac{\sigma_{AB}^2(\boldsymbol{x})}{(\epsilon + \sigma_{AA}(\boldsymbol{x}))(\epsilon + \sigma_{BB}(\boldsymbol{x}))} \tag{9.49}$$

$$g(\boldsymbol{x}, \boldsymbol{y}) = e^{-\|\boldsymbol{x}-\boldsymbol{y}\|^2/2\sigma^2} \tag{9.50}$$

$$n(\boldsymbol{x}) = \int_{\boldsymbol{y}\in\Omega} g(\boldsymbol{x}, \boldsymbol{y})\, d\boldsymbol{y} \tag{9.51}$$

$$\bar{A}(\boldsymbol{x}) = \frac{1}{n(\boldsymbol{x})} \int_{\boldsymbol{y}\in\Omega} A(\boldsymbol{y}) g(\boldsymbol{x}, \boldsymbol{y})\, d\boldsymbol{y} \tag{9.52}$$

$$\sigma_{AB}(\boldsymbol{x}) = \frac{1}{n(\boldsymbol{x})} \int_{\Omega} (A(\boldsymbol{y}) - \bar{A}(\boldsymbol{x}))(B(\boldsymbol{y}) - \bar{B}(\boldsymbol{x})) g(\boldsymbol{x}, \boldsymbol{y})\, d\boldsymbol{y} \tag{9.53}$$

where $\epsilon > 0$ ensures that $\Delta(A, B; \boldsymbol{x})$ is always defined. The computation of the deformation field morphing $A$ to $B$ is then formalized as solving

$$\hat{\boldsymbol{u}} = \operatorname*{argmin}_{\boldsymbol{u}} \int_{\Omega} \Delta(A \circ \boldsymbol{u}, B; \boldsymbol{x})\, d\boldsymbol{x} + \Delta(\boldsymbol{u}). \tag{9.54}$$

The solution can be found by gradient descent employing a coarse to fine strategy. If image $B$ is properly sampled it is also possible to get an improved estimate by resampling it to higher resolution, obtaining $B'$ and matching the windowed data from $A$ to a subsampled version of $B'$ matching the resolution of $A$ but located in between the original sampling provided by $B$.

The availability of a dense, pixel-to-pixel, correspondence between images opens the way to many interesting applications which go beyond template matching. A first one is to use the Jacobian determinant $|J_{\boldsymbol{u}}|$ of the deformation field to characterize the local behavior of the mapping:

1. A continuously differentiable mapping $\boldsymbol{u}$ is invertible if $|J_{\boldsymbol{u}}| \neq 0$.
2. If $|J_{\boldsymbol{u}}| > 0$ the mapping preserves orientation.
3. The absolute value of $|J_{\boldsymbol{u}}|$ represents the factor by which the mapping expands or shrinks the volume (or area).

The last property is particularly important in the analysis of medical images as it allows us to characterize the evolution of anatomical structures during therapy. The first property permits the identification of mapping singularities, such as the (dis)appearance of a structure, possibly computing two deformation fields, from $A$ to $B$ and from $B$ to $A$.

The deformation field $\boldsymbol{u}$ allows us to vectorize image sets in such a way that pixel-by-pixel operations become more meaningful, increasing the advantage of using linear space methods. Let us reconsider the problem of computing the average face from a set of images. Intuitively, it should be like a real face minimizing the dissimilarity from all faces in the set, without exhibiting artifact features. Usually the average face is computed (from a set of frontal, geometrically normalized images) by pixel averaging. However, the resulting face is not what we would call a true average face, even if it is the average of the face vectors. A better, more meaningful, average face is obtained by introducing an average morphed face $\bar{F}$

$$\bar{F} = \frac{1}{N} \sum_{i=1}^{N} F_I \circ \boldsymbol{u}_i \tag{9.55}$$

for appropriately chosen fields $\boldsymbol{u}_i$. In order to find the deformation fields we need to change perspective, thinking only in terms of deformation fields: the average face will be recovered afterwards, as a by-product of the computation of the fields. In order to give all images the same status, we compute the fields in such a way that every mapped image $F_I \circ \boldsymbol{u}_i$ is as similar as possible to all other images, minimizing an average global error $\bar{\Delta}$

$$\bar{\Delta} = \frac{1}{n-1} \sum_{i \neq j} \int_{\Omega} \Delta(F_i \circ \boldsymbol{u}_i, F_j \circ \boldsymbol{u}_j; x)\, d\boldsymbol{x} + \sum_k \Delta(\boldsymbol{u}_k). \tag{9.56}$$

The fields are initialized to the identity and the minimization process carried over iteratively in a coarse to fine framework. Convergence can be improved by imposing $\sum_k \boldsymbol{u}_k = 0$ after each time step by subtracting the mean field $N^{-1} \sum_k \boldsymbol{u}_k$ from each single field.

The information on the geometry of faces is contained in the deformation fields mapping them to the mean face and we can try to capture it using PCA. The deformation fields are vector fields and a simple way to capture their correlation is to compute their scalar product

$$\Sigma^{\boldsymbol{u}}_{ij} = \boldsymbol{u}_i \cdot \boldsymbol{u}_j = \frac{1}{|\Omega|} \int_{\Omega} \boldsymbol{u}_i(\boldsymbol{x}) \cdot \boldsymbol{u}_j(\boldsymbol{x})\, d\boldsymbol{x} \tag{9.57}$$

which defines the correlation matrix, due to the fact that we previously imposed $\sum_i \boldsymbol{u}_i = 0$. The dimension of the resulting correlation matrix $\Sigma^{\boldsymbol{u}}$ is $N \times N$ and we can extract its eigenvalues $\lambda^{\boldsymbol{u}}_k$ and eigenvectors $\boldsymbol{e}^{\boldsymbol{u}}_k$ from which we obtain the shape principal directions or modes

$$s_k = \sum_i (\boldsymbol{e}^{\boldsymbol{u}}_k)_i \boldsymbol{u}_i \tag{9.58}$$

in a way completely analogous to the PCA technique described in Section 8.1. The average image can then be animated by applying to it a linear combination of shape modes: the way we constructed the deformation fields and computed their correlation matrix ensures that all resulting images will possess a correct face geometry (see also Intermezzo 9.3). Incorporating textural information is straightforward: we apply PCA to the images morphed towards the average face

$$F'_i = F_i \circ \boldsymbol{u}_i - \bar{F}. \tag{9.59}$$

We can then work separately on shape and texture or we can compute a single correlation matrix

$$\Sigma_{ij} = \frac{1}{\sigma_s^2} \Sigma^{\boldsymbol{u}}_{ij} + \frac{1}{\sigma_t^2} F'_i \cdot F'_j. \tag{9.60}$$

where

$$\sigma_s^2 = \frac{1}{N} \sum_{i=1}^{N} \|\boldsymbol{u}_i\|^2, \quad \sigma_t^2 = \frac{1}{N} \sum_{i=1}^{N} \|F'_i\|^2.$$

**Intermezzo 9.3.** FACS: Facial Action Coding System

Facial expressions are an important, non-verbal, language that people use for communication. As such, a compact code to describe their many nuances is useful to both psychologists that need to decipher it and to animators that use it to transmit specific information. Starting from a detailed investigation of the facial muscles whose activity is associated to facial expressions, a compact coding system, FACS, was developed in the mid 1970s and used in psychological studies and animation. There are 46 action units (AUs) to which an intensity score can be associated based on a five degree scale of evidence: trace (A), slight (B), marked/pronounced (C), severe/extreme (D), and maximum (E). A few examples are

**AU10** upper lip raiser;

**AU11** nasolabial deepener;

**AU12** lip corner puller;

**AU13** cheek puffer;

**AU25** lips part;

**AU27** mouth stretch.

An emotion like happiness corresponds to a complex expression of basic AUs, one of which is

$$\text{happiness} = (\text{A12} + \text{A13}) \wedge (\text{A25} + \text{A27}) \wedge (\text{A10}) \wedge (\text{A11}). \tag{9.61}$$

A major limitation of FACS is that it does not consider the detailed time evolution of expressions, a factor currently believed to be of significant importance. The techniques discussed in the chapter have been used to perform automatic recognition of FACS AUs, supplementing them with detailed dynamical information (Donato *et al.* 1999; Ekman and Friesen 1978; Essa and Pentland 1997; Tian *et al.* 2001).

The use of a joint correlation matrix allows us to capture the dependencies of shape and texture due, for example, to racial characteristics.

The possibilities offered by modeling deformation fields do not end here. A particularly interesting one is related to image space modeling of object variations, a couple of examples being facial expressions and face rotations. When we image the appearance of object $O_1$ undergoing a transformation controlled by a limited number of parameters $\boldsymbol{\alpha}$, leading to $O_1(\boldsymbol{\alpha})$, we are exploring a low-dimensionality manifold that can be captured well, or at least locally, by means of PCA. A different object $O_2$ of the same class undergoing the same transformation will span a different, but close manifold. If we make the additional assumption that the two manifolds are locally parallel we can try to approximate the changes of $O_2$ by means of the known spatial changes $\boldsymbol{u}_1$ of $O_1$ and of a mapping $\boldsymbol{u}_{1,2}$ from $O_1$ to $O_2$

$$\boldsymbol{u}_2(\boldsymbol{\alpha}) \approx (\boldsymbol{u}_{2,1} + \boldsymbol{u}_1(\boldsymbol{\alpha})) + \boldsymbol{u}_{1,2} \tag{9.62}$$

and similarly for the textural components

$$I_2(\boldsymbol{x}; \boldsymbol{\alpha}) \approx I_2(\boldsymbol{x}) + [I_1(\boldsymbol{u}_1(\boldsymbol{u}_{2,1}(\boldsymbol{x}); \boldsymbol{\alpha})) - I_1(\boldsymbol{u}_{2,1}(\boldsymbol{x}))]. \tag{9.63}$$

The resulting transformation corresponds to displacing object $O_2(\boldsymbol{\alpha})$ in the joint shape texture manifold by the vector describing the change of $O_1$ to $O_1(\boldsymbol{\alpha})$. This allows us to modify an image depicting person $A$ with a neutral expression into an image depicting $A$ smiling once we have two images of $B$, one neutral and one smiling (see Figure 9.5). Another interesting application based on the same approach is that of illumination compensation (see Figures 9.6 and 9.7). Shape space PCA can be exploited to classify deformation fields. As an example let us consider facial expressions. The studies presented in Intermezzo 9.3 suggest that a low-dimensional representation of facial expression is appropriate. PCA naturally leads to

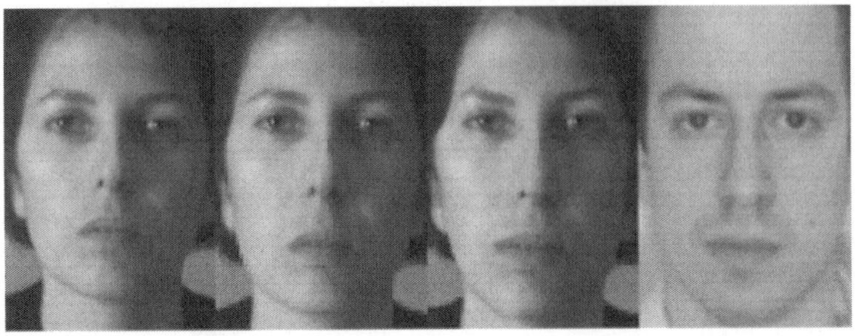

**Figure 9.6.** Optical flow techniques, such as the one described in Intermezzo 9.2, can be used to map images of different subjects under different illumination by first applying a normalizing preprocessing such as the local contrast operator defined in Equation 14.8. Source: Brunelli (1997).

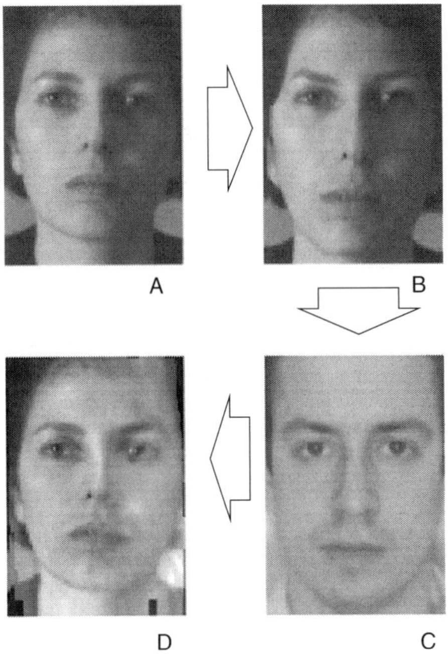


**Figure 9.7.** Illumination compensation is another application of the dense correspondence maps between images of different faces. Source: Brunelli (1997).

such a representation of the deformation fields underlying the variety of facial expressions and provides a sound basis on how complex expression can be represented by means of a (linear) combination of more basic elements, the principal directions. Expression, or, more generally, deformation dynamics, i.e. the temporal evolution of the deformation field, can be incorporated by normalizing the time span over which the deformation develops, and concatenating the fields corresponding to a subset of instants. PCA can then be performed as before but taking into consideration the extended vectors. As discussed in Chapter 8, an alternative approach is to introduce a set of subspaces, one per expression or deformation class, separately modeling variability and dynamics.

## 9.5. Bibliographical Remarks

The dynamical approach to the Hough transform used to introduce deformable templates is based on the paper by Gyulassy and Harlander (1991). A different perspective is presented in the papers by Yuille *et al.* (1991) and by Ohlsson *et al.* (1992). The step from the Hough transform to deformable models, using the former to trigger the latter, is clearly presented in the papers by Garrido and Perez De La Blanca (1998) and Ecabert and Thiran (2004).

Deformable models of the type described in Section 9.2 were proposed by Yuille *et al.* (1992) and extended in Epstein and Yuille (1994) and Coughlan *et al.* (2000). Active contour models, also known as snakes, are a less constrained approach to boundary detection introduced by Kass *et al.* (1988). A significant extension that allows for the simultaneous detection of several objects and inner and exterior boundaries can be found in Caselles *et al.* (1997).

A rather complete presentation of active shape models can be found in Cootes *et al.* (1995), on which Section 9.3 is based. Their application to medical images is considered in Cootes *et al.* (1994), where some additional examples are provided. An application to face analysis can be found in Luettin *et al.* (1996), where active shape models are used to track the shape of lips for an automatic lip-reading system.

Research on diffeomorphic matching and its application in computational anatomy is vast. The presentation of diffeomorphic matching is essentially the one presented by Charpiat *et al.* (2005). The invertibility required by diffeomorphic matching can be made stronger by requiring that forward and reverse deformations be the inverse of one another. The resulting problem of consistent image registration is analyzed in Christensen and Johnson (2001). A good reference to elastic matching is the paper by Bajcsy and Kovačič (1989). The paper by Grenander and Miller (1998) can be considered as a manifesto of computational anatomy. Another extensive presentation, with many references, can be found in the paper by Thompson and Toga (2002). An interesting approach to dense matching for large deformations based on viscous fluid dynamics was proposed by Christensen *et al.* (1996). In this paper the template is considered as a viscous fluid whose motion, controlled by the Navier–Stokes equation, attempts to minimize the difference among the templates. A more efficient solver for the partial differential equations governing fluid dynamics is presented in Bro-Nielsen and Gramkow (1996).

The idea of using a parallel deformation of templates coupled to approximation network 'in-betweening' for graphical animation was introduced by Poggio and Brunelli (1992) and further elaborated in Beymer *et al.* (1993) and Beymer and Poggio (1996). The application of dense correspondences to illumination compensation is considered in Brunelli (1997).

## References

Bajcsy R and Kovačič S 1989 Multiresolution elastic matching. *Computer Vision, Graphics and Image Processing* **46**, 1–21.

Barron J, Fleet D and Beauchemin S 1994 Performance of optical flow techniques. *International Journal of Computer Vision* **12**, 43–77.

Beymer D and Poggio T 1996 Image representation for visual learning. *Science* **272**, 1905–1909.

Beymer D, Shashua A and Poggio T 1993 Example based image analysis and synthesis. Technical Report AIM-1431; CBCL-080, MIT.

Bro-Nielsen M and Gramkow C 1996 Fast fluid registration of medical images. *Proceedings of the 4th International Conference on Visualization in Biomedical Computing*, Lecture Notes in Computer Science, vol. 1131, pp. 267–276. Springer.

Brunelli R 1994 Training neural nets through stochastic minimization. *Neural Networks* **7**(9), 1405–1412.

Brunelli R 1997 Estimation of pose and illuminant direction for face processing. *Image and Vision Computing* **15**, 741–748.

Brunelli R and Tecchiolli G 1995 Stochastic minimization with adaptive memory. *Journal of Computational and Applied Mathematics* **57**, 329–343.

Caselles V, Kimmel R and Sapiro G 1997 Geodesic active contours. *International Journal of Computer Vision* **22**, 61–79.

Charpiat G, Faugeras O and Keriven R 2005 Image statistics based on diffeomorphic matching. *Proceedings of the 10th International Conference on Computer Vision and Pattern Recognition (ICCV'05)*, vol. 1, pp. 852–857.

Christensen G and Johnson H 2001 Consistent image registration. *IEEE Transactions on Medical Imaging* **20**, 568–582.

Christensen G, Rabbitt R and Miller M 1996 Deformable templates using large deformation kinematics. *IEEE Transactions on Image Processing* **5**, 1435–1447.

Cootes T, Hill A, Taylor C and Haslam J 1994 The use of active shape models for locating structures in medical images. *Image and Vision Computing* **12**, 355–366.

Cootes T, Taylor C, Cooper D and Graham J 1995 Active shape models: their training and application. *Computer Vision and Image Understanding* **61**, 38–59.

Coughlan J, Yuille A, English C and Snow D 2000 Efficient deformable template detection and localization without user initialization. *Computer Vision and Image Understanding* **78**, 303–319.

Donato G, Bartlett M, Hager J, Ekman P and Sejnowski T 1999 Classifying facial actions. *IEEE Transactions on Pattern Analysis and Machine Intelligence* **21**, 974–989.

Ecabert O and Thiran J 2004 Adaptive Hough transform for the detection of natural shapes under weak affine transformations. *Pattern Recognition Letters* **25**, 1411–1419.

Ekman P and Friesen W 1978 *Facial Action Coding System*. Consulting Psychologists Press Inc.

Epstein R and Yuille A 1994 Training a general purpose deformable template. *Proceedings of the International Conference on Image Processing (ICIP'94)*, vol. 1, pp. 203–207.

Essa I and Pentland A 1997 Coding, analysis, interpretation, and recognition of facial expressions. *IEEE Transactions on Pattern Analysis and Machine Intelligence* **19**, 757–763.

Garrido A and Perez De La Blanca N 1998 Physically-based active shape models initialization and optimization. *Pattern Recognition* **31**, 1003–1017.

Grenander U and Miller M 1998 Computational anatomy: an emerging discipline. *Quarterly of Applied Mathematics* **LVI**, 617–694.

Gyulassy M and Harlander M 1991 Elastic tracking and neural network algorithms for complex pattern recognition. *Computer Physics Communications* **66**, 31–46.

Horn B and Schunk B 1981 Determining optical flow. *Artificial Intelligence* **17**, 185–203.

Kass M, Witkin A and Terzopoulos D 1988 Snakes: active contour models. *International Journal of Computer Vision* **1**, 321–331.

Luettin J, Thacker N and Beet S 1996 Locating and tracking facial speech features. *Proceedings of the 13th IAPR International Conference on Pattern Recognition (ICPR'96)*, vol. 1, pp. 652–656.

Ohlsson M, Peterson C and Yuille A 1992 Track finding with deformable templates – the elastic arms approach. *Computer Physics Communications* **71**, 77–98.

Poggio T and Brunelli R 1992 A novel approach to graphics. A.I. Memo No. 1354, MIT.

Thompson P and Toga A 2002 A framework for computational anatomy. *Computing and Visualization in Science* **5**, 13–34.

Tian Y, Kanade T and Cohn J 2001 Recognizing action units for facial expression analysis. *IEEE Transactions on Pattern Analysis and Machine Intelligence* **23**, 97–115.

Yuille A, Hallinan P and Cohen D 1992 Feature extraction from faces using deformable templates. *International Journal of Computer Vision* **8**, 99–111.

Yuille A, Honda K and Peterson C 1991 Particle tracking by deformable templates. *Proceedings of the International Joint Conference on Neural Networks*, vol. 1, pp. 7–12.

# 10 COMPUTATIONAL ASPECTS OF TEMPLATE MATCHING

Fetch us in fuel; and be quick, thou'rt best,
To answer other business. Shrug'st thou, malice?

*The Tempest*
WILLIAM SHAKESPEARE

The drawback of template matching is its high computational cost, which has two distinct origins. The first source of complexity is the necessity of using multiple templates to accommodate the variability exhibited by the appearance of complex objects. The second source of complexity is related to the representation of the templates: the higher the resolution, i.e. the number of pixels, the heavier the computational requirements. Besides some computational tricks, this chapter presents more organized, structural ways to improve the speed at which template matching can be performed.

## 10.1. Speed

One of the major drawbacks of template matching techniques, at least when using a straightforward implementation of the relevant formulas, is their high computational cost. The computation of the $L_2$ or $L_1$ distance between two $m \times m$ templates requires $3m^2 - 1$ operations. In the general case, the detection of a single template $T_{m\times m}$ within a high-resolution image $I_{n\times n}$ by means of a sliding way approach requires a significant amount of operations $\mathcal{C}$ even when matching is performed by computing the $L_1$ distance without normalization:

$$\mathcal{C}_{L_1} = (3m^2 - 1)(n - m + 1)^2 = O(m^2(n - m + 1)^2). \tag{10.1}$$

The possibility of speeding up the computation depends on many factors, including the way used to quantify the similarity of patterns, e.g. whether the matching relies on normalized correlation or on the computation of an $L_p$ distance, and the requirements of the application, e.g. whether a single, best match must be found or multiple instances should be detected.

### 10.1.1. EARLY JUMP-OUT

We start our review of speeding up techniques from the simple case of $L_p$ matching between patterns searching for the best matching position $(\hat{u}_1, \hat{u}_2)$ of the template within the image:

$$(\hat{u}_1, \hat{u}_2) = \underset{(u_1,u_2)}{\operatorname{argmin}}\, [d_p(T, I(u_1, u_2))]^p = \underset{(u_1,u_2)}{\operatorname{argmin}} \sum_{i_1,i_2} |T(i_1, i_2) - I(u_1 + i_1, u_2 + i_2)|^p. \tag{10.2}$$

*Template Matching Techniques in Computer Vision: Theory and Practice* Roberto Brunelli

In order to find the required position, we need to move $\boldsymbol{u} = (u_1, u_2)$ over the complete image while keeping track of the minimum distance value. Let us introduce a scanning function $S_A : t \in \mathbb{Z} \rightarrow (x_1, x_2) \in \mathbb{Z}^2$ for image $A$ and the partial matching function

$$\Delta(T, I, \boldsymbol{u}, k_T) = \sum_{t=1}^{k_T} |T(S_T(t)) - I(S_T(t) + \boldsymbol{u})|^p. \tag{10.3}$$

The image of the scanning function $S_A$ for an image $A_{l \times l}$ is the set of points belonging to the associated image, and its domain is the set of the integers $\{1, \ldots, l^2\}$. The computation of the best matching position requires two scanning functions: one for the image, giving the position $\boldsymbol{u}$ at which the template is positioned for the matching; and one for the template, giving the sequence necessary for the computation of $\Delta(T, I, \boldsymbol{u}, k_T)$. In order to find the best matching position we do not necessarily need to compute $\Delta(T, I, S_I(k_I), m^2)$ for every point in the image: we may jump out of the computation whenever

$$\Delta(T, I, S_I(k_I), k_T) > \Delta_{\max} \tag{10.4}$$

where $\Delta_{\max}$ represents the maximum acceptable mismatch as the comparison at the given position will fail to satisfy the required constraint due to the fact that $\Delta$ is non-decreasing in $k_T$. As we want to compute as few values of $\Delta(T, I, S_I(k_I), k_T)$ as possible, we may choose the scanning function in such a way that template pixels that are likely to be very different from those not belonging to the template are considered first. A scanning function based on this principle can be found by computing the expected difference $H[l]$ for each possible pixel intensity $l$

$$H[l] = \sum_q p_q |l - q| \tag{10.5}$$

where $p_q$ is the probability that an image pixel has value $q$. The pixels of the template can then be sorted by decreasing expected difference of their value and considered by the scanning function in the resulting order. In some applications we are not interested in all possible matches whose distance is less than $\Delta_{\max}$, but only in the best one. An important case is block matching for video compression. The video frame is subdivided into non-overlapping blocks of homogeneous size: the best match of each of them in the successive frame must be found because it is advantageous to code the differences between the two blocks, as they exhibit a lower entropy, instead of coding the new block from scratch. In these cases we can adaptively redefine $\Delta_{\max}$ as the minimum computed value of $\Delta(T, I, S_I(k_I), m^2)$. The scanning function $S_I$ controlling the positioning of the template over the image also deserves some attention. Let us consider the task of tracking a single object in multiple subsequent video frames: the already computed positions can be used to forecast the position of the template in the next frame. A scanning strategy that starts from the expected position of the target, spiraling away from it, can be advantageous, providing at early stages a low $\Delta_{\max}$ value. A further improvement of the early jump-out strategy is to use a sequence of thresholds $\Delta_j$ that are dynamically updated during the matching process. For each match, we keep track of the values $\Delta(T, I, S_I(k_I), k_T)$: if we match the whole template and the final mismatch is lower than $\Delta_{\max}$ we create a sequence of partial thresholds $\Delta_j$

$$\Delta_j = \frac{1}{\alpha}[(\alpha - 1)\Delta(T, I, S_I(k_I), j) + \Delta(T, I, S_I(k_I), m^2)] \quad j < m^2 \tag{10.6}$$

$$\Delta_{m^2} = \Delta(T, I, S_I(k_I), m^2) \tag{10.7}$$

with $\alpha \geq 1$, and let $\Delta_{\max} = \Delta_{m^2}$. If a sequence of thresholds is available we change the jump-out test of Equation 10.4 with the following:

$$\Delta(T, I, S_I(k_I), k_T) > \Delta_{k_T}. \tag{10.8}$$

The larger the value of $\alpha$, the closer $\Delta_j$ is to $\Delta(T, I, S_I(k_I), j)$, and the greedier the strategy.

### 10.1.2. THE USE OF SUM TABLES

The computation of the normalized correlation coefficient $r_P$ is more complex than the computation of a sum of pixel-wise differences due to the centering and scaling of the values. Let us rewrite $\rho$ as

$$\rho(\boldsymbol{u}) = \frac{\sum_{i_1=0}^{m-1} \sum_{i_2=0}^{m-1} I(u_1 + i_1, u_2 + i_2) T(i_1, i_2) - m^2 \mu_I(\boldsymbol{u}) \mu_T}{\|I'(\boldsymbol{u})\| \|T'\|} \tag{10.9}$$

$$\mu_I(\boldsymbol{u}) = \frac{1}{m^2} \sum_{i_1=0}^{m-1} \sum_{i_2=0}^{m-1} I(u_1 + i_1, u_2 + i_2) \tag{10.10}$$

$$\|I'(\boldsymbol{u})\| = \left[ \sum_{i_1=0}^{m-1} \sum_{i_2=0}^{m-1} I^2(u_1 + i_1, u_2 + i_2) - m^2 \mu_I^2(\boldsymbol{u}) \right]^{1/2} \tag{10.11}$$

$$\mu_T = \frac{1}{m^2} \sum_{i_1=0}^{m-1} \sum_{i_2=0}^{m-1} T(i_1, i_2) \tag{10.12}$$

$$\|T'\| = \left[ \sum_{i_1=0}^{m-1} \sum_{i_2=0}^{m-1} T^2(i_1, i_2) - m^2 \mu_T^2 \right]^{1/2}. \tag{10.13}$$

The computations required by Equations 10.12–10.13 do not depend on the position in the image and can be performed at the beginning of the matching process. The computations required by Equations 10.10–10.11 can be performed efficiently by means of sum tables (see also Figure 10.1). The sum table associated to an image $A(x_1, x_2)$ is a function $S(A; x_1, x_2)$ defined as

$$S(A; x_1, x_2) = A(x_1, x_2) + S(A; x_1 - 1, x_2) + S(A; x_1, x_2 - 1) - S(A; x_1 - 1, x_2 - 1) \tag{10.14}$$

when the arguments are non-negative, and 0 otherwise. If we compute $S(A; x_1, x_2)$ scanning $A$ in a top-down, left–right order, the computation required by Equation 10.14 is limited to three sums. The sum table allows the computation of the sum of the values of $A$ over an arbitrary rectangular region using only three operations:

$$\begin{aligned} \sum_{i_1=0}^{m_1-1} \sum_{i_2=0}^{m_2-1} A(x_1 + i_1, x_2 + i_2) = {} & S(A; x_1 + m_1 - 1, x_2 + m_2 - 1) \\ & - S(A; x_1 - 1, x_2 + m_2 - 1) \\ & - S(A; x_1 + m_1 - 1, x_2 - 1) \\ & + S(A; x_1 - 1, x_2 - 1). \end{aligned} \tag{10.15}$$

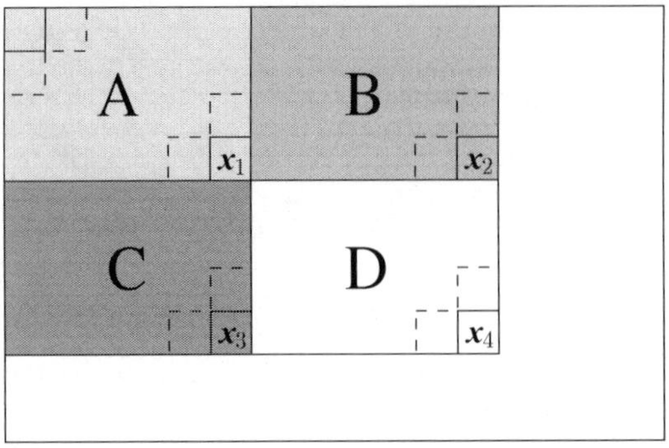


**Figure 10.1.** The sum table of an image allows the efficient computation of its integral over an arbitrary rectangular region. The values of the sum table at positions $\boldsymbol{x}_1$, $\boldsymbol{x}_2$, $\boldsymbol{x}_3$, $\boldsymbol{x}_4$ correspond respectively to the sum of image values over rectangles A, A + B, A + C, and A + B + C + D. The sum over D then corresponds to $S(I; \boldsymbol{x}_4) + S(I; \boldsymbol{x}_1) - S(I; \boldsymbol{x}_2) - S(I; \boldsymbol{x}_3)$.

In order to compute $r_P(\boldsymbol{u})$ we precompute the sum tables

$$S(I; x_1, x_2), \quad S(I^2; x_1, x_2)$$

that can be used for the rapid computation of the quantities in Equations 10.10–10.11 exploiting Equation 10.15. The only computation that cannot take advantage of sum tables is that involving the product of the image and the template. There are, however, situations where even this computation can take advantage of a sum table, and an important case is that of defect detection. In this case, we have two aligned images, one of which must be compared to the other to spot significant differences. The comparison is performed at each pixel by considering a suitably sized neighborhood of it. In this case the role of the template image $T$ changes and it is completely equivalent to that of $I$ so that Equations 10.12–10.13 must be modified, including a spatial dependence, mimicking Equations 10.10–10.11, and similarly in Equation 10.9. The computation of Equation 10.9 now benefits from the computation of another sum table, $S(IT; x_1, x_2)$. In this case both images have dimensions $n \times n$ while $m \times m$ identifies the size of the window $W$ used at each position $\boldsymbol{u}$ to compare $I$ and $T$. The final result will then be an image of size $(n - m + 1)^2$ where at each position we have the correlation value of $I_W(\boldsymbol{u})$ with $T_W(\boldsymbol{u})$, the subscript meaning that we compare a window $W$ located at $\boldsymbol{u}$.

The computation based on the use of sum tables is $O(n^2)$ while the direct method is $O(n^2 m^2)$ giving the former an advantage of some orders of magnitude. When the image alignment is not perfect, it is necessary to consider different translations of the reference image, recomputing for each translation only $S(IT; x_1, x_2)$.

### 10.1.3. HIERARCHICAL TEMPLATE MATCHING

An approach that works well for objects with significant low spatial frequency content is that of employing a coarse to fine strategy. The image and the template are progressively decimated after being properly filtered to avoid aliasing (see Chapter 2) creating a layered image representation, the resolution pyramid (see Algorithm 10.1 and Figure 10.2), repeating

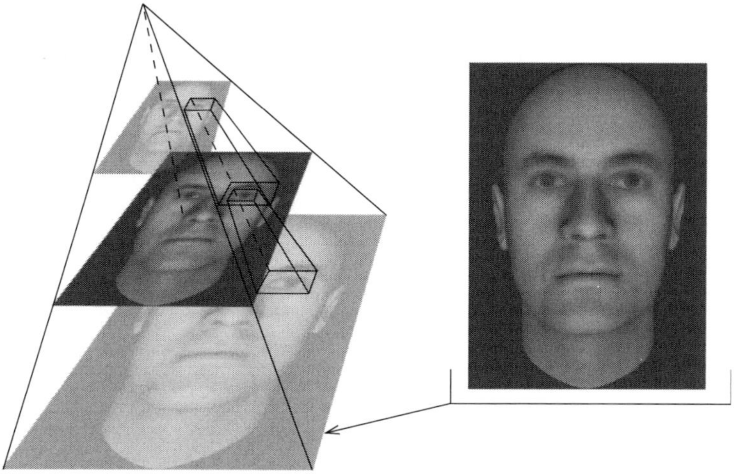

**Figure 10.2.** Hierarchical template matching relies on the resolution pyramid, obtained by progressive decimation of the image (and templates) at full resolution.

the procedure for the template. The number of levels $l_M$ depends on the templates and on the context where they must be found. If the templates depict objects whose fine details need to be checked in order to distinguish them, the number of levels will be limited, as the objects become progressively harder to discern as the resolution decreases.

---

**Algorithm 10.1**: Resolution pyramid

---

**Data**: The number of pyramid levels $l_M$, the image at the original resolution $I_{l_M}$, the required subsampling spacing $\mathbf{\Delta}$.
**Initialization**: Set current level $l = l_M - 1$.
**while** $l > 0$ **do**
  limit frequency content of image $I_{l+1}$ so that it can be subsampled with spacing $\mathbf{\Delta}$;
  subsample $I_{l+1}$ obtaining its decimated version $I_l$;
  $l \rightarrow l - 1$
**end**
**return** *The set* $\{I_l\}_{l=1}^{l_M}$ *representing the resolution pyramid.*

---

The improvement in speed derives from the possibility of progressively focusing the search area at each level around the most promising locations spotted at the previous, lower resolution one. Algorithm 10.2 illustrates hierarchical matching for the commonly used subsampling spacing $\mathbf{\Delta} = (2, 2)$.

The threshold $\theta_l$ used to select the most promising locations should depend on the pyramid resolution level as the distribution of correlation values depends on the resolution at which they are computed. An alternative strategy is to project at the next level a fraction $\alpha(l)$ of the positions for which $r_P$ has been computed, corresponding to the highest values. However, resampling effects may result in losing the correct template location, making hierarchical correlation not completely equivalent to an exhaustive search at the highest resolution. Similarity measures other than the correlation coefficient can be used, making obvious

**Algorithm 10.2**: Hierarchical matching

**Data**: A resolution pyramid $\{I_l\}_{l=1}^{l_M}$ and a set of level specific matching thresholds $\{\theta_l\}$ to filter promising positions.

**Initialization**: Set current level $l = 1$ and initial set of image positions at which matching must be performed $L = \{\boldsymbol{u} : u_1 < n_{l1},\ u_2 < n_{l2}\}$ where $\boldsymbol{n}_l$ represents image dimensions at level $l$.

**while** $l \leq l_M$ **do**

compute $\{r_P(\boldsymbol{u})\}_{\boldsymbol{u} \in L}$;

extract the set of best matching locations $\{\boldsymbol{u} : r_P(\boldsymbol{u}) \geq \theta_l\} \to L$;

project $L$ at the higher resolution level to control the next matching step

$$\bigcap_{\boldsymbol{u} \in L} \{(2u_1, 2u_2) \oplus \mathcal{N}_8\} \to L$$

where $\mathcal{N}_8$ represents the eight-connected neighborhood of a pixel.

**end**

**return** *The set L of most promising template positions.*

modifications to the previous workflow. In some cases, when matching and downsampling are carefully tuned to each other, it is possible to have provable equivalence to a full search.

### 10.1.4. METRIC INEQUALITIES

Metrics enjoy properties, such as the triangular inequality, that can be used to speed up the matching process without losing any correct match. Let us consider the case of matching based on the $L_p$ norm of Equation 10.2 and let us denote by $I_T(\boldsymbol{u})$ the region of image $I$ located at $\boldsymbol{u}$ and congruent to $T$. The $L_p$ norm satisfies the triangular inequality

$$(L_p(I_T(\boldsymbol{u}), T))^p = (\|I_T(\boldsymbol{u}) - T\|_p)^p \geq |\|I_T(\boldsymbol{u})\|_p - \|T\|_p|^p. \tag{10.16}$$

If we denote by $|_S$ the restriction to a subset of the template, or corresponding image window, a similar bound applies

$$\|I_T(\boldsymbol{u}) - T\|_{p,S}^p \geq |\|I_T(\boldsymbol{u})\|_{p,S} - \|T\|_{p,S}|^p. \tag{10.17}$$

Let us partition the set of rows of the template (and the corresponding image window) into $k$ disjoint subsets $R_i$: for simplicity we may think of $k - 1$ subsets each composed of $m/k$ consecutive rows and a last subset comprising the remaining rows. Exploiting the triangular inequality for the subsets we may write

$$\|I_T(\boldsymbol{u}) - T\|_p^p \geq \sum_{r=1}^{k} |\|I_T(\boldsymbol{u})\|_{p,R_r} - \|T\|_{p,R_r}|^p. \tag{10.18}$$

The inequality gives a sufficient condition for pruning positions at which the match cannot possibly satisfy a constraint based on a maximum allowed distance $\Delta_{\max}$. In fact, whenever

$$\sum_{r=1}^{k} |\|I_T(\boldsymbol{u})\|_{p,R_r} - \|T\|_{p,R_r}|^p > \Delta_{\max} \tag{10.19}$$

$I_T(\boldsymbol{u})$ cannot satisfy

$$\|I_T(\boldsymbol{u}) - T\|_p^p \leq \Delta_{\max} \tag{10.20}$$

because of Equation 10.18. If Equation 10.19 does not hold, instead of computing the complete distance we substitute one of the addenda in the sum with the corresponding computed image template distance (restricted to the corresponding set of rows). This provides a tighter lower bound, because we substituted one of the terms with a greater one (see Equation 10.17), and we check again to see if we can prune the current position

$$\|I_T(\boldsymbol{u}) - T\|_{p,R_i}^p + \sum_{r=1,r\neq i}^{k} |\|I_T(\boldsymbol{u})\|_{p,S} - \|T\|_{p,S}|^p > \Delta_{\max}. \tag{10.21}$$

The procedure can be repeated until we exhaust the subsets, having had the possibility to prune with tighter and tighter bounds, progressively computing the actual distance between the image and the template. The speed-up derives from the possibility of computing the RHS of Equation 10.17 in a very efficient way using sum tables as described in Section 10.1.2. The approach can be easily modified to find only the best matching instance by substituting $\Delta_{\max}$ with the lowest $\|I_T(\boldsymbol{u}) - T\|_p^p$ value found so far. The resulting speed-up is very significant, reaching orders of magnitude with respect to the direct approach and outperforming Fourier-based approaches (see Section 10.1.5) when the level of noise is limited.

A somewhat similar approach can be used in the case of normalized correlation by exploiting the Cauchy–Schwartz inequality, the $p = q = 2$ case of the more general inequality

$$\left[\sum_{i=1}^{n} a_i^p\right]^{1/p} \left[\sum_{i=1}^{n} b_i^q\right]^{1/q} \geq \sum_{i=1}^{n} a_i b_i \tag{10.22}$$

where all values must be positive and $p^{-1} + q^{-1} = 1$. Denoting by $|_a^b$ the restriction to rows $a, a+1, \ldots, b$, we may introduce the auxiliary term $\psi_Z|_a^b$

$$\psi_Z(\boldsymbol{u})|_a^b = \sum_{i_2=a}^{b} \sum_{i_1=0}^{m-1} (I(u_1 + i_1, u_2 + i_2) - \mu_I(\boldsymbol{u}))(T(i_1, i_2) - \mu_T), \tag{10.23}$$

the operator $\Sigma(A)|_a^b$ that represents the sum of the values of image $A$ restricted to the range of rows $[a, b]$, and two additional expressions

$$\beta'(\boldsymbol{u})|_s^{m-1} = \frac{\sqrt{(\|T\||_s^{m-1})^2 + m(m-s)\mu_T^2 - 2\mu_T\,\Sigma(T)|_s^{m-1}}}{\sqrt{(\|I_T\||_s^{m-1})^2 + m(m-s)\mu_I(\boldsymbol{u})^2 - 2\mu_I(\boldsymbol{u})\,\Sigma(I_T)|_s^{m-1}}} \tag{10.24}$$

$$\begin{aligned}\beta''(\boldsymbol{u})|_s^{m-1} &= \|I_T\||_s^{m-1}\|T\||_s^{m-1} - \mu_T \cdot \Sigma(I_T)|_s^{m-1} \\ &\quad - \mu_I(\boldsymbol{u}) \cdot \Sigma(T)|_s^{m-1} + m(m-s)\mu_T\mu_{I_T}(\boldsymbol{u})\end{aligned} \tag{10.25}$$

that majorate $\psi_Z(\boldsymbol{u})|_s^{m-1}$ so that

$$r_P(\boldsymbol{u}) \leq \frac{\psi_Z(\boldsymbol{u})|_0^{s-1} + \beta'(\boldsymbol{u})|_s^{m-1}}{\|I'(\boldsymbol{u})\|\|T'\|} \tag{10.26}$$

$$r_P(\boldsymbol{u}) \leq \frac{\psi_Z(\boldsymbol{u})|_0^{s-1} + \beta''(\boldsymbol{u})|_s^{m-1}}{\|I'(\boldsymbol{u})\|\|T'\|}. \tag{10.27}$$

The speed-up derives from the possibility of computing $\beta'$ and $\beta''$ and the denominator using sum tables, reducing direct computation to $\psi_Z(\boldsymbol{u})|_0^{s-1}$, a subset of the whole template. The inequalities are used to prune the position whenever

$$\frac{\psi_Z(\boldsymbol{u})|_0^{s-1} + \beta'(\boldsymbol{u})|_s^{m-1}}{\|I'(\boldsymbol{u})\| \|T'\|} \leq r_{\min} \tag{10.28}$$

and similarly for $\beta''$. The minimum acceptable correlation value $r_{\min}$ can be provided or can be estimated using a fraction $\alpha < 1$ of the best correlation value $r_H$ found with a hierarchical approach: $r_{\min} = \alpha r_H$. A reasonable range for $\alpha$ is the interval [0.9, 0.95]. The speed-up resulting from the use of Equation 10.28 is more limited than the one achievable in the distance-based comparison, with typical values in the range [2, 5]. The approaches presented in this section are equivalent to exhaustive search: they are guaranteed to find the best matching position.

### 10.1.5. THE FFT ADVANTAGE

The discrete Fourier correlation theorem (see Intermezzo 3.2) states that the Fourier transform of the discrete, unnormalized, circular (cross-)correlation of two real functions (images) $I_A$ and $I_B$ is

$$I_A \otimes I_B = \mathcal{F}^{-1}\{\mathcal{F}\{I_A\}\mathcal{F}^*\{I_B\}\} \tag{10.29}$$

so that the spatial domain values can be recovered via inverse Fourier transforms. The theorem is based on two important hypotheses:

1. The images have the same size.
2. The images are periodic.

The first requirement can be easily satisfied by padding the two images with zero so that they have the same dimensions. The algorithms of modern fast Fourier transforms (FFTs) are not significantly penalized by using image dimensions that are not of the form $b^k$. The effects due to the second requirement not being satisfied can be cured, if needed, by windowing (see Equation 10.77). The arithmetic complexity of the computation required by Equation 10.29 is roughly $30n^2 \log_2 n$, which must be compared to the complexity $(3m^2 - 1)(n - m + 1)^2$ of the direct computation of the unnormalized correlation. The relative speed-up of the Fourier approach over the direct one is presented graphically in Figure 10.3. The relative efficiency of the transform method increases when $m$ approaches $n/2$ and when they both become large. The possibility of taking advantage of Fourier space computation even when computing normalized cross-correlation is discussed in Section 10.1.7.

### 10.1.6. PCA-BASED SPEED-UP

In the previous sections we addressed the problem of finding a single template within an image. In many situations, such as face detection, we are not searching for a specific template, but rather for any good enough match with an entire class of templates. The idea of progressive pruning can be easily applied to PCA-based detection. Let $\{\boldsymbol{e}_i\}_{i=1,\ldots,k}$ be the set of first $k$ eigenvectors sorted by decreasing eigenvalues $\lambda_1 \geq \lambda_2 \geq \cdots \geq \lambda_k$ (see Equation 8.16).

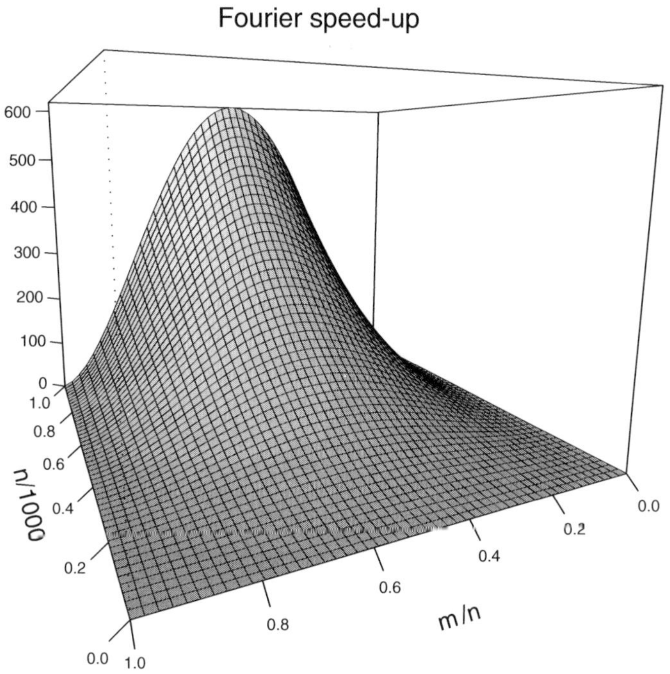


**Figure 10.3.** The relative speed-up of Fourier-based correlation as a function of image ($n$) and template ($m$) sizes.

The idea is to progressively project $I_T(\boldsymbol{u})$ over the space spanned by $\{\boldsymbol{e}_i\}_{i=1,\dots,k}$

$$c_i = I_T(\boldsymbol{u}) \cdot \boldsymbol{e}_i \tag{10.30}$$

$$\hat{I}_{T,i}(\boldsymbol{u}) = \hat{I}_{T,i-1}(\boldsymbol{u}) + c_i \boldsymbol{e}_i \tag{10.31}$$

$$\delta_i = \|\hat{I}_{T,i}(\boldsymbol{u}) - I_T(\boldsymbol{u})\|_2 \tag{10.32}$$

checking the following pruning conditions:

$$P_\delta(\delta_i) \leq \theta_1 \tag{10.33}$$

$$P_c(|c_i|) \leq \theta_2 \tag{10.34}$$

where the two distributions $P_\delta$ and $P_c$ are estimated when PCA is performed. As we saw in Section 8.1.1, they are usually modeled as normal distributions whose parameters can be derived by maximum likelihood estimation. If the conditions reported in Equations 10.33–10.34 are not satisfied, the projection process continues, otherwise position $\boldsymbol{u}$ is no longer considered because it does not provide (probabilistically) enough evidence for the template.

### 10.1.7. A COMBINED APPROACH

Transform-based computation, sum-table-based efficiency, and principal subspace projection can be combined to obtain a fast pattern matching algorithm when large sets of templates must be considered. The algorithm is best described as a sequence of an off-line phase followed by an on-line processing step and exploits many of the ideas discussed in previous sections.

**Off-line** Let $\boldsymbol{x}_i, i = 1, \ldots, N$, be the set of templates. The following steps can be performed off-line, prior to the detection process, as they only operate on the training database.

1. Normalize data so that the sum of the coordinates of each vector is 0:
$$x'_{i,j} = x_{i,j} - \sum_{l=1}^{m^2} x_{i,l}. \tag{10.35}$$
This normalization is the one required for the computation of $r_P$ and is different from the one used in PCA where averaging is performed across corresponding pixels of different images.
2. Make each sample an $L_2$ unit norm vector
$$\boldsymbol{x}''_i = \frac{\boldsymbol{x}'_i}{\|\boldsymbol{x}'_i\|_2} \quad \forall i = 1, \ldots, N. \tag{10.36}$$
3. Compute the arithmetic average of the $N$ normalized vectors
$$\boldsymbol{\mu} = \frac{1}{N}\sum_{i=1}^{N} \boldsymbol{x}''_i. \tag{10.37}$$
4. Compute the first $k$ principal directions $\boldsymbol{e}_i$ for the normalized set $\{\boldsymbol{x}''_i - \boldsymbol{\mu}\}$ (see Section 8.1.2 on how to choose an appropriate value of $k$).
5. Compute and normalize $\boldsymbol{e}_0$, the projection of $\boldsymbol{\mu}$ onto the orthogonal complement of the space spanned by the eigenvectors
$$\boldsymbol{e}_0 = \left(\boldsymbol{\mu} - \sum_{l=1}^{k}(\boldsymbol{\mu}\cdot\boldsymbol{e}_l)\boldsymbol{e}_l\right) \Big/ \left\|\boldsymbol{\mu} - \sum_{l=1}^{k}(\boldsymbol{\mu}\cdot\boldsymbol{e}_l)\boldsymbol{e}_l\right\|_2. \tag{10.38}$$
Note that $\boldsymbol{e}_0$ satisfies Equation 10.35, being the average of vectors that already satisfy it.
6. Compute the Fourier transforms $\mathcal{F}\{\boldsymbol{e}_l\}$, $l = 0, \ldots, k$, of the eigenvectors applying the appropriate zero padding to them based on the image to be searched.

**On-line** The following steps are applied to each image where the templates must be located.

1. At each position $\boldsymbol{u}$ in the image we need to compute
$$r_P(\boldsymbol{u}) \approx \frac{\boldsymbol{x}\cdot I'_T(\boldsymbol{u})}{\|I'_T(\boldsymbol{u})\|^2} \tag{10.39}$$
where $I'_T(\boldsymbol{u})$ is the image window normalized according to Equation 10.35 and
$$\boldsymbol{x} = \sum_{l=0}^{k}(I'_T(\boldsymbol{u})\cdot\boldsymbol{e}_l)\boldsymbol{e}_l \tag{10.40}$$

is the projection of $I'_T(\boldsymbol{u})$ onto the principal subspace (see Step 4 of the off-line phase), identifying its closest approximation. This leads to

$$r_P(\boldsymbol{u}) \approx \frac{\sum_{l=0}^{k} (I'_T(\boldsymbol{u}) \cdot \boldsymbol{e}_l)^2}{\left\| I'_T(\boldsymbol{u}) \right\|^2}. \tag{10.41}$$

The (approximate) normalization uses $\left\| I'_T(\boldsymbol{u}) \right\|^2$ instead of $\|\boldsymbol{x}\| \| I'_T(\boldsymbol{u}) \|$ because the former can be computed efficiently and $\|\boldsymbol{x}\| \leq \left\| I'_T(\boldsymbol{u}) \right\|$. This inequality implies that when the approximation is good, the result is approximately the same, while when the approximation is bad the value of the correlation coefficient will be lower than the correct one, a fact that does not usually cause problems. The denominator of Equation 10.39 can be computed efficiently using sum tables (see Equations 10.10–10.11).

2. The sum of the coordinates of each eigenvector is 0 so that

$$\boldsymbol{e}_l \cdot I'_T(\boldsymbol{u}) = \sum_{i_1=0}^{m-1} \sum_{i_2=0}^{m-1} e_l(i_1, i_2)(I^2(u_1 + i_1, u_2 + i_2) - \mu_I(\boldsymbol{u})) = \boldsymbol{e}_l \cdot I_T(\boldsymbol{u}). \tag{10.42}$$

The numerator of Equation 10.41 can then be computed efficiently in the Fourier domain and back transformed to the spatial domain:

$$(\boldsymbol{e}_l \otimes I)(\boldsymbol{u}) = \mathcal{F}^{-1}\{\mathcal{F}\{\boldsymbol{e}_l\}\mathcal{F}^*\{I\}\} \; \forall \, l = 0, \ldots, k \tag{10.43}$$

with appropriate zero padding of $\boldsymbol{e}_l$. We can then efficiently obtain the normalized correlation coefficient at every position

$$r_P(\boldsymbol{u}) = \frac{\sum_{l=0}^{k} [(\boldsymbol{e}_l \otimes I)(\boldsymbol{u})]^2}{\left\| I'_T(\boldsymbol{u}) \right\|^2}. \tag{10.44}$$

This algorithm provides not only the matching score at each image position, but also the coefficients of the projection onto the principal subspace from Equation 10.43.

## 10.2. Precision

Another very important aspect of template matching is the (measurement) precision with which the location of templates can be computed. Let us recall that, for measurements, accuracy quantifies closeness to the true value, while precision measures reproducibility: that is, how close to each other measurements are. A valid detector should be both accurate and precise. Several applications, such as super resolution imaging (which derives a high-resolution image from a set of lower resolution ones) or measurements in 3D, require subpixel accuracy in the detection of landmark features. At first sight it might seem that the precision of localization cannot exceed the discreteness that characterizes the digital images considered. However, in the case of properly sampled signals, there are several opportunities for reliable computation of template position at subpixel accuracy, often at the level of 0.1 pixels, and sometimes even more, up to 0.01 pixels.

In order to get an appreciation of the obtainable precision, let us consider a one-dimensional template $T$ of length $m$ and its $2n_\epsilon$ linearly resampled versions $T_r$

$$T\left(i_1 + \frac{r}{n_\epsilon}\right) \approx T_r(i_1) = \begin{cases} \frac{r}{n_\epsilon}[T(i_1+1) - T(i_1)] + T(i_1) & r > 0 \\ \frac{r}{n_\epsilon}[-T(i_1-1) + T(i_1)] + T(i_1) & r < 0 \end{cases} \tag{10.45}$$

where $|r| = 1, \ldots, n_\epsilon$ (when $r = 0$ we have the original template). The average difference $\Delta_T$ between successive template samples is given by

$$\Delta_T = \frac{1}{m} \sum_{i=1}^{m} |T(i+1) - T(i)|. \tag{10.46}$$

If we assume that image $I$ exactly corresponds to the untranslated template, its distance from the translated versions is given by

$$\Delta(I, T_r) = \frac{r}{n_\epsilon} m \Delta_T \tag{10.47}$$

and

$$\Delta(T_{r+1}, T_r) = \frac{1}{n_\epsilon} m \Delta_T. \tag{10.48}$$

The standard deviation of $\Delta(I, T_r)$ due to noise $\eta$ is $\sigma_\Delta = \sqrt{m\sigma_\eta}$ and the standard deviation of the difference of two such values, such as those due to two adjacent translated templates, is then $\sqrt{2m\sigma_\eta}$. In order for the displacement $n_\epsilon^{-1}$ to be detectable, we must have

$$\Delta(T_{r+1}, T_r) > \sqrt{2m\sigma_\eta} \tag{10.49}$$

and

$$n_\epsilon < \sqrt{\frac{m}{2\sigma_\eta}} \Delta_T \tag{10.50}$$

which for $\Delta_T \approx 5$, $\sigma_\eta = 1$, and $m = 100$ results in 0.028 pixels; that is, a 35-fold increase in precision.

In order to limit the amount of computation, an algorithm for subpixel template detection would use a standard approach to get within 1 pixel of the actual position, then switch to the match of the resampled templates. Referring to the previous one-dimensional case, let us assume that the best position is found at $i_0$ so we need to compute

$$\hat{r} = \operatorname*{argmin}_{r \in \{-n_\epsilon, \ldots, n_\epsilon\}} \Delta(T_r, I, i_0) \tag{10.51}$$

that locates the template at $i_0 - \hat{r}$. If we plot $\Delta(T_r, I, i_0)$ as a function of $r$ we will see that the resulting function (approximately) represents a sampled parabola. The equation of the parabola can be estimated by the least squares method to get the position of the minimum value, thereby obtaining a refined estimate for the template location. The strategy outlined for the one-dimensional case generalizes directly to the two-dimensional case. The template must be resampled on a regular grid and the refined location is obtained by interpolating the resulting paraboloid (see Intermezzo 10.1). The refinement based on the interpolation of $\Delta(T_r, I, i_0)$ often results in a five-fold increase of subpixel accuracy so that a smaller number of resampled templates is necessary to get the required precision.

**Intermezzo 10.1.** Fitting a paraboloid

Let $\{x_i, y_i, z_i\}_{i=1}^{s}$ be a set of $s$ samples assumed to lie on a paraboloid

$$z = ax^2 + bxy + cy^2 + dx + ey + f = \boldsymbol{P} \cdot \boldsymbol{Q}(x, y) \tag{10.52}$$

where $\boldsymbol{P} = (a, b, c, d, e, f)$ and $\boldsymbol{Q}(x, y) = (x^2, xy, y^2, x, y, 1)$. We want to find the point $\boldsymbol{P}$ which minimizes the sum of squared errors

$$E(\boldsymbol{P}) = \sum_{i=1}^{s} (\boldsymbol{P} \cdot \boldsymbol{Q}_i - z_i)^2 \tag{10.53}$$

where $\boldsymbol{Q}_i = \boldsymbol{Q}(x, y)$. The minimum is attained when the gradient of $E$ is the zero vector

$$\nabla E = 2 \sum_{i=1}^{s} (\boldsymbol{P} \cdot \boldsymbol{Q}_i - z_i) \boldsymbol{Q}_i = \boldsymbol{0} \tag{10.54}$$

leading to the system

$$\left( \sum_{i=1}^{s} \boldsymbol{Q}_i \boldsymbol{Q}_i^T \right) \boldsymbol{P} = \sum_{i=1}^{s} z_i \boldsymbol{Q}_i \tag{10.55}$$

from which $\boldsymbol{P}$ can be derived. The location of the minimum of the paraboloid is then given by

$$(x_0, y_0) = \left( \frac{2cd - be}{b^2 - 4ac}, \frac{bd - 2ae}{b^2 - 4ac} \right). \tag{10.56}$$

### 10.2.1. A PERTURBATIVE APPROACH

Template resampling is not the only way to achieve subpixel accuracy. An interesting approach is based on the possibility of approximating signal translation by means of convolution

$$f(x + t) = f * \delta_{-t}(x) = \int f(u) \delta_{-t}(x - u)\, du \tag{10.57}$$

where $\delta_{-t}(x) = \delta(x + t)$ is the Dirac delta function. In the discrete case we obtain

$$f(i + t) = (f * \delta_{-t})(i) = \sum_j f(j) \delta_{-t}(i - j). \tag{10.58}$$

When the translation is not represented by an integer value, the data of $f$ are missing due to sampling, but, under the assumption that the sampling frequency is significantly larger than the signal frequency, it is possible to resort to the linear approximation

$$f(i + \epsilon) \approx (1 - \epsilon) f(i) + \epsilon f(i + 1) \tag{10.59}$$

which amounts to a convolution with $(1 - \epsilon)\delta_0 + \epsilon \delta_{-1}$, an approximation of $\delta_{-\epsilon}$. The above equation generalizes, for a smooth function $f$, to

$$\sum_i b_i f(x + \lambda_i) \approx f\left(x + \sum_i \lambda_i b_i\right) \tag{10.60}$$

given that $\sum_i b_i = 1$ and $b_i > 0$. Let us now revert to the problem of matching template $T$ in image $I$. Let us consider the case of precise matching: we have already obtained the position of the template to within 1 pixel. In this case $t \in [-1, 1]$. We further assume that the template appearing in the image is subject to noise $\eta$, arbitrary (intensity) offset $O$, and intensity scaling

$$I(u_1 + i_1, u_2 + i_2) = \sum_{k \in \mathcal{N}} a_k T(i_1 + k_1, i_2 + k_2) + O + \eta \tag{10.61}$$

where $\mathcal{N}$ represents the points with integer coordinates in a neighborhood of $\mathbf{0}$, e.g. its four-connected neighborhood

$$\mathcal{N} = \{(i, j) : |i| + |j| \leq 1\} = \{(0, 0), (-1, 0), (1, 0), (0, 1), (0, -1)\}.$$

The parameters $a_k$ and $O$ can be found by means of least squares minimization

$$(\hat{a}_k, \hat{O}) = \underset{a_k, O}{\operatorname{argmin}} \sum_{(i_1, i_2) \in T} \left( I(u_1 + i_1, u_2 + i_2) - \sum_{k \in \mathcal{N}} a_k T(i_1 + k_1, i_2 + k_2) - O \right)^2. \tag{10.62}$$

The quality of the approximation can be quantified with

$$\sigma_T = \frac{1}{m} \sqrt{\sum_{i \in T} \left( I(\boldsymbol{u} + \boldsymbol{i}) - \sum_{k \in \mathcal{N}} \hat{a}_k T(\boldsymbol{i} + \boldsymbol{k}) - \hat{O} \right)^2}. \tag{10.63}$$

The coefficients $\hat{a}_k$ take care of the rescaling and of the translation of the template according to the two-dimensional generalization of Equation 10.60. The translation between the image and the template can be estimated also by translating the image window, obtaining an alternative estimate

$$T(i_1, i_2) = \sum_{k \in \mathcal{N}} a'_k I(u_1 + i_1 + k_1, u_2 + i_2 + k_2) + O' + \eta. \tag{10.64}$$

The intensity scaling factor is given by $\hat{s} = \sum_k \hat{a}_k$ and the coefficients for the estimation of the translation are given by $\hat{b}_k = \hat{a}_k \hat{s}^{-1}$. The generalization of Equation 10.60 results in

$$I(\boldsymbol{u} + \boldsymbol{i}) \approx \hat{s} \sum_{k \in \mathcal{N}} \hat{b}_k T(\boldsymbol{i} + \boldsymbol{k}) + \hat{O} \tag{10.65}$$

$$\approx \hat{s} T(\boldsymbol{i} + \sum_{k \in \mathcal{N}} b_k \boldsymbol{k}) + \hat{O} \tag{10.66}$$

finally providing an estimate $\hat{\boldsymbol{t}}$ for the translation

$$\hat{\boldsymbol{t}}_T = \sum_{k \in \mathcal{N}} \hat{b}_k \boldsymbol{k}. \tag{10.67}$$

In the same way from Equation 10.64 we can obtain an alternative estimate $\hat{\boldsymbol{t}}_I$. The final estimate is obtained by combining the estimates, weighting them with the inverse of the corresponding variances:

$$\hat{\boldsymbol{t}} = \left( \frac{1}{\sigma_T^2} \hat{\boldsymbol{t}}_T + \frac{1}{\sigma_I^2} \hat{\boldsymbol{t}}_I \right) \left( \frac{1}{\sigma_T^2} + \frac{1}{\sigma_I^2} \right)^{-1}. \tag{10.68}$$

### 10.2.2. PHASE CORRELATION

We saw in Section 6.3 that it is possible to get a very sharp correlation peak with a suitably designed matched filter. We want to gain additional insights in order to exploit the result for subpixel localization.

Let us consider two images $I_a$ and $I_b$ and the corresponding discrete Fourier transforms $\tilde{I}_a$ and $\tilde{I}_b$. The phase-only circular correlation of the two images is defined in Fourier space as

$$\tilde{R}(k_1, k_2) = \frac{\tilde{I}_a(k_1, k_2)\tilde{I}_b^*(k_1, k_2)}{|\tilde{I}_a(k_1, k_2)\tilde{I}_b^*(k_1, k_2)|} = e^{i(\theta_a(k_1,k_2)-\theta_b(k_1,k_2))} \tag{10.69}$$

and in the spatial domain by the inverse transform of $\tilde{R}(k_1, k_2)$

$$R(i_1, i_2) = \frac{1}{n_1 n_2} \sum_{k_1,k_2} \tilde{R}(k_1, k_2) e^{+i2\pi k_1 i_1/n_1} e^{+i2\pi k_2 i_2/n_2} \tag{10.70}$$

where $k_1 = -m_1, \ldots, +m_1$, $n_1 = 2m_1 + 1$, and similarly for the other coordinate. If image $I_b$ is a circular translation of $I_a$ by the integer displacement $(\Delta_1, \Delta_2)$, its Fourier transform is given by

$$\tilde{I}_b = \tilde{I}_a e^{-2\pi i(k_1(\Delta_1/n_1)+k_2(\Delta_2/n_2))} \tag{10.71}$$

leading to

$$\tilde{R}(k_1, k_2) = \frac{\tilde{I}_a(k_1, k_2)\tilde{I}_a^*(k_1, k_2)e^{+2\pi i(k_1(\Delta_1/n_1)+k_2(\Delta_2/n_2))}}{|\tilde{I}_a(k_1, k_2)\tilde{I}_a^*(k_1, k_2)e^{+2\pi i(k_1(\Delta_1/n_1)+k_2(\Delta_2/n_2))}|} = e^{+2\pi i(k_1(\Delta_1/n_1)+k_2(\Delta_2/n_2))} \tag{10.72}$$

whose inverse transform is a delta function

$$R(i_1, i_2) = \delta(i_1 + \Delta_1, i_2 + \Delta_2). \tag{10.73}$$

If the translation does not correspond to an integer number of samples but is small, say $(\delta_1, \delta_2)$, the following approximation holds:

$$\tilde{R}(k_1, k_2) \approx e^{+2\pi i(k_1(\delta_1/n_1)+k_2(\delta_2/n_2))} \tag{10.74}$$

due to the shift theorem of the continuous Fourier transform and the fact that the DFT is an approximation of the continuous one. We are then led to the following spatial domain phase-only correlation response:

$$R(i_1, i_2) \approx \frac{\alpha}{n_1 n_2} \frac{\sin[\pi(i_1 + \delta_1)]}{\sin[(\pi/n_1)(i_1 + \delta_1)]} \frac{\sin[\pi(i_2 + \delta_2)]}{\sin[(\pi/n_2)(i_2 + \delta_2)]} \tag{10.75}$$

where $\alpha$ models the effect of small noise components: $\alpha = 1$ when there is no noise and $\alpha < 1$ when noise is present. The translation that optimally aligns the template with the image is then recovered by a least squares fit of the peak shape. In order to apply these results successfully to subpixel accuracy template detection, attention must be paid to a few details. The derivation is based on the assumption of periodic images: their wrapping around at the edges introduces discontinuities that do not actually exist in the patterns we are interested in. A way to reduce the effect is to apply a windowing function to the images before computing the DFT

$$I(i_1, i_2) \to W(i_1, i_2)I(i_1, i_2) \tag{10.76}$$

a common choice being the Hann window

$$W(i_1, i_2) = \frac{1}{4}\left[1 + \cos\left(\frac{\pi i_1}{n_1}\right)\right]\left[1 + \cos\left(\frac{\pi i_2}{n_2}\right)\right]. \tag{10.77}$$

The peak is very sharp so that a limited number of points is needed to get a good estimate of the parameters.

Phase-only correlation (Figure 10.4) emphasizes high frequencies due to the division by the magnitude of the Fourier transform appearing in Equation 10.69. However, high frequencies may have a lower signal to noise ratio, therefore leading to unreliable estimates. A solution is to apply a low-pass filter in the Fourier domain, preserving the low spatial frequency components where most of the energy of natural images is located. The shape of the peak in the spatial domain changes and must be computed specifically for the filter used. A practical choice is a Gaussian kernel

$$\tilde{H}(k_1, k_2) \approx e^{-2\pi^2\sigma^2(k_1^2+k_2^2)} \tag{10.78}$$

whose smoothing is controlled by $\sigma$

$$R(i_1, i_2) = \frac{1}{n_1 n_2}\sum_{k_1,k_2} \tilde{R}(k_1, k_2)\tilde{H}(k_1, k_2)e^{+i2\pi k_1 i_1/n_1}e^{+i2\pi k_2 i_2/n_2} \tag{10.79}$$

$$\approx \frac{1}{n_1 n_2}\sum_{k_1,k_2} e^{+2\pi i(k_1(\delta_1/n_1)+k_2(\delta_2/n_2))}\tilde{H}(k_1, k_2)e^{+i2\pi(k_1(i_1/n_1)+k_2(i_2/n_2))} \tag{10.80}$$

$$= H(i_1 + \delta_1, i_2 + \delta_2) \tag{10.81}$$

$$\approx \frac{1}{2\pi\sigma^2}e^{-[(i_1+\delta_1)^2+(i_2+\delta_2)^2]/(2\sigma^2)}. \tag{10.82}$$

The technique can be applied also to the case of rotated and scaled patterns using the Fourier–Mellin transform (see Intermezzo 10.2).

## 10.3. Bibliographical Remarks

The literature on how to speed up template matching computation is vast and the present chapter is based on a few papers that address the problem from the perspective adopted by this book. The idea of using feature probabilities to reduce the expected computational cost of template matching is described by Margalit and Rosenfeld (1990). A short review of jump-out techniques for fast-motion estimation in video coding is reported by Huang *et al.* (1996) who also describe the adaptive jump-out technique presented in this chapter. The possibility of speeding up some of the computations needed to obtain the normalized correlation coefficient is widely appreciated and described by, among others, Tsai and Lin (2003) who also present the case for its use in the specific application of defect detection.

The use of the Cauchy–Schwartz inequality in deriving useful distance bounds for the normalized correlation coefficient is described in Di Stefano *et al.* (2005) while the idea of

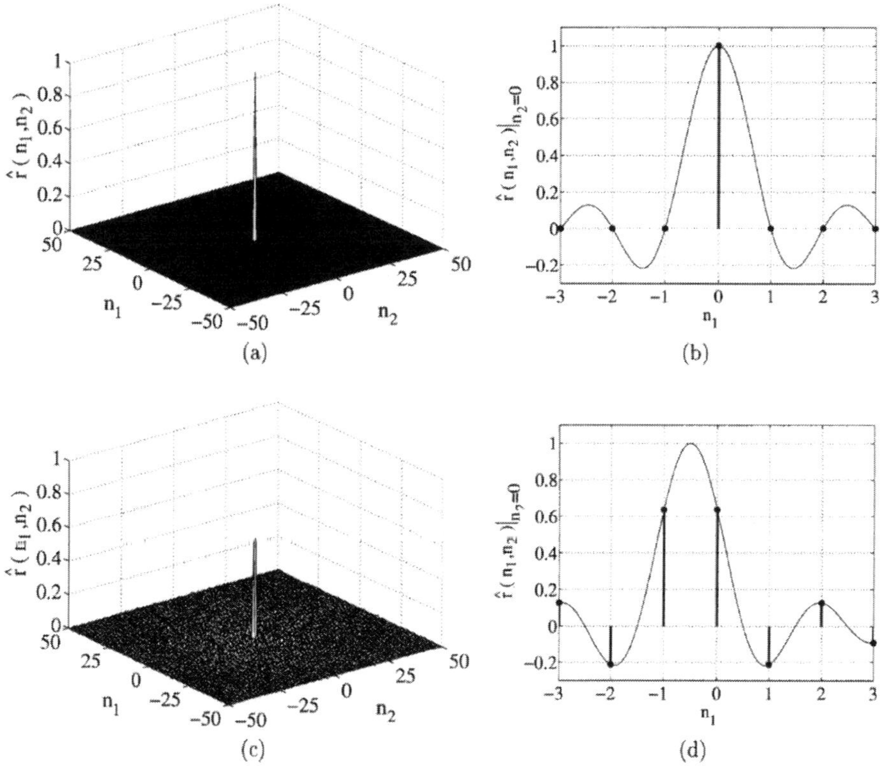


**Figure 10.4.** The upper plot shows the results of phase-only correlation for a null relative displacement while the lower plot shows the result for a displacement $\delta = (0.5, 0)$. The plots on the right show an enlarged view of the profile at $n_2 = 0$. Source: Takita *et al.* (2003).

getting progressively tighter bounds is described in Tombari *et al.* (2009). The possibility of improving template matching speed by hierarchical strategies has also been known for a long time and is exploited in many approaches. A coarse to fine algorithm equivalent to full search and exploiting properties of $L_p$ norms is described by Gharavi-Alkhansari (2001). The triangular inequality among subtemplates is exploited in the paper by Kawanishi *et al.* (2004). The possibility of using projection kernels to speed up template matching, particularly in the low-noise limit, is addressed in full detail in Hel-Or and Hel-Or (2005) and Ben-Artzi *et al.* (2007).

Using the fast Fourier transform to speed up the computation of correlation in the non-normalized case is a basic result of Fourier theory. The possibility of using it even for the computation of the normalized coefficient is addressed by Uenohara and Kanade (1997) together with the possibility of matching to a whole subspace of templates without requiring the explicit computation of the best matching projection. The computational advantages of the FFT approach carry over to the robust estimation of correlation: it is possible to expand a robust kernel in terms of trigonometric functions and to compute a robust correlation, performing pixel-level weighting by convolving with the kernel, by repeated application of the FFT theorem (Fitch *et al.* 2005).

**Intermezzo 10.2.** The Fourier–Mellin transform

The shift theorem of the Fourier transform, exploited in Equation 10.71 and Equation 10.74, tells us that circular image translation only affects the phase component of the transform: if we restrict consideration to the magnitude of the transform, we lose information (and invertibility) but we gain translational invariance. The Fourier transform has two additional properties regarding the way it transforms under scaling and rotation in the original spatial domain:

1. Scaling by $\lambda$:

$$\mathcal{F}\{I(\lambda \boldsymbol{x})\}(\boldsymbol{u}) = \frac{1}{\lambda^2}\mathcal{F}\{I(\boldsymbol{x})\}\left(\frac{1}{\lambda}\boldsymbol{u}\right). \tag{10.83}$$

2. Rotation by $\theta$:

$$\begin{aligned}\mathcal{F}\{I(x_1 \cos\theta - x_2 \sin\theta, x_1 \sin\theta + x_2 \cos\theta)\}(\boldsymbol{u}) = \mathcal{F}\{I(\boldsymbol{x})\}(&u_1 \cos\theta \\ &- u_2 \sin\theta, u_1 \sin\theta + u_2 \cos\theta).\end{aligned} \tag{10.84}$$

Thus scaling and rotations map from the spatial to the frequency domain and the result holds (approximately) for the DFT. If we are able to map rotations and scaling into translations, we can add yet another Fourier transform: its magnitude is then invariant to translations (of the original image), scaling, and rotations (by virtue of their being transformed into translations). The log-polar transform discussed in Section 2.4.3 transforms rotations and scaling into translations. In the current case the choice of the coordinate origin is quite natural: the point $(0, 0)$ in Fourier space. The sequence of the above steps is known as the Fourier–Mellin transform. Appropriate frequency-variant smoothing must be applied in Fourier space before applying the log-polar transform. The technique presented in the main text for high-accuracy localization based on phase correlation can be applied to determine the scaling and rotation parameters $(\lambda, \theta)$ in the log-polar transformed Fourier space. Using the estimated parameters, the template is normalized to match the scale and orientation of the image, and its accurate position can then be estimated using the original phase-only correlation approach. The Fourier–Mellin transform has found many applications, including invariant object recognition, watermarking and image registration (Derrode and Ghorbel 2001; Milanese *et al.* 1998; Reddy and Chatterji 1996; Zhang *et al.* 2006).

The description of subpixel accuracy template matching is derived from three papers. The paper by Frischholz and Spinnler (1993) is based on the idea of template resampling and presents the argument supporting the computation of the maximum attainable precision. The idea of expanding a translated template as a linear combination of convolutions with translated Dirac functions is presented in Lan and Mohr (1998). The presentation of phase correlation and its potential in subpixel accuracy matching is based on the papers by Takita *et al.* (2003) and Nagashima *et al.* (2006).

## References

Ben-Artzi G, Hel-Or H and Hel-Or Y 2007 The Gray-code filter kernels. *IEEE Transactions on Pattern Analysis and Machine Intelligence* **29**, 382–393.

Derrode S and Ghorbel F 2001 Robust and efficient Fourier–Mellin transform approximations for gray-level image reconstruction and complete invariant description. *Computer Vision and Image Understanding* **83**, 57–78.

Di Stefano L, Mattoccia S and Tombari F 2005 ZNCC-based template matching using bounded partial correlation. *Pattern Recognition Letters* **26**, 2129–2134.

Fitch A, Kadyrov A, Christmas W and Kittler J 2005 Fast robust correlation. *IEEE Transactions on Image Processing* **14**, 1063–1073.

Frischholz R and Spinnler K 1993 A class of algorithms for real-time subpixel registration. *Proceedings of SPIE*, vol. 1989, pp. 50–59.

Gharavi-Alkhansari M 2001 A fast globally optimal algorithm for template matching using low-resolution pruning. *IEEE Transactions on Image Processing* **10**, 526–533.

Hel-Or Y and Hel-Or H 2005 Real-time pattern matching using projection kernels. *IEEE Transactions on Pattern Analysis and Machine Intelligence* **27**, 1430–1445.

Huang HC, Hung YP and Hwang WL 1996 Adaptive early jump-out technique for fast motion estimation in video coding. *Proceedings of the 13th IAPR International Conference on Pattern Recognition (ICPR'96)*, vol. 2, pp. 864–868.

Kawanishi T, Kurozumi T, Kashino K and Takagi S 2004 A fast template matching algorithm with adaptive skipping using inner-subtemplates' distances. *Proceedings of the IEEE Conference on Computer Vision and Pattern Recognition (CVPR'04)*, pp. 654–657.

Lan Z and Mohr R 1998 Direct linear sub-pixel correlation by incorporation of neighbor pixels' information and robust estimation of window transformation. *Machine Vision and Applications* **10**, 256–268.

Margalit A and Rosenfeld A 1990 Using feature probabilities to reduce the expected computational cost of template matching. *Computer Vision, Graphics and Image Processing* **52**, 110–123.

Milanese R, Cherbuliez M and Pun T 1998 Invariant content-based image retrieval using the Fourier-Mellin transform. *Proceedings of the International Conference on Advances in Pattern Recognition*, pp. 73–82.

Nagashima S, Aoki T, Higuchi T and Kobayashi K 2006 A subpixel image matching technique using phase-only correlation. *Proceedings of the International Symposium on Intelligent Signal Processing and Communications Systems (ISPACS2006)*, pp. 701–704.

Reddy B and Chatterji B 1996 An FFT-based technique for translation, rotation, and scale-invariant image registration. *IEEE Transactions on Image Processing* **5**, 1266–1271.

Takita K, Aoki T, Sasaki Y, Higuchi T and Kobayashi K 2003 High-accuracy subpixel image registration based on phase-only correlation. *IEICE Transactions on Fundamentals of Electronics, Communications and Computer Sciences* **E86A**, 1925–1934.

Tombari F, Mattoccia S and Di Stefano L 2009 Full search-equivalent pattern matching with incremental dissimilarity approximations. *IEEE Transactions on Pattern Analysis and Machine Intelligence* **31**, 129–141.

Tsai DM and Lin CT 2003 Fast normalized cross correlation for defect detection. *Pattern Recognition Letters* **24**, 2625–2631.

Uenohara M and Kanade T 1997 Use of Fourier and Karhunen-Loeve decomposition for fast pattern matching with a large set of templates. *IEEE Transactions on Pattern Analysis and Machine Intelligence* **19**, 891–898.

Zhang J, Ou Z and Wei H 2006 Fingerprint matching using phase-only correlation and Fourier–Mellin transforms. *Proceedings of the Sixth International Conference on Intelligent Systems Designs and Applications*, vol. 2, pp. 379–383.

# 11 MATCHING POINT SETS: THE HAUSDORFF DISTANCE

Nor yet the other's distance comfort me.

*Pericles*
WILLIAM SHAKESPEARE

Matching sets of points using techniques targeted at area matching is far from optimal, regarding both efficiency and effectiveness. This chapter shows how to compare sparse templates, composed by points with no textural properties, using an appropriate distance. Robustness to noise and template deformation as well as computational efficiency are analyzed.

## 11.1. Metric Pattern Spaces

In previous chapters we have mainly considered templates not as geometrical entities but rather as pictorial patterns. Even when we neglected all textural information, as in Chapter 7 where we considered curvilinear binary patterns, the way we detected them using the Hough transform was not fully geometric in nature. We now switch to a fully geometrical perspective by considering all pixels representing a binary pattern as points in image space solely characterized by their position (see Figure 11.1). The discrete nature of images is then naturally transferred to point sets, leading to a unified treatment. We need to find a useful way to quantify the similarity of sets of points by introducing an appropriate metric so that we can compare them and detect specific arrangements. The ideas of invariance to pattern transformations and of robustness of the comparison measure to noise and outlying structures preserve their importance in the new framework. A generally useful invariance is that to affine transformations, which represent a good approximation to weak perspective projections of a nearly coplanar set of points. However, the amount of invariance required is heavily dependent on the specific task. Before introducing a specific shape metric, let us recall the conditions under which a function identifies a metric and some properties it should possess to be a useful one. A metric on a set $S$ is a function $d : S \times S \to \mathbb{R}$ that satisfies the following conditions for any element $x, y, z \in S$:

1. $d(x, y) = 0$ if and only if $x = y$.
2. $d(y, z) \leq d(x, y) + d(x, z)$, the strong triangle inequality.

The other conditions usually associated with a metric, namely symmetry and non-negativity, follow from the stated ones. The usual triangle inequality

$$d(x, z) \leq d(x, y) + d(y, z) \tag{11.1}$$

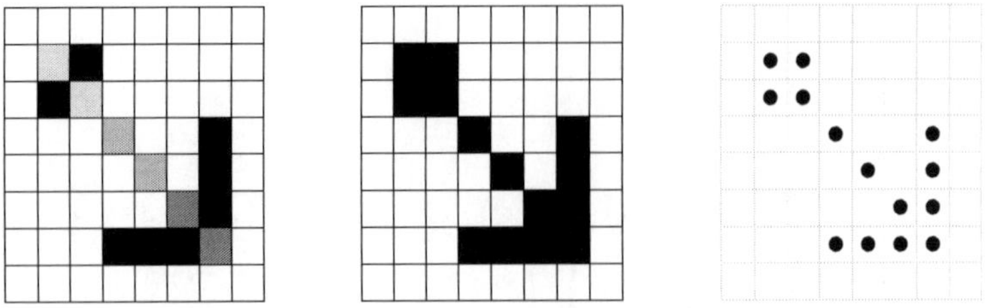

**Figure 11.1.** Three different ways of considering the same pattern: using textural information, as a binary template, and as a set of points.

follows from the above one plus symmetry. There are applications, partial shape matching being one of them, where the triangle inequality is not a meaningful requirement. A clarifying example is when $x$, $y$, $z$ correspond respectively to a man, a horse, and a centaur. Under partial matching the distances $d(\text{man, centaur})$ and $d(\text{centaur, horse})$ are close to zero, but $d(\text{man, horse})$ is high. We can now formalize the space of shapes we will be considering in this chapter.

**Intermezzo 11.1.** What a topological space is

**Definition 11.1.1.** *A topological space is a set $X$, together with a collection $\mathfrak{T}$ of subsets of $X$, satisfying the following axioms:*

1. *The empty set and $X$ are in it.*
2. *The union of any collection of sets in $\mathfrak{T}$ is also in $\mathfrak{T}$.*
3. *The intersection of any finite collection of sets in $\mathfrak{T}$ is also in $\mathfrak{T}$.*

*The collection $\mathfrak{T}$ is usually called a topology.*

**Definition 11.1.2.** *Let $X$ be a topological space (see Intermezzo 11.1), $\mathfrak{S}$ a collection of subsets of $X$, and $d(\cdot, \cdot)$ a metric on $\mathfrak{S}$. The triple $(X, \mathfrak{S}, d)$ is called a metric pattern space.*

We may formalize the notion of pattern deformation by the introduction of a transformation group $T$ such that $t(S) \in \mathfrak{S}$ and $t^{-1}(S) \in \mathfrak{S}$ for each $t \in T$ and $S \in \mathfrak{S}$. There are four features that an ideal metric should possess from the point of view of template matching and we will state them formally.

**Definition 11.1.3 (Deformation robustness).** *A distance function $d(\cdot, \cdot)$ is called deformation robust if for each $S \in \mathfrak{S}$ and $\epsilon > 0$, an open neighborhood $I \subseteq T$ of the identity transformation exists such that $d(S, t(S)) < \epsilon$, $\forall t \in I$.*

The definition formalizes the idea of robustness to perturbations, i.e. small deformations, such as those due to small affine distortion. If we denote by $\mathrm{Cl}(S)$ the closure of $S \in X$, the boundary is defined as $\mathrm{Bd}(S) = \mathrm{Cl}(S) \cap \mathrm{Cl}(X - S)$. We can then introduce the notion of blur robustness: the insensitivity to additions near the boundary of the shape, such as the appearance of detail due to different degrees of image blurring.

**Definition 11.1.4 (Blur robustness).** *A distance function $d(\cdot, \cdot)$ is called blur robust if for each $S \in \mathfrak{S}$ and $\epsilon > 0$ there exists an open neighborhood $U$ of $\mathrm{Bd}(S)$ such that $d(S, B) < \epsilon$, $\forall B \in \mathfrak{S}$ for which $B - U = S - U$ and $\mathrm{Bd}(S) \subseteq \mathrm{Bd}(B)$.*

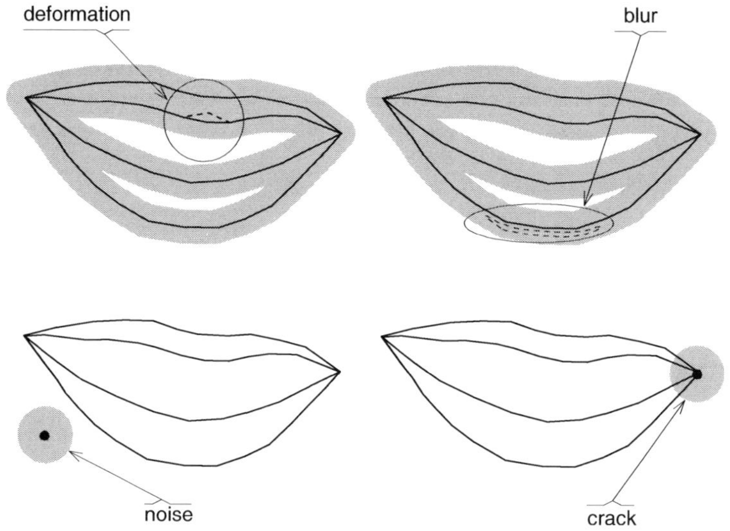


**Figure 11.2.** Examples of the four different types of robustness described in the text for a shape comparison metric.

A commonly experienced deformation of a shape is the formation of small cracks in its boundary that may change its connectivity. While the topological impact of such deformations is major, we would like them to have a limited impact on the distance between the patterns. A crack can be formalized as a closed subset $C \in \mathrm{Bd}(S)$ consisting of limit points of $\mathrm{Bd}(S) - C$. Therefore all open neighborhoods of a point $\boldsymbol{x} \in C$ intersect $\mathrm{Bd}(S) - C$: cracks represent parts of the boundary that can be reconstructed by taking the closure of the boundary remaining after their removal. Note that this crack robustness is not covered by the previous blur robustness, which only addresses the possibility of adding detail to the boundary.

**Definition 11.1.5 (Crack robustness).** *A distance function $d(\cdot, \cdot)$ is called crack robust if for each $S \in \mathfrak{S}$, each crack $C$ of $S$, and $\epsilon > 0$ there exists an open neighborhood $U$ of $C$ such that $S - U = B - U$ implies $d(S, B) < \epsilon$ for all $B \in \mathfrak{S}$.*

Therefore, if crack robustness applies, changing a shape in a small enough neighborhood of a crack results in patterns that are close, with respect to the chosen metric, to the original shape even if connectedness is not preserved. The last definition addresses noise robustness.

**Definition 11.1.6 (Noise robustness).** *A distance function $d(\cdot, \cdot)$ is called noise robust if for each $S \in \mathfrak{S}$, $x \in X$, and $\epsilon > 0$ there exists an open neighborhood $U$ of $x$ such that $S - U = B - U$ implies $d(S, B) < \epsilon$ for all $B \in \mathfrak{S}$.*

If a metric satisfies noise robustness changes of the patterns that occur within small regions do not cause discontinuities in their mutual distance. The different kinds of robustness are illustrated in Figure 11.2.

## 11.2. Hausdorff Matching

Given a metric space $X$, the Hausdorff metric $d_H$ can be defined for any collection of closed, bounded subsets: we will consider it specifically for the Euclidean space $X = \mathbb{R}^n$ denoting by $\mathfrak{R}^n$ the collection of all compact, non-empty subsets of $\mathbb{R}^n$. Let us denote the $\epsilon$ neighborhood of a subset $S \in \mathbb{R}^n$ by

$$S^\epsilon = \bigcup_{x \in S} B^\epsilon(\boldsymbol{x}) \tag{11.2}$$

where $B^\epsilon(\boldsymbol{x})$ denotes the Euclidean ball with center $\boldsymbol{x} \in \mathbb{R}^n$ and radius $\epsilon$.

**Definition 11.2.1.** *The Hausdorff metric $d_H : \mathfrak{R}^n \times \mathfrak{R}^n \to \mathbb{R}$ is given by*

$$d_H(S_1, S_2) = \inf\{\epsilon > 0 \mid S_1 \subseteq S_2^\epsilon \textit{ and } S_2 \subseteq S_1^\epsilon\}. \tag{11.3}$$

Less formally, it is given by the minimum amount of thickening that we must apply in turn to each of the two sets in order to include the other one (with no thickening).

It is not difficult to prove that the Hausdorff metric satisfies the two metric requirements. The metric $d_H$ is invariant under transformations belonging to $\mathrm{Iso}(\mathbb{R}^n)$, the group of Euclidean isometries, i.e. rigid motions combined with reflection:

$$d_H(S_1, S_2) = d(g(S_1), g(S_2)) \quad g \in \mathrm{Iso}(\mathbb{R}^n). \tag{11.4}$$

The Hausdorff metric can also be proved to be deformation, blur, and crack robust but it is not noise robust. In the case of discrete sets of points, as when digital images are considered, the Hausdorff metric can be defined as follows.

**Definition 11.2.2.** *The Hausdorff metric between two finite sets $E = \{\boldsymbol{e}_1, \ldots, \boldsymbol{e}_q\}$ and $F = \{f_1, \ldots, f_r\}$ is given by*

$$d_H(E, F) = \max(d_h(E, F), d_h(F, E)) \tag{11.5}$$

*where*

$$d_h(E, F) = \max_{\boldsymbol{e} \in E} \left[ \min_{f \in f}(\|\boldsymbol{e} - f\|) \right] \tag{11.6}$$

*and $\|\cdot\|$ is the $L_2$ (Euclidean) norm and $d_h(\cdot)$ is called the directed Hausdorff distance.*

In the following we will be using the specialization of the Hausdorff metric given by Equation 11.5 as it is the most appropriate one for the case of discrete images. The directed Hausdorff distance is not symmetric and it is not a metric. Its value is found by computing for each point in set $E$ the distance from its nearest neighbor in $F$, returning the highest among the values so computed. When $d_H(E, F) = d$, every point in $E$ is within a distance $d$ of some point of $F$ and vice versa. No explicit pairing of points between the two sets is required and, as an example, many points from $E$ could match a single point of $F$. As $d_H$ constrains every point of each set to be close enough to a point of the other set, it is not appropriate for partial matching tasks: the points belonging to the complement of the matched portion within each of the sets would in fact result in a large distance. In shape matching applications, we are interested in comparing two given sets for all possible relative positions

$$d_H^m(E, F) = \min_{g_1, g_2 \in G} d_H(g_1(E), g_2(F)) \tag{11.7}$$

where $G$ is a suitable transformation group. Whenever the distance between two points is preserved by the action of $G$, we have that

$$d_H^m(I, T) = \min_{g_1, g_2 \in G} d_H(I, g_1^{-1} g_2(T)) = \min_{g \in G} d_H(I, g(T)) \tag{11.8}$$

where sets $E$ ($F$) have been substituted by $I$ ($T$) making the notation reminiscent of image (template): the template set is moved over the image in order to find the best matching position (see Figure 11.5). The transformation group we are more interested in is that of translations.

## 11.3. Efficient Computation of the Hausdorff Distance

The computation of the Hausdorff distance in the discrete space characteristic of raster images relies on the efficient computation of the minimum distance of every pixel position from the set of pixels representing the shape. The resulting function $D_S(\boldsymbol{x})$ is usually called the distance map of $S$ and gives the minimum distance of point $\boldsymbol{x}$ from the shape set (see also Intermezzo 11.2 and Figure 11.5b). The directed Hausdorff distance $d_h$ can then be computed as

$$d_h(I, T) = \max_{\boldsymbol{a} \in I} D_T(\boldsymbol{a}) \tag{11.9}$$

while the complete Hausdorff distance, for a fixed relative position of the two sets, is given by

$$d_H(I, T) = \max\Big(\max_{\boldsymbol{a} \in I} D_T(\boldsymbol{a}), \max_{\boldsymbol{b} \in T} D_I(\boldsymbol{b})\Big). \tag{11.10}$$

If we need to compute $d_H^m$ over the set of possible translations of the two sets, according to Equation 11.8 we may restrict ourselves to translating only one of them, i.e. $T$, by the (discrete) translation vector $\boldsymbol{x}$, obtaining

$$d_H(I, T \oplus \boldsymbol{x}) = \max\Big(\max_{\boldsymbol{a} \in I} D_T(\boldsymbol{a} - \boldsymbol{x}), \max_{\boldsymbol{b} \in T} D_I(\boldsymbol{b} + \boldsymbol{x})\Big) \tag{11.11}$$

which represents the Hausdorff distance of the two sets for a relative displacement $\boldsymbol{x}$ and we used the Minkowski sum notation to represent the translation of a set: $T \oplus \boldsymbol{x} = \{\boldsymbol{t} + \boldsymbol{x} \mid \boldsymbol{t} \in T\}$. The optimal alignment of the two sets is obtained at

$$\boldsymbol{x}_0 = \operatorname*{argmin}_{\boldsymbol{x}} d_H(I, T \oplus \boldsymbol{x}) \tag{11.12}$$

and the corresponding distance value is

$$d_H^m(I, T) = \min_{\boldsymbol{x}} d_H(I, T \oplus \boldsymbol{x}). \tag{11.13}$$

In many image matching applications the use of the directed Hausdorff distance is more appropriate. The reason is that when we want to detect a model $T$ in an image $I$ that contains many additional shapes, we want to check that every point of the model is close to some image detail but not the other way round, because this would require that unrelated image detail be matched by the model structure. The computation of the directed distance is then

$$d_h(T \oplus \boldsymbol{x}) = \max_{\boldsymbol{b} \in T} D_I(\boldsymbol{b} + \boldsymbol{x}). \tag{11.14}$$

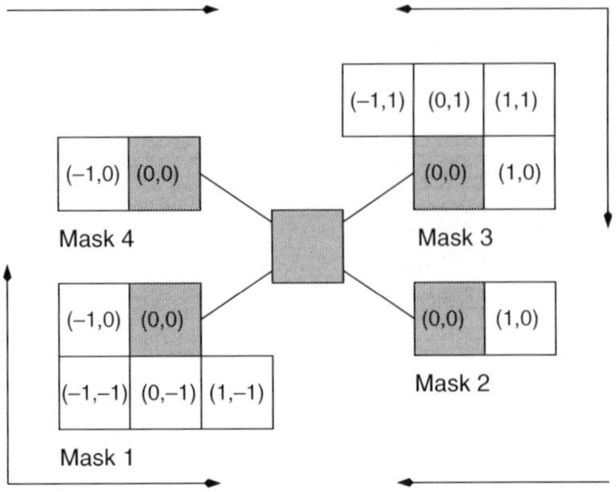


**Figure 11.3.** The masks and scanning directions employed by the algorithm eight-point sequential signed Euclidean transform. The diagram uses a right-handed coordinate system with $x$ increasing to the right and $y$ increasing upwards.

In spite of the profound conceptual differences, the actual computation of $d_H$ presents a formal structure resembling the computation of correlation. The similarity can be readily appreciated when the above computations are rewritten exploiting the matrix structure of the discrete, binary images representing our shapes. Considering, for simplicity, the computation of $d_h$, and adopting the notation $T(\boldsymbol{x}) = T[x_1, x_2]$ for integer point coordinates, we may write

$$\begin{aligned} d_h^T[x_1, x_2] &= d_h(T \oplus \boldsymbol{x}) \\ &= \max_{\boldsymbol{b} \in T} D_I(\boldsymbol{b} + \boldsymbol{x}) \\ &= \max_{\boldsymbol{b} \in T} D_I[b_1 + x_1, b_2 + x_2] \\ &= \max_{(b_1, b_2): T[b_1, b_2] = 1} D_I[b_1 + x_1, b_2 + x_2] \\ &= \max_{(b_1, b_2)} T[b_1, b_2]\, D_I[b_1 + x_1, b_2 + x_2] \end{aligned} \tag{11.15}$$

and compare it to the binary correlation $C[x_1, x_2]$ of the arrays (corresponding to) $T$ and $I$

$$C[x_1, x_2] = \sum_{b_1} \sum_{b_2} T[b_1, b_2] I[b_1 + x_1, b_2 + x_2]. \tag{11.16}$$

The only difference is given by the substitution of image $I$ with the corresponding distance transform $D_I$ and by the replacement of summation by maximization. It is important to observe that small errors in pixel locations have a limited impact on Hausdorff matching: the value of the distance does not change by more than the maximum pixel displacement. In the case of correlation the effect of even a single pixel displacement may have a dramatic impact on the final result: in the worst case, the value of correlation can drop from its possible maximal value to 0.

As the distance map is computed only at points of integer coordinates, the above procedure provides an approximation of the true minimum value of the Hausdorff distances. The error of the approximation is bounded by 1 under any $L_p$ norm used for the computation of point distances. The bound derives from the relation

$$\max_{\epsilon} |d_H(I, T \oplus (\boldsymbol{x} + \epsilon)) - d_H(I, T \oplus \boldsymbol{x})| \leq \|\epsilon\| \tag{11.17}$$

**Intermezzo 11.2.** The signed Euclidean transform

The term *distance transform* identifies a class of algorithms acting on binary images and producing an aligned, real (or integer) image reporting for each pixel its minimum distance from a set of zero-distance pixels $D_0$. Depending on the application, the latter can be identified with the background pixels, often characterized by a 0 value, or with its complement set, i.e. the so-called foreground pixels.

The computation of the Hausdorff distance requires the latter type of zero-distance pixels. While the Hausdorff distance can be defined in terms of any point distance, we will focus our attention on the Euclidean distance $L_2$. The eight-point sequential signed Euclidean distance transform not only associates to each pixel $\boldsymbol{x}$ its distance from the zero-distance pixels but also identifies among them the closest one

$$\boldsymbol{x}_0(\boldsymbol{x}) = \operatorname*{argmin}_{\boldsymbol{y} \in D_0} \|\boldsymbol{x} - \boldsymbol{y}\|$$

providing the offset vector $\Delta_0(\boldsymbol{x}) = \boldsymbol{x} - \boldsymbol{x}_0$ from which the actual distance can be computed. Let us denote by $O_1$ and $O_2$ the matrices containing respectively the $x_1$ and $x_2$ components of the offset vector and let $Z$ be a value greater than the maximum possible distance (e.g. the diagonal of the image considered). The algorithm (see Figure 11.3) is based on the following three phases:

1. Initialize the offset matrices $O_1(x_1, x_2) = O_2(x_1, x_2) = 0$ for points belonging to the zero-distance set $D_0$ and $O_1(x_1, x_2) = O_2(x_1, x_2) = Z$ for points in the complement $D_0^C$ of $D_0$.
2. Backward scan, for each image line in turn, starting from the bottom line:
   (a) move mask 1 from left to right aligning in turn the element containing (0, 0) to each pixel of the binary image belonging to $D_0^C$:
      (i) add the vector of each element of the mask to the corresponding vectors built from $O_1$ and $O_2$ and determine the vector with minimum length;
      (ii) if the minimum length vector has size less than $Z$, update $O_1$ and $O_2$ with the coordinates of the minimum size vector just computed;
   (b) repeat the update phase moving mask 2 from right to left.
3. Forward scan, for each image line in turn, starting from the first line:
   (a) repeat the update phase moving mask 3 from right to left, updating;
   (b) repeat the update phase moving mask 4 from left to right.

The minimum distance map (see Figure 11.4) is then derived by the computed offset vectors: $D(x_1, x_2) = \|(O_1(x_1, x_2), O_2(x_1, x_2))\|$ (Ye 1988).

| | | | | |
|---|---|---|---|---|
| (4,0) | (3,0) | (2,0) | (1,0) | **[(0,0)]** |
| (0,-4) | (3,1) | (2,1) | (1,1) | (0,1) |
| (0,-3) | (-1,-3) | (2,2) | (1,2) | (0,2) |
| (0,-2) | (-1,-2) | (-2,-2) | (1,3) | (0,3) |
| (0,-1) | (-1,-1) | (-2,-1) | (-3,-1) | (0,4) |
| **[(0,0)]** | (-1,0) | (-2,0) | (-3,0) | (-4,0) |

**Figure 11.4.** The two-component result obtained using the signed Euclidean transform: each pixel contains its offset from the nearest zero-distance point. The points belonging to the zero-distance set $D_0$ are boxed and located at the lower left and upper right corners.

so that the maximum possible change undergone by $d_H(I, T \oplus (\boldsymbol{x} + \boldsymbol{\epsilon}))$, when $\boldsymbol{x} + \boldsymbol{\epsilon}$ is limited to the pixel whose center is $\boldsymbol{x}$, is $\|\boldsymbol{\epsilon}\|_p \leq 1$. This simple fact can be exploited to improve significantly the computation of $d_H$ by introducing a hierarchical strategy. Let us introduce a set of regular cells $\{C_i\}$ (with centers $\boldsymbol{x}_i$) whose union covers the whole image and let

$$\Delta(C_i) = \max_{\boldsymbol{\epsilon}+\boldsymbol{x}_i \in C_i} \|\boldsymbol{\epsilon}\|. \tag{11.18}$$

Let us now impose a threshold $\theta$ on the maximum allowed (directed) Hausdorff distance: $d_h(T, I) \leq \theta$. By Equation 11.17 we have

$$d_h(I, T \oplus \boldsymbol{x}_i + \boldsymbol{\epsilon}) \geq d_h(I, T \oplus \boldsymbol{x}_i) - \Delta(C_i) \quad \forall \boldsymbol{x}_i + \boldsymbol{\epsilon} \in C_i \tag{11.19}$$

so that we do not need to further consider cell $C_i$ whenever

$$d_h(I, T \oplus \boldsymbol{x}_i) > \theta + \Delta(C_i) \tag{11.20}$$

because for every point within the cell the constraint $d_h(T, I) \leq \theta$ is violated. All remaining cells can be divided into smaller subcells and the procedure is applied recursively to each of them until all cells have been considered.

## 11.4. Partial Hausdorff Matching

We have seen that by using the directed Hausdorff distance it is possible to match a (whole) template to a portion of a whole image. A simple modification in the definition of $d_h$ allows us to find the best match among all possible subsets of given size $k$ of a template $T$ with image $I$. We merely replace the selection of the maximum distance value with that of the $k$th ranked one:

$$d_h^{(k)}(T, I) = \Big( \underset{\boldsymbol{b}\in T}{D_I(\boldsymbol{b})} \Big)_{(k)} \tag{11.21}$$

where ranking is by ascending value. This simple modification automatically considers the subset of $k$ points of $T$ that are closest to $I$ in the computation of the matching value. If the number of points of $T$ is $n_T$ we have that

$$d_h^{(n_T)}(T, I) = d_h(T, I). \tag{11.22}$$

The computation of the $k$th ranked value does not require full sorting of the data and can be accomplished in $O(n_T)$ time, the same complexity required by the computation of the maximum value (with a relative constant of 2). A partial bidirectional Hausdorff pseudo-distance is defined as

$$d_H^{(kl)}(E, F) = \max(d_h^{(k)}(E, F), d_h^{(l)}(E, F)). \tag{11.23}$$

It is possible to prove that there exist two sets $E_k$ and $F_l$ of size not less than $k$ and $l$ respectively, for which $d_H(E_k, F_l) = d_H^{(kl)}(E, F)$. The triangle inequality holds only if the two subsets are compared to the same portion of a third set.

Let us now consider the computation of $d_h^{(k)}$ from the perspective of binary images, as we did before when comparing Hausdorff matching to correlation. Instead of fixing $k$, or, equivalently, the fraction $f_k = k/n_T$, and assessing the matching quality by the value of $d_h^{(k)}$, let us fix a distance value $d$ and assess the matching by the largest fraction $f$ such that $d_h^{(k)} \leq d$. We may associate to each compared set an image $F$ where the pixel corresponding to the points of the set have value 1, and the remaining ones 0. Let us dilate image $F$ replacing each of its pixels of value 1 with a disk of radius $\delta$ and let $F_\delta$ be the result. If we linearize the binary images we may rewrite the Hausdorff fraction as

$$f_h^\delta(E, F) = \frac{\boldsymbol{E}^T \boldsymbol{F}_\delta}{\boldsymbol{E}^T \boldsymbol{E}^T} \tag{11.24}$$

where $E$ is our template image. When we need to detect many different shapes $\boldsymbol{E}_i$ we may employ a subspace approach computing the principal components of our model set and approximate the value of the Hausdorff fraction using the projection of the models onto a subspace. Before proceeding, let us observe that PCA is not particularly good at representing linear structures and, by Equation 11.24, we see that the templates and the image to which they are matched undergo different processings. Both problems can be solved by augmenting the model set with the $d$-dilated versions of the templates, so that the resulting principal components are equally well suited for the models and for the dilated image. If we denote by $\boldsymbol{e}_i \in \mathbb{R}^m$ the coefficients of the expansion of $\boldsymbol{E}$ onto the first $m$ eigenvectors, $\boldsymbol{f}$ the expansion of the unknown image, and $\boldsymbol{\mu}_E$ the average model, we have the following approximation of the Hausdorff fraction:

$$f_h^\delta(E_i, F) \approx \frac{\boldsymbol{e}_i^T \boldsymbol{f} + \boldsymbol{e}_i^T \boldsymbol{\mu}_E + \boldsymbol{f}^T \boldsymbol{\mu}_E - \|\boldsymbol{\mu}_E\|^2}{\|\boldsymbol{e}_i\|^2} \tag{11.25}$$

where we used $E = [(E - \boldsymbol{\mu}_E) + \boldsymbol{\mu}_E]$ and the fact that orthonormal transformations preserve dot products. The only quantity that needs to be computed for each model is $\boldsymbol{e}_i^T \boldsymbol{f}$; the other values can be precomputed when the eigenvectors are built, or need to be computed only once (e.g. $\boldsymbol{e}_i^T \boldsymbol{\mu}_E$). The computation of a bidirectional Hausdorff fraction

$$f_H^\delta(E_i, F) = \min(f_h^\delta(E_i, F^\delta, ) f_h^\delta(F, E_i^\delta)) \tag{11.26}$$

does not present any additional difficulty.

## 11.5. Robustness Aspects

The Hausdorff distance defined in Equation 11.5 is not robust to outliers. In a sense, we may even consider it to be based on them because it is defined by the maximum among the matching distance values. The introduction of the partial Hausdorff distance in Equation 11.21 achieves robustness by selecting the value at a predefined rank $k$: greater distances have no impact on the final value and robustness is achieved up to a fraction of outliers equal to $(n_E - k)/n_E$ where $n_E$ is the number of points in $E$.

Several other modifications of $d_H$ can be devised that provide reduced sensitiveness to outliers. The partial directed Hausdorff distance can be further modified by replacing the min operation with the selection of a ranked value:

$$d_h^{(P,Q)} = P^{\text{th}}_{\boldsymbol{e} \in E}(Q^{\text{th}}_{\boldsymbol{f} \in F}(\|\boldsymbol{e} - \boldsymbol{f}\|)) \tag{11.27}$$

where $P^{\text{th}}_{\boldsymbol{e} \in E}$ represents the $P$th ranked value among the set of $n_E$, $Q$th ranked Euclidean distance values associated to each element $\boldsymbol{e} \in E$ from the elements of set $F$. The standard definition of the partial directed Hausdorff distance is recovered when $Q = 1$. Good matching

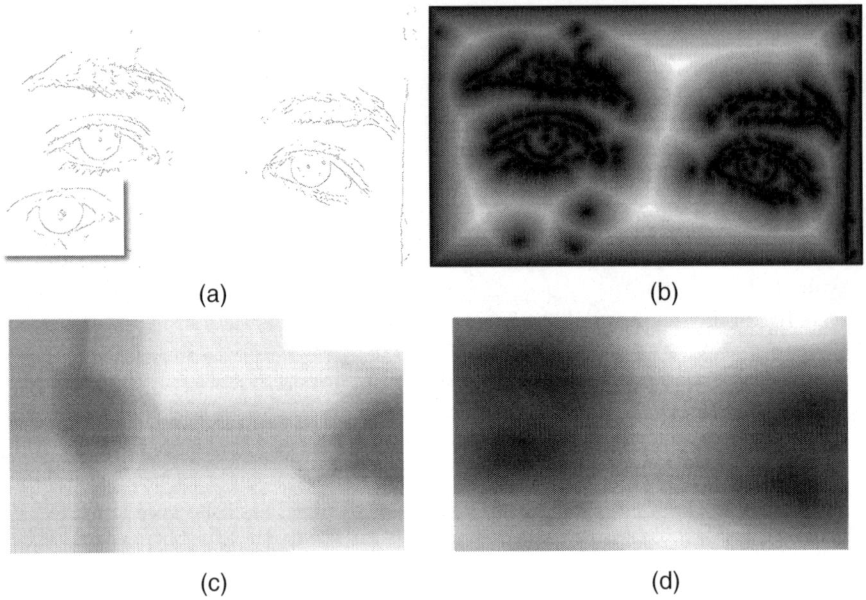


**Figure 11.5.** An eye template is matched to the edge map (a) of another person's face using the Hausdorff distance: (b) reports the Euclidean distance map, (c) the Hausdorff distance for different positions of the template, and (d) a map of $d_h^{\mathrm{MHD}}$, introduced in Equation 11.28.

results are reported in the literature for continuous patterns with $P \in [0.8, 0.9]$ and $Q \in [0.01, 0.05]$. If the patterns are sparse, replacing the min operator with the selection of a ranked value is detrimental; if the point set is sparse, the next best matching point might be quite distant and all distances would be greater than expected even when the models match perfectly. If the patterns are continuous, e.g. when they are represented by long edges or compact areas, a low $Q$th ranked value is an indicator that local continuity is preserved and more selective matching can be achieved in low-noise situations.

A simple modification of $d_h$ that results in a measure with reduced sensitivity to outliers is the replacement of the max operator with an averaging operation:

$$d_h^{\mathrm{MHD}} = \frac{1}{n_E} \sum_{\boldsymbol{e} \in E} D_F(\boldsymbol{e}). \tag{11.28}$$

This new measure is not technically robust and its reduced sensitivity to outliers derives from spreading out their effect over all the values (see Figure 11.5).

A general way of introducing robustness into the computation of $d_h$ is to replace the Euclidean distance with a different function that can limit (or eliminate) the effect of outliers following the strategy upon which $M$-estimators are based:

$$d_h^M = \frac{1}{n_E} \sum_{\boldsymbol{e} \in E} \rho(D_F(\boldsymbol{e})) \tag{11.29}$$

where function $\rho$ is required to be convex, symmetric, and with a unique minimum at zero. A simple $\rho$ that results in a bounded influence of outliers is

$$\rho(x) = \begin{cases} |x| & |x| \leq \tau \\ \tau & |x| > \tau. \end{cases} \tag{11.30}$$

When $\tau = \infty$ we recover the usual directed Hausdorff distance while $\tau = 0$ leaves us with no information: the value of the distance is identically 0. The value of $\tau$ must be chosen experimentally small enough to achieve good outlier resilience but at the same time large enough so that the range of values assumed by $d_h^\rho$ is large enough to discriminate good matches from bad ones, and, for the good ones, to provide enough information in order to locate the optimal matching position with good accuracy. An alternative modification of $d_h$ based on ideas from robust statistics is that of using a linear combination of order statistics:

$$d_h^{\mathrm{LTS},k} = \frac{1}{k} \sum_{i=1}^{k} d_h(\boldsymbol{e})_{(i)} \tag{11.31}$$

where the parenthesized index identifies the order statistics. When $k = n_E$ we recover $d_h^{\mathrm{MHD}}$. The two estimators $d^\rho$ and $d^{\mathrm{LTS},k}$, $k < n_E$, are technically robust as they respectively bound and eliminate the effect of a non-null number of outliers, yet they achieve a good efficiency due to their averaging of multiple values. The latter feature is important when the position of the template must be located with accuracy. The computational complexity of $d_h$, $d_h^\rho$, and $d_h^{\mathrm{LTS},k}$ are similar and higher than $d_h^M$ but significantly lower than $d_h^{(P,Q)}$.

## 11.6. A Probabilistic Perspective

Hausdorff distance matching can be cast in a probabilistic framework, namely that of maximum likelihood estimation, that allows us to further tune it to specific application needs by modeling errors and data outliers.

In order to cast our matching problem in a maximum likelihood setting, let our model $T$ be characterized by $n_T$ features $\{\boldsymbol{t}_i\}$ (its points) and let us associate to each possible location $\boldsymbol{x}$ of the model a set of $n_T$ measurements $D_I(\boldsymbol{t}_i(\boldsymbol{x}))$ representing the distance of each feature point from the closest feature in image $I$. As correlation between the measurements decreases quickly with separation of the points, we consider them as (approximately) independent so that their joint probability density function can be approximated by the product of the distributions of each single measurement leading to the following expression for the likelihood at position $\boldsymbol{x}$:

$$p(D_I(\boldsymbol{t}_1(\boldsymbol{x})), \ldots, D_I(\boldsymbol{t}_{n_T}(\boldsymbol{x}))|\boldsymbol{x}) \approx \prod_{i=1}^{n_T} p(D_I(\boldsymbol{t}_i(\boldsymbol{x}))) = L(\boldsymbol{x}) \tag{11.32}$$

which can be transformed to the more convenient form

$$\ln L(\boldsymbol{x}) = \sum_{i=1}^{n_T} \ln p(D_I(\boldsymbol{t}_i(\boldsymbol{x}))) \tag{11.33}$$

which preserves the relative ordering of the likelihood values. The Hausdorff fraction of Equation 11.24 corresponds to

$$\ln p(D_I(\boldsymbol{t}_i(\boldsymbol{x}))) = \begin{cases} 1 & \text{if } D_I(\boldsymbol{t}_i(\boldsymbol{x})) < \delta \\ 0 & \text{otherwise.} \end{cases} \tag{11.34}$$

Each point in the template contributes a (constant) value if it is within distance $\delta$ of an image feature, and a different (smaller) value if it is not. Note, however, that Equation 11.34 does not correspond to a real probability as it is not normalized.

The hierarchical strategy to speed up the matching process based on Equation 11.18 can be adapted to the probabilistic formulation. The first step is to generate a probability map $P_I$ from the distance map to avoid repeated computation of the terms $\ln p(D_I(\boldsymbol{t}_i(\boldsymbol{x})))$. Let us again consider a cell $C_i$ centered at $\boldsymbol{x}_i$. Assuming that $p(D_I(\boldsymbol{t}_i(\boldsymbol{x})))$ is a decreasing function of $D_I$, we may bound the maximum possible probability value in the following way:

$$p(D_I(\boldsymbol{t}_i(\boldsymbol{x}))) \leq p[\max(0, D_I(\boldsymbol{x}_{C(\boldsymbol{t}_i(\boldsymbol{x}))}) - \Delta(C(\boldsymbol{t}_i(\boldsymbol{x}))))] \tag{11.35}$$

where $C(\boldsymbol{t}_i(\boldsymbol{x}))$ identifies the cell to which $\boldsymbol{t}_i(\boldsymbol{x})$ belongs. We consider the translation of our template at a given position, obtain a bound for its $p$ by summing the above bounds for each of its points and discard the cell from further analysis if the following relation holds:

$$\sum_i \ln p[\max(0, D_I(\boldsymbol{x}_{C(\boldsymbol{t}_i(\boldsymbol{x}))}) - \Delta(C(\boldsymbol{t}_i(\boldsymbol{x}))))] < \theta_p. \tag{11.36}$$

Knowledge of the a priori probability of model positions $p(\boldsymbol{x})$ can be easily incorporated

$$\ln L(\boldsymbol{x}) = \ln p(\boldsymbol{x}) + \sum_{i=1}^{N_T} \ln p(D_I(\boldsymbol{t}_i(\boldsymbol{x}))) \tag{11.37}$$

and exploited to improve search efficiency. Depending on the structure of the a priori probability $p$ for a given template position, a similar bound might be efficiently computable also for Equation 11.37, and the hierarchical strategy extended to it. The probability reported in Equation 11.34 does not model accurately the process of edge pixel matching. A more correct way of assigning the probability is that of considering $p(\cdot)$ as due to two different situations:

1. The distance is related to a pixel in the image that effectively corresponds to the pixel of the template.

2. The distance comes from matching an extraneous pixel.

This can be formalized by expressing $p$ as a convex linear combination of two distributions $p_1$ and $p_2$

$$p(d) = \lambda p_1(d) + (1 - \lambda) p_2(d), \quad \lambda \in [0, 1]. \tag{11.38}$$

A practical choice for $p_1$ would be a normal distribution while $p_2$ can be replaced by the (constant) expected probability for a random outlier. The effect of a constant $p_2$ is to provide a non-null lower bound on the probability associated to each template feature, effectively preventing large distance from driving the overall probability of the pattern to zero.

The parameters of $p_1$, $p_2$, and $\lambda$ can also be estimated from data if the matching ground truth is available (see Appendix C for a description of the cross-validation technique). It is also possible to localize templates to subpixel accuracy by fitting a second-order polynomial to the log-likelihood data obtained from Hausdorff matching. When the likelihood is Gaussian, the log-likelihood is a second-order polynomial: in the other cases, the approximation is usually good enough in the proximity of the best matching position.

## 11.7. Invariant Moments

Comparing binary images pixel by pixel and comparing shapes by means of the Hausdorff distance do not exhaust the possibilities for matching arbitrary shapes. A simple alternative, when partial matching is not required, is based on the comparison of invariant moments.

Let $f(x_1, x_2)$ be a real, positive function with bounded support on a region $R \subset \mathbb{R}^2$. Its moments $m_{q,r}$ of order $q + r$ are defined as

$$m_{q,r} = \int_R f(x_1, x_2) x_1^q x_2^r dx_1\, dx_2. \tag{11.39}$$

If we let $f(x_1, x_2) = 1$, the moments characterize the shape of region $R$ neglecting any additional textural information. The Fourier transform is

$$f(\omega_1, \omega_2) = \int_R f(x_1, x_2) e^{-2\pi i (x_1\omega_1 + x_2\omega_2)}\, dx_1\, dx_2 \tag{11.40}$$

$$= \int_R f(x_1, x_2) \sum_{q=0}^{\infty} \sum_{r=0}^{\infty} \frac{(-2\pi i)^{q+r}}{q!r!} x_1^q x_2^r \omega_1^q \omega_2^r\, dx_1\, dx_2 \tag{11.41}$$

$$= \sum_{q=0}^{\infty} \sum_{r=0}^{\infty} \frac{(-2\pi i)^{q+r}}{q!r!} m_{q,r} \omega_1^q \omega_2^r \tag{11.42}$$

and the original function can be recovered via the inverse Fourier transform of the moment-based expansion, justifying the claim that moments characterize the shape of $R$. Designing moment-like quantities that exhibit translational and scale invariance is easy. The expressions

$$\mu_{q,r} = \int_R (x_1 - \bar{x}_1)^q (x_2 - \bar{x}_2)^r\, dx_1\, dx_2 \tag{11.43}$$

where barred symbols represent averages, are known as centered moments and are translationally invariant. Under a scale change $(x_1, x_2) \to (\alpha x_1, \alpha x_2)$ the centered moments transform as

$$\mu'_{q,r} = \alpha^{q+r+2} \mu_{q,r} \tag{11.44}$$

so that if we pose $\gamma = (q + r + 2)/2$ the quantities

$$\eta_{q,r} = \frac{\mu_{q,r}}{\eta_{0,0}^{\gamma}} \tag{11.45}$$

are invariant to scaling and translation. It is possible to find quantities, the so-called Hu's moments, that are also rotation invariant:

$$\phi_1 = \eta_{2,0} + \eta_{0,2} \tag{11.46}$$

$$\phi_2 = (\eta_{2,0} - \eta_{0,2})^2 + 4\eta_{1,1} \tag{11.47}$$

$$\phi_3 = (\eta_{3,0} - 3\eta_{1,2})^2 + (3\eta_{2,1} - \eta_{0,3})^2 \tag{11.48}$$

$$\phi_4 = (\eta_{3,0} + \eta_{1,2})^2 + (\eta_{2,1} + \eta_{0,3})^2 \tag{11.49}$$

$$\begin{aligned}\phi_5 = {} & (\eta_{3,0} - 3\eta_{1,2})(\eta_{3,0} + \eta_{1,2})[(\eta_{3,0} - 3\eta_{1,2})^2 - 3(\eta_{2,1} + \eta_{0,3})^2] \\ & + (3\eta_{2,1} - \eta_{0,3})(\eta_{2,1} + \eta_{0,3})[3(\eta_{3,0} - 3\eta_{1,2})^2 - (\eta_{2,1} + \eta_{0,3})^2]\end{aligned} \tag{11.50}$$

$$\begin{aligned}\phi_6 = {} & (\eta_{2,0} - \eta_{0,2})[(\eta_{3,0} + \eta_{1,2})^2 - (\eta_{2,1} + \eta_{0,3})]^2 \\ & + 4\eta_{1,1}(\eta_{3,0} + \eta_{1,2})(\eta_{2,1} + \eta_{0,3})\end{aligned} \tag{11.51}$$

$$\begin{aligned}\phi_7 = {} & (3\eta_{2,1} - \eta_{3,0})(\eta_{3,0} + \eta_{1,2})[(\eta_{3,0} - 3\eta_{1,2})^2 - 3(\eta_{2,1} + \eta_{0,3})^2] \\ & + (3\eta_{2,1} - \eta_{0,3})(\eta_{2,1} + \eta_{0,3})[3(\eta_{3,0} + \eta_{1,2})^2 - (\eta_{2,1} + \eta_{0,3})^2].\end{aligned} \tag{11.52}$$

A generic shape can then be mapped into $\mathbb{R}^7$ and compared to other shapes by means of the distance of its invariant moments set. It is not possible to discriminate rotationally invariant objects because the associated Hu's moments are equally zero. The invariance of the moments is not perfect in the discrete space of images and precision issues may also arise in the computation of higher order moments.

## 11.8. Bibliographical Remarks

The formalization of the four different robustness aspects for distances among point sets is based on the report by Hagedoorn and Veltkamp (1999). Our treatment of Hausdorff matching relies heavily on the paper by Huttenlocher and Rucklidge (1993). A good introduction to the computation of distance maps and to their efficient use for edge matching using a hierarchical structure can be found in Borgefors (1988). The possibility of using a PCA-based approach to improve efficiency when matching a large class of models is considered by Huttenlocher *et al.* (1992). A probabilistic formulation of Hausdorff matching was introduced by Olson (2002).

Robustness of variations of Hausdorff matching is considered in several papers among which we may cite Sim *et al.* (1999). The introduction of neighborhood limited point matching with a penalty when no such match can be found, as proposed by Takács (1998), also provides robustness in terms of a bounded influence of missing matches.

A spatially weighted version for face recognition is introduced in Guo *et al.* (2003) while the possibility of reducing the number of points in the templates by restricting consideration to high-curvature points on edge curves, weighting them by their structural distinctiveness, is investigated by Gao (2003). The use of a normalized version of the first eigenface to weight the contribution of the different points in the computation of the directed Hausdorff distance is considered in Lin *et al.* (2003). An interesting variation of Hausdorff matching is proposed in the paper by Vivek and Sudha (2007). Instead of relying on simple edges, the intensity image is first transformed in a similar way to that proposed by the census transform: pixels whose transformed values match contribute in the same way as edge pixels in the standard formulation, while pixels whose transformed values differ contribute a large penalty. A confidence measure for the matching of fiducial points is used also by Wang and Chua (2006) to improve Hausdorff matching for gray-level images.

## References

Borgefors G 1988 Hierarchical chamfer matching: a parametric edge matching algorithm. *IEEE Transactions on Pattern Analysis and Machine Intelligence* **10**, 849–865.

Gao Y 2003 Efficiently comparing face images using a modified Hausdorff distance. *IEE Proceedings – Vision, Image and Signal Processing* **150**, 346–350.

Guo B, Lam K, Lin K and Siu W 2003 Human face recognition based on spatially weighted Hausdorff distance. *Pattern Recognition Letters* **24**, 499–507.

Hagedoorn M and Veltkamp R 1999 Metric pattern spaces. Technical Report 1999-03, Utrecht University, Information and Computing Sciences.

Huttenlocher D and Rucklidge W 1993 Multi-resolution technique for comparing images using the Hausdorff distance. *Proceedings of the IEEE Conference on Computer Vision and Pattern Recognition (CVPR'93)*, pp. 705–706.

Huttenlocher D, Klanderman G and Rucklidge W 1992 Comparing images using the Hausdorff distance under translation. *Proceedings of the IEEE Conference on Computer Vision and Pattern Recognition (CVPR'92)*, pp. 654–656.

Lin K, Lam K and Siu W 2003 Spatially eigen-weighted Hausdorff distances for human face recognition. *Pattern Recognition* **36**, 1827–1834.

Olson C 2002 Maximum-likelihood image matching. *IEEE Transactions on Pattern Analysis and Machine Intelligence* **24**, 853–857.

Sim D, Kwon O and Park R 1999 Object matching algorithms using robust Hausdorff distance measures. *IEEE Transactions on Image Processing* **8**, 425–429.

Takács B 1998 Comparing face images using the modified Hausdorff distance. *Pattern Recognition* **31**, 1873–1881.

Vivek E and Sudha N 2007 Robust Hausdorff distance measure for face recognition. *Pattern Recognition* **40**, 431–442.

Wang Y and Chua C 2006 Robust face recognition from 2D and 3D images using structural Hausdorff distance. *Image and Vision Computing* **24**, 176–185.

Ye Q 1988 The signed Euclidean transform and its applications. *Proceedings of the 19th IAPR International Conference on Pattern Recognition (ICPR'88)*, vol. 1, pp. 495–499.

# 12 SUPPORT VECTOR MACHINES AND REGULARIZATION NETWORKS

But we in it shall be remember'd;
We few, we happy few, we band of brothers;

*Henry V*
WILLIAM SHAKESPEARE

Template matching can be considered a classification task and the corresponding probabilistic concepts have already been investigated. When the probability distribution of the templates is unknown, the design of a classifier becomes more complex and many critical estimation issues surface. This chapter presents basic results upon which two interrelated, powerful classifier design paradigms stand.

## 12.1. Learning and Regularization

In many cases we do not know the probability distribution of the patterns that we want to detect or recognize and the techniques presented in Chapter 3 cannot be applied directly. We must learn all the probabilistic dependencies from a finite set of examples $\{(\boldsymbol{x}_i, y_i)\}_{i=1}^N$ of pattern instances $\boldsymbol{x}_i \in \mathbb{R}^{n_d}$ and corresponding class labels $y_i \in L$. Formally, we must learn a function $f$

$$f : \mathbb{R}^{n_d} \to L \tag{12.1}$$

and, most importantly, we want to learn a function that not only summarizes well the data we are given but also correctly labels data to be seen: it should generalize well. We start our journey into the field of learning (and regularization as an associated necessity) by revisiting the solution presented in Chapter 6 for the problem of matching variable patterns. The synthetic discriminant functions, addressed in Chapter 6, can be considered as a linear classifier, attempting to condense all the information necessary for the discrimination of patterns into a single direction:

$$\boldsymbol{h} = \sum_i c_i \boldsymbol{x}_i. \tag{12.2}$$

Assuming that $\boldsymbol{h}$ is built by requiring it to output 1 on the patterns of a given class and 0 on the remaining ones, classification is achieved by considering the projection of the unknown pattern onto $\boldsymbol{h}$

$$\boldsymbol{h} \cdot \boldsymbol{x} = \left(\sum_i c_i \boldsymbol{x}_i\right) \cdot \boldsymbol{x} = \sum_i c_i (\boldsymbol{x}_i \cdot \boldsymbol{x}) \tag{12.3}$$

*Template Matching Techniques in Computer Vision: Theory and Practice* Roberto Brunelli

and thresholding the result

$$\boldsymbol{h} \cdot \boldsymbol{x} > \theta, \quad \theta \in [0, 1]. \tag{12.4}$$

If we consider Equation 12.3, we see that it has the structure of a linear combination of scalar products. In Chapter 8, we were able to introduce a powerful generalization of PCA, namely kernel PCA, with a simple trick, preserving the structure of the computations but moving to a different feature space. It is then natural to ask ourselves if a similar trick can be performed in this case. More specifically we are interested in exploring a generalization of the type

$$\boldsymbol{h}' \cdot \boldsymbol{x}' = f(\boldsymbol{x}) = \sum_i c_i(\phi(\boldsymbol{x}_i) \cdot \phi(\boldsymbol{x})) = \sum_i c_i K_\phi(\boldsymbol{x}_i, \boldsymbol{x}) \tag{12.5}$$

following closely the strategy of kernel PCA. The computation of the advanced synthetic discrimination functions based on Equation 6.31 further required the minimization of an additional term based on the spectral envelope of the training samples. The function of this regularizing term was to keep under control the behavior of the filter acting on patterns that were not used in the construction of the filter. In a sense, it provided controlled smoothing of filter response. Interestingly enough, the structure of this regularizing term can also be related to the computation of scalar products

$$E_M = \boldsymbol{h}^\dagger T \boldsymbol{h} = (T^{1/2}\boldsymbol{h})^\dagger (T^{1/2}\boldsymbol{h}) = (T^{1/2}\boldsymbol{h}) \cdot (T^{1/2}\boldsymbol{h}) = \|\boldsymbol{h}\|_T^2 \tag{12.6}$$

because $T$ is a real diagonal matrix. The minimized energy $E_M$ can be considered as the norm $\|\boldsymbol{h}\|^2$ of the filter computed in a different space, a norm that is large for functions with high frequency content.

Let us continue the analysis of the procedure leading to synthetic discriminant functions. The filter was required to have a predefined response $\boldsymbol{u} = (u_1, u_2, \ldots, u_N)$ on a given set of (training) patterns. We saw that the linear structure of the filter allowed us to impose a number of constraints up to the dimension of the pattern space. However, the number of patterns to be discriminated is, in general, higher than the dimension of the space where they are described. Should many training samples be available a least squares approach would be required as it would be impossible to satisfy all the resulting constraints. With a slight change of notation ($\boldsymbol{u}$ becomes $\{y_i\}$), we can then reformulate the problem solved with advanced synthetic discriminant functions as

$$\hat{\boldsymbol{h}} = \operatorname*{argmin}_{\boldsymbol{h}} \left( \sum_i (y_i - \boldsymbol{h} \cdot \boldsymbol{x})^2 + \lambda \|\boldsymbol{h}\|_T^2 \right) \tag{12.7}$$

leading to a constrained minimization problem.

The idea of not limiting ourselves to reduce the approximation error but to impose some additional constraints to better shape the solution is a very useful one and, in many cases, a necessity. Let us reconsider the problem of template matching in its general form. Given a pattern $\boldsymbol{x}$ we want to find the optimal classifier such that $\phi(\boldsymbol{x}) = 1$ when $\boldsymbol{x}$ is an instance of our template class, and $\phi(\boldsymbol{x}) = -1$ when it is not. We saw in Chapter 3 that the problem can be solved directly when we know the distribution of the data $p(\boldsymbol{x})$. When the latter is not known, the problem is transformed into an optimization one: we select a parametric class of functions, and we choose the best among them. Unfortunately, the most useful optimality criterion, i.e. the minimum probability of error, cannot be computed because we do not know the probability density of our pattern class. We must then resort to some suboptimal, but computable criterion for selecting our classifier, such as the minimum error on an available, finite size, training set. Unfortunately, even the minimization of an apparently simple criterion such as the classification error on a finite set of samples poses exceptional computational complexities: the resulting error function is piecewise flat leading to a combinatorial problem. We must then find a mathematically tractable approximation of the optimal criterion, facing a sequence of critical choices:

1. Select the optimality criterion.
2. Select a computationally tractable approximation.
3. Select an appropriate parametric class of functions.
4. Find a solution.

Each of the above steps causes some departure from the optimal solution and we need to consider all of them with care. Let us now depart slightly from the perspective of classification by adopting an approximation stance. Given a set $\{(\boldsymbol{x}_i, y_i)\}_i$, we search for a function $\hat{f}$ minimizing the empirical (approximation) squared error

$$E_{\text{emp}}^{\text{MSE}} = \frac{1}{N} \sum_i (y_i - f(\boldsymbol{x}_i))^2 \tag{12.8}$$

$$\hat{f}(\boldsymbol{x}) = \operatorname*{argmin}_f \ E_{\text{emp}}^{\text{MSE}}(f; \{(\boldsymbol{x}_i, y_i)\}_i). \tag{12.9}$$

$E_{\text{emp}}^{\text{MSE}}$ is then our computationally tractable approximation to the real optimality criterion, namely minimum probability of error. If we do not somehow restrict the choice of $f$ the problem is ill posed: it has many possible solutions and many of them useless. An example of the latter class of solutions is

$$\hat{f}(\boldsymbol{x}) = \sum_i y_i \delta(\boldsymbol{x} - \boldsymbol{x}_i) \tag{12.10}$$

a function for which $E_{\text{emp}}^{\text{MSE}} = 0$ but that will classify incorrectly all future instances of our template class that do not match exactly one of the training samples. Generalizing from the example of Equation 12.7, we employ a regularization procedure, turning the optimization problem of Equation 12.9 into

$$\hat{f}(\lambda) = \operatorname*{argmin}_{f \in \mathfrak{H}} \frac{1}{N} \sum_i (y_i - f(\boldsymbol{x}_i))^2 + \lambda \|f\|_{\mathfrak{H}}^2 \tag{12.11}$$

where $\|f\|_{\mathfrak{H}}$ is the norm of $f$ in the (function) space $\mathfrak{H}$ to which we restrict our quest for a solution and $\lambda > 0$ controls the amount of regularization. Note the explicit dependence of $\hat{f}$ on the regularization parameter $\lambda$. What we are trying to obtain by solving the minimization problem of Equation 12.11 is the function $f_0$ describing the distribution of $(\boldsymbol{x}, y)$, from which our samples are drawn. It can be shown that there is a unique value $\hat{\lambda}$ that minimizes the error between $f_0$ and $\hat{f}$ for a given number of samples and given function space. The situation is somewhat similar to that considered when shrinkage estimators were introduced. If $\lambda$ is very large, the largest contribution is from the function norm: $\hat{f}$ will be shrunk towards the null function, resulting in a solution which is oblivious to available data, and corresponding to high bias and low variance. On the contrary, if $\lambda$ is small, available data will lead the solution, potentially resulting in a low-bias, high-variance situation. If space $\mathfrak{H}$ is not rich enough to approximate $f_0$ well, even the best achievable solution will be affected by a significant bias, usually called approximation error. If space $\mathfrak{H}$ is too rich, it will be easy to reduce the empirical error to zero for any given, arbitrary, dataset leading to potential high-variance problems such as those mentioned for Equation 12.10. The optimal value $\hat{\lambda}$ can be determined using cross-validation techniques, and in particular leave-one-out error estimation (see Theorem C.2.1).

We now need to select a useful space and an appropriate norm. Following the steps that lead to kernel PCA, we first select a positive definite kernel function $K$

$$\boldsymbol{x} \rightarrow K(\cdot, \boldsymbol{x}) : \phi(\boldsymbol{x}') = K(\boldsymbol{x}'; \boldsymbol{x}) \tag{12.12}$$

that associates to each point in our pattern space a function for which

$$\sum_{i,j=1}^{n} c_i c_j K(\boldsymbol{x}_i, \boldsymbol{x}_j) > 0 \quad \forall \boldsymbol{c} \neq \boldsymbol{0} \in \mathbb{R}^n \quad \text{and} \quad \boldsymbol{x}_i \neq \boldsymbol{x}_j, \ \text{for } i \neq j. \tag{12.13}$$

We construct our function space $\mathfrak{H}$ as the space spanned by the following linear combinations:

$$f(\cdot) = \sum_{i=1}^{n} \alpha_i K(\cdot, \boldsymbol{x}_i) \tag{12.14}$$

with arbitrary $\boldsymbol{x}_i$, $\alpha_i$, and $n$, resembling the structure of Equation 12.5. In order to define a norm in this function space we introduce a dot product for the functions that belong to it, and, given $g(\cdot) = \sum_{j=1}^{m} \beta_j K(\cdot, \boldsymbol{x}')$,

$$\langle f, g \rangle = \sum_{i=1}^{n} \sum_{j=1}^{n'} \alpha_i \beta_j K(\boldsymbol{x}_i, \boldsymbol{x}'_j) \tag{12.15}$$

from which we may compute the norm of $f$

$$\langle f, f \rangle = \sum_{ij} \alpha_i \alpha_j K(\boldsymbol{x}_i, \boldsymbol{x}'_j) \geq 0. \tag{12.16}$$

From the action of the dot product we derive two important equations that justify the name reproducing kernel Hilbert space (RKHS) associated with Equation 12.14:

$$\langle K(\cdot, x), f \rangle = f(x) \tag{12.17}$$

$$\langle K(\cdot, x), K(\cdot, x') \rangle = K(x, x'). \tag{12.18}$$

We are now ready to solve the minimization problem of Equation 12.11 using the calculus of variations. We need to find the function (instead of a point) minimizing the functional (instead of a function). The technique of finding extremal points by setting the derivative of a function to zero can be transferred to the minimization of a functional with the introduction of the concept of functional derivative. At an extremal function the derivative of the functional must be zero as a functional, so that

$$\frac{\delta E\{f\}}{\delta f}[g] = 0 = \frac{d}{d\epsilon} E\{f + \epsilon g\}|_{\epsilon=0} \quad \forall g \tag{12.19}$$

where the square bracket notation identifies the application of the functional to a function. As we want to minimize with respect to a function, we need to compute the functional derivative with respect to $f$, apply it to an element $g$ of the RKHS, and set it equal to zero:

$$\frac{\delta}{\delta f}\left(\frac{1}{n}\sum_i (y_i - f(\boldsymbol{x}_i))^2 + \lambda \|f\|_{\mathfrak{H}}^2\right)[g] = -\frac{2}{n}\sum_i (y_i - f(\boldsymbol{x}_i))g(\boldsymbol{x}_i) + 2\lambda\langle f, g\rangle = 0. \tag{12.20}$$

The above equation must hold for any $g$ and we choose $g = K(\cdot, \boldsymbol{x})$. From the reproducing feature of the kernel in Equation 12.17 we have $\langle f, g\rangle = f(\boldsymbol{x})$ so that we may rewrite Equation 12.20 as

$$f(\boldsymbol{x}) = \sum_i \left[\frac{y_i - f(\boldsymbol{x}_i)}{n\lambda}\right] K(\boldsymbol{x}, \boldsymbol{x}_i). \tag{12.21}$$

Introducing

$$c_i = \frac{y_i - f(\boldsymbol{x}_i)}{n\lambda} \tag{12.22}$$

we are led to

$$f(\boldsymbol{x}) = \sum_i c_i K(\boldsymbol{x}, \boldsymbol{x}_i) \tag{12.23}$$

and inserting Equation 12.23 into Equation 12.22 gives

$$y_i = n\lambda c_i + \sum_j c_j K(\boldsymbol{x}_j, \boldsymbol{x}_i) \tag{12.24}$$

which, letting $K_{ij} = K(\boldsymbol{x}_i, \boldsymbol{x}_j)$, can be conveniently rewritten in matrix form

$$(n\lambda + K)\boldsymbol{c} = \boldsymbol{y}. \tag{12.25}$$

As $K$ is positive definite as well as $n\lambda I$, the matrix $(n\lambda I + K)$ is itself positive definite and therefore invertible. Equation 12.25 then has a unique solution

$$\boldsymbol{c} = (n\lambda I + K)^{-1}\boldsymbol{y}. \tag{12.26}$$

The solution is then expressed in terms of the available samples in spite of the more general norm used:

$$f(\boldsymbol{x}) = \sum_i c_i K(\boldsymbol{x}, \boldsymbol{x}_i) \tag{12.27}$$

where the sum runs over the available examples (see also Figure 12.1 for a graphical representation of Equation 12.27). So far we have worked only at the kernel level, but in order to better understand the function norm used to regularize the problem we need to understand how the dot product works on the functions belonging to our space. To gain some understanding, we focus on a specific class of kernels, namely translation invariant kernels:

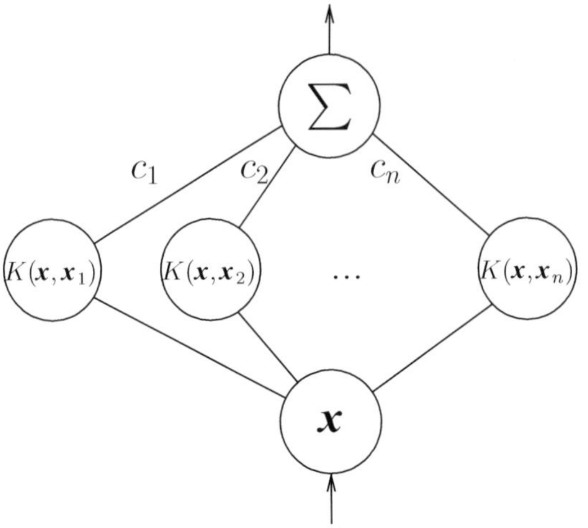


**Figure 12.1.** The solution to the regularized minimization problem of Equation 12.11 reported in Equation 12.27 can be represented by a network structure.

$K(\boldsymbol{x}, \boldsymbol{x}') = f(\boldsymbol{x} - \boldsymbol{x}')$ with a continuous, strictly positive definite function $f \in L_1(\mathbb{R}^d)$. Under these assumptions, it is possible to prove that

$$K(\boldsymbol{x}, \boldsymbol{x}') = \int e^{-i\langle \boldsymbol{x}-\boldsymbol{x}', \boldsymbol{\omega}\rangle} \upsilon(\boldsymbol{\omega})\, d\boldsymbol{\omega} = \int e^{-i\langle \boldsymbol{x}, \boldsymbol{\omega}\rangle} \upsilon(\boldsymbol{\omega}) \overline{e^{-i\langle \boldsymbol{x}', \boldsymbol{\omega}\rangle}}\, d\boldsymbol{\omega} \tag{12.28}$$

where $\upsilon(\boldsymbol{\omega})$ is a symmetric non-vanishing function. Recalling that the $d$-dimensional Fourier transform and its inverse are given by

$$F[f](\boldsymbol{\omega}) = (2\pi)^{-d/2} \int f(\boldsymbol{x}) e^{-i\langle \boldsymbol{x}, \boldsymbol{\omega}\rangle}\, d\boldsymbol{x}$$

$$F^{-1}[F[f]](\boldsymbol{x}) = (2\pi)^{-d/2} \int f(\boldsymbol{x}) e^{i\langle \boldsymbol{x}, \boldsymbol{\omega}\rangle}\, d\boldsymbol{\omega} = f(\boldsymbol{x})$$

we may compute the Fourier transform of $K(\boldsymbol{x}, \cdot)$

$$F[K(\boldsymbol{x}, \cdot)](\boldsymbol{\omega}) = (2\pi)^{-d/2} \iint (\upsilon(\boldsymbol{\omega}') e^{-i\langle \boldsymbol{x}, \boldsymbol{\omega}'\rangle}) e^{i\langle \boldsymbol{x}', \boldsymbol{\omega}'\rangle}\, d\boldsymbol{\omega}' e^{-i\langle \boldsymbol{x}', \boldsymbol{\omega}\rangle}\, d\boldsymbol{x}' \tag{12.29}$$

$$= (2\pi)^{d/2} \upsilon(\boldsymbol{\omega}) e^{-i\langle \boldsymbol{x}, \boldsymbol{\omega}\rangle}. \tag{12.30}$$

This allows us to rewrite Equation 12.28 as

$$K(\boldsymbol{x}, \boldsymbol{x}') = (2\pi)^{-d} \int \frac{F[K(\boldsymbol{x}, \cdot)](\boldsymbol{\omega}) \overline{F[K(\boldsymbol{x}', \cdot)](\boldsymbol{\omega})}}{\upsilon(\boldsymbol{\omega})}\, d\boldsymbol{\omega} \tag{12.31}$$

which, by linearity, allows us to express in a similar way all functions in our RKHS. The interesting thing is that now we may introduce a so-called regularization operator $\mathfrak{Y}$ mapping the functions residing in our RKHS into a dot product space so that the natural dot space of the mapped functions is given by our kernel function

$$\mathfrak{Y} : f \to (2\pi)^{-d/2} v^{-1/2} F[f] \tag{12.32}$$

$$K(\boldsymbol{x}, \boldsymbol{x}') = \int (\mathfrak{Y} K(\boldsymbol{x}, \cdot))(\boldsymbol{\omega}) \overline{(\mathfrak{Y} K(\boldsymbol{x}, \cdot))(\boldsymbol{\omega})}\, d\boldsymbol{\omega}. \tag{12.33}$$

This is important for two reasons:

1. We can characterize our kernel in Fourier space.
2. We have an explicit expression for a regularization operator whose action is to extract from a function the part that is effectively considered by the kernel-controlled regularization.

In particular, we see that small values of $v(\omega)$ amplify the corresponding frequencies in Equation 12.32 so that penalizing $\|f\|_K$ results in a strong attenuation of them. The Gaussian kernel is a widely used one and its regularization effects can be easily appreciated in the Fourier domain:

$$k(r) = \sigma^{-d} e^{-r^2/2\sigma^2} \tag{12.34}$$

$$F[k](\omega) = e^{-\omega^2\sigma^2/2} \tag{12.35}$$

implying that the larger the value of $\sigma$, the faster $v(\omega) = F[k](\omega)$ goes to zero, and the stronger the smoothness imposed by penalizing the RKHS norm. In a way not dissimilar from the design of a matched filter, if we know the power spectrum of the function to be estimated, we can choose $K(\boldsymbol{x}, \cdot)$ in such a way that $F[k]$ matches the expected value of the power spectrum of $f$.

## 12.2. RBF Networks

The structure of Equation 12.27 can be interpreted in terms of a network structure with a single layer of hidden units, the basis functions, corresponding to different kernel instances located at the samples (see Figure 12.1). An important result states that the networks derived from a regularization principle can approximate arbitrarily well continuous functions defined on a compact subset of $\mathbb{R}^d$ provided that enough units are used. Regularization networks can then be considered as universal approximators like feedforward neural networks with a single hidden layer of sigmoidal units.

We have seen that an important class of kernels used in RKHS computations is that of radial kernels, those depending only on $\|\boldsymbol{x} - \boldsymbol{y}\|$, and, consequently, translation invariant. These kernels can be characterized in Fourier space and their regularization effect made explicit. Among them the Gaussian kernel is widely used and can be used to provide a class of regularized solutions for the problem described by Equation 12.11 depending on the value of $\sigma$. If we consider the situation for $\lambda = 0$, thereby removing regularization, the solution will be an exact interpolant of the available data but, depending on the width of the Gaussian kernel used, it may exhibit oscillatory behavior. Another significant problem is that the complexity of the interpolant solution is linear in the size of the dataset: the linear expansion has a number of terms equal to the number of points. This implies that the evaluation of $f(\boldsymbol{x})$ may be extremely costly for large datasets.

Appreciation of these practical limitations led to the introduction of a class of approximation networks which are based on a set of modifications to the basic regularization approach:

**Intermezzo 12.1.** K-means: a simple clustering algorithm

The k-means algorithm clusters $n$ points $\mathbf{x}_i$ into $k < n$ partitions $C_j$, $j = 1, \ldots, k$, trying to minimize the following square error function:

$$\sum_{j=1}^{k} \sum_{\mathbf{x}_i \in C_j} (\mathbf{x}_i - \boldsymbol{\mu}_j)^2 \tag{12.36}$$

where $\boldsymbol{\mu}_j$ is the centroid of the points belonging to $C_j$.

The most common form of the algorithm uses an iterative refinement known as Lloyd's algorithm. Lloyd's algorithm starts by partitioning the input points into $k$ initial sets, either at random or using some heuristic knowledge. It then calculates the mean point, or centroid, of each set. It constructs a new partition by associating each point with the closest centroid. The centroids are then recalculated for the new clusters, and the algorithm repeated by alternate application of these two steps until convergence, which is obtained when the points no longer switch clusters (or, alternatively, centroids are no longer changed). This algorithm is essentially equivalent to a gradient descent performed using Newton's method and can be proved to converge to a local minimum. The efficiency of Lloyd's algorithm is due to its relation with Newton's method. As it converges to a local minimum, several different initial conditions should be tried and the solution leading to the lowest value of Equation 12.36 should be retained (Bottou and Bengio 1994).

1. The number of basis functions is no longer (necessarily) equal to the number of points $\mathbf{x}_i$.
2. The centers $\mathbf{t}_j$ of the basis functions are not required to be at the input vectors and they are computed during the training process to optimize the performance of the network.
3. Each basis function has its own regularizing parameters $\mathbf{q}_j$ (for a Gaussian kernel, $\sigma_j$, or even $\Sigma_j$) that may also be determined by the training process.

These modifications can be formalized in the approximation structure

$$f(\mathbf{x}) = \sum_{j=1}^{n} c_j K(\mathbf{x} - \mathbf{t}_j; \mathbf{q}_j) + b \tag{12.37}$$

with $n$ centers, $n < N$, the number of data samples, $b$ is an offset value, and the translational invariance has been made explicit using $\mathbf{x} - \mathbf{t}_j$ as kernel argument. These networks are also known as radial basis function (RBF) networks when the kernel is radial and as GBF networks when the kernel is Gaussian but not necessarily radial. The increased flexibility, however, does not come for free: the computation in the most general case, when centers and kernel parameters are optimized during the training process, is complex and plagued by the presence of multiple local minima. A partial solution is to first estimate the position of the centers (e.g. using a clustering algorithm such as the one presented in Intermezzo 12.1) and the kernel parameters (using heuristics or cross-validation techniques), and to finally compute the coefficients of the expansion using a pseudo-inverse approach:

$$\mathbf{c} = K^{+}\mathbf{y} \quad \text{where} \quad K = K_{ji} = K(\mathbf{x}_j; \mathbf{t}_i, \mathbf{q}_i). \tag{12.38}$$

In many cases we can avoid the computation of $b$ by prenormalizing our data to zero average.

### 12.2.1. RBF NETWORKS FOR GENDER RECOGNITION

An important characteristic of RBF networks is their descriptive capacity, especially when all of their parameters are optimized during the training phase. This can be appreciated in a simple, but not trivial task: gender discrimination using a geometrical feature vector. Using simple image processing techniques, including template matching, we can reliably extract from a geometrically normalized frontal face image an 18-dimensional feature vector:

- pupils to nose, mouth, and chin vertical distances;
- nose and mouth width;
- zygomatic and bigonial breadth;
- six radii describing chin shape;
- mouth height;
- upper and lower lip thickness;
- pupil to eyebrow separation and eyebrow thickness.

The features are presented graphically in Figure 12.2. Let us consider a two-unit GBF network required to output 1 (−1) for a vector corresponding to a male (female):

$$f_G(\boldsymbol{x}) = \sum_{j=1}^{2} c_j G(\|\boldsymbol{x} - \boldsymbol{t}_j\|_W) \tag{12.39}$$

where the norm is a weighted norm

$$\|\boldsymbol{x} - \boldsymbol{t}\|_W = (\boldsymbol{x} - \boldsymbol{t})^T W^T W (\boldsymbol{x} - \boldsymbol{t}) \tag{12.40}$$

$W$ being a square matrix, in our case constrained to be diagonal. Strictly speaking the basis functions are no longer radial, but, as $W$ is shared by all units, we can transform our data $\boldsymbol{x} \to \boldsymbol{x}' = W^{-1}\boldsymbol{x}$, and in the new data space, after the computation of $W$, the units will be radial. Classification is then performed based on the sign of network output. However, we would like to make the magnitude of network response correspond to the confidence in the classification, following a fuzzy classification paradigm. We may then restrict the range of network output by applying a sigmoid (logistic) nonlinearity

$$\lambda(x) = \frac{1 - e^{-2x}}{1 + e^{-2x}} \tag{12.41}$$

on network output

$$f_{G,1}(\boldsymbol{x}) = \lambda(f_G(\boldsymbol{x})). \tag{12.42}$$

The centers of the networks are initialized with the average male/female vectors and the corresponding coefficients with 1/−1. The network is trained by minimizing the following error:

$$\Delta_G = \Delta_c + \frac{1}{N}\sqrt{\sum_{i=1}^{N}(y_i - \lambda(f_G(\boldsymbol{x}_i)))^2} \tag{12.43}$$

where $\Delta_c$ represents the percent classification error. The training process modifies the coefficients, the centers, and the metrics $W$. The male/female prototypes are displayed in

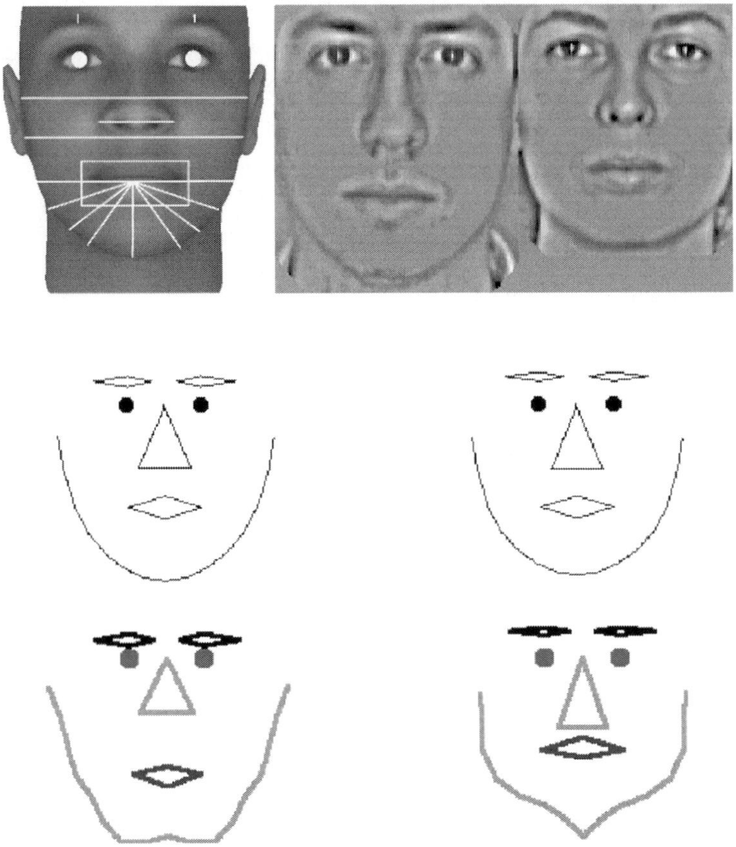

**Figure 12.2.** Gender classification using a Gaussian basis function network. The drawings in the middle row represent the centers of the network before training and those in the bottom row show the effects of the training process. Source: Brunelli and Poggio (1993).

Figure 12.2. The elements $w_i$ representing the diagonal of $W$ are modified significantly by the training process resulting in a major contribution of eyebrow thickness, followed by upper lip thickness, pupil to eyebrow separation, and lower lip thickness. The database used for classification comprised 168 vectors equally distributed over 21 males and 21 females. The classification on the training set was 92% correct while the actual performance measured using a leave-one-out strategy was 87.5%. Human performance is often considered to be close to perfect in this kind of task, so a comparative test was done. The face images from which the vectors were computed were normalized by dividing them by the local average, cropping out hair, and masking any residual facial hair. The resulting images were then used for an informal psychophysical experiment: 17 subjects were required to press M (F) for male (female) without any time constraint upon being presented with a sequence of images on a computer screen. Some of the subjects were familiar with a large subset of the people represented in the database. The result was surprising: an average score of 90% correct classification was achieved, performance being unaffected by familiarity with the database people.

## 12.3. Support Vector Machines

We have seen in Section 12.1 that if we gauge approximation quality by its average squared error over a training set, it is possible to get a solution to the regularized problem in closed form as a linear combination of kernel functions located on the sample points. While for a generic error measure it is not always possible to find a closed form solution, it is possible to prove that the special form of the regularized solution found for the quadratic error holds in general.

**Theorem 12.3.1 (Representer Theorem).** *Denote by $\Omega : [0, \infty) \to \mathbb{R}$ a strictly monotonic increasing function, by $\mathfrak{X}$ a set, and by $c : (\mathfrak{X} \times \mathbb{R}^2)^N \to \mathbb{R} \cup \{\infty\}$ an arbitrary loss function. Then each minimizer $f \in \mathfrak{H}$ of the regularized risk*

$$c((x_1, y_1, f(x_1)), \ldots, (x_N, y_N, f(x_N))) + \Omega(\|f\|_{\mathfrak{H}}) \tag{12.44}$$

*admits a representation of the form*

$$f(x) - \sum_{i=1}^{N} \alpha_i K(x_i, x). \tag{12.45}$$

In spite of the possibility of choosing the solution as a linear combination of kernel functions applied to an arbitrary number of arbitrarily chosen points (the functions in the space identified by the regularization kernel of Equation 12.14), the solution is always expressed by Equation 12.45 even if an efficient solution for the expansion coefficients is rather exceptional.

It is sometimes desirable to exclude part of the solution from regularization. A very common example is when we search a solution of the form $f(x) = g(x) + b$ where $b$ is a constant. The representer theorem is flexible enough to accommodate this type of requirement, leading to a variant called the semiparametric representer theorem.

**Theorem 12.3.2 (Semiparametric Representer Theorem).** *Under the hypothesis of the representer theorem, let $\{\psi_p\}_{p=1}^{M} : \mathfrak{X} \to \mathbb{R}$ be a set of $M$ real-valued functions such that the $N \times M$ matrix $[\psi_p(\boldsymbol{x}_i)]_{ip}$ has rank $M$. Then any $\tilde{f} : f + h$ with $f \in \mathfrak{H}$ and $h \in \mathrm{span}\{\psi_p\}$ minimizing the regularized risk*

$$c((x_1, y_1, \tilde{f}(x_1)), \ldots, (x_N, y_N, \tilde{f}(x_N))) + \Omega(\|\tilde{f}\|_{\mathfrak{H}}) \tag{12.46}$$

*admits a representation of the form*

$$\tilde{f}(x) = \sum_{i=1}^{N} \alpha_i K(x_i, x) + \sum_{p=1}^{M} \beta_p \psi_p(x) \tag{12.47}$$

*with $\beta_p \in \mathbb{R}$ for all $p$.*

The main reason for introducing the above theorem is that it allows us to approach an important learning paradigm, namely that of support vector machines, from a regularization perspective.

The squared error loss function is very convenient from a computational point of view in that it often allows us to find a closed form for the solution of related problems, as we saw in Section 12.1, but it is not the most natural choice for classification problems. In these problems the most appropriate loss would be the classification error on the training set which, due to its piecewise flatness, unfortunately would make the optimization problem very hard. A major problem of the square loss is that it gives too much unwanted weight to gross errors. One such error could weight as many not so evident errors and we would rather prefer to get rid of the latter than the former. Another important point is that in a situation of attained perfect classification, we would like the classification boundary to provide as stable a performance as possible, being as far as possible from the points that are closest to it. The idea in this case is to have some margin for unexpected data close to one of the two classes but more extreme in membership than those available in the training set. Let us consider as usual a binary classification problem with class labels $\pm 1$ and a decision function $\text{sign}(f(\boldsymbol{x}))$. The soft margin loss function

$$c((x_1, y_1, f(x_1)), \ldots, (x_N, y_N, f(x_N))) = \frac{1}{N} \sum_{i=1}^{N} \max(0, 1 - y_i f(\boldsymbol{x}_i)) \tag{12.48}$$

addresses these issues resulting in the minimization of the regularized risk

$$\operatorname*{argmin}_f \frac{1}{N} \sum_{i=1}^{N} \max(0, 1 - y_i f(\boldsymbol{x}_i)) + \frac{\lambda}{2} \|f\|^2 \tag{12.49}$$

with $\lambda > 0$ or

$$\operatorname*{argmin}_f \frac{1}{\lambda N} \sum_{i=1}^{N} \max(0, 1 - y_i f(\boldsymbol{x}_i)) + \frac{1}{2} \|f\|^2 \tag{12.50}$$

so that for $\lambda \to 0$ the dominating part is the classification error. In this case, if the problem is separable, the solution will be forced to correctly classify all of the data while at the same time minimizing the function norm.

When classification for point $\boldsymbol{x}_i$ is correct, it will still contribute to the loss value unless the value $|f(\boldsymbol{x}_i)| \geq 1$, so that among all the solutions that classify the data correctly, the one providing the largest classification margin will be favored. Let us note, however, that if the classes are not separable a single gross error will still have a major impact on the solution. Let us consider the solution offered by the semiparametric representer theorem when the kernel function corresponds to the usual dot product and $\psi_1 = 1$:

$$f(x) = \sum_{i=1}^{N} \alpha_i \boldsymbol{x}_i \cdot \boldsymbol{x} + b = \boldsymbol{w} \cdot \boldsymbol{x} + b \tag{12.51}$$

where

$$\boldsymbol{w} = \sum_{i=1}^{N} \alpha_i \boldsymbol{x}_i \tag{12.52}$$

and the regularizing norm entering the minimization problem is $\|f\| = \|\boldsymbol{w}\|_{L_2}$. The classification surface is then given by a hyperplane and classification is obtained by projecting onto the plane normal and thresholding according to the value of $b$ (see Figure 12.3):

$$\text{sign}\left(\sum_{i=1}^{N} \alpha_i \boldsymbol{x}_i \cdot \boldsymbol{x} + b\right) \underset{y=-1}{\overset{y=1}{\gtrless}} 0. \tag{12.53}$$

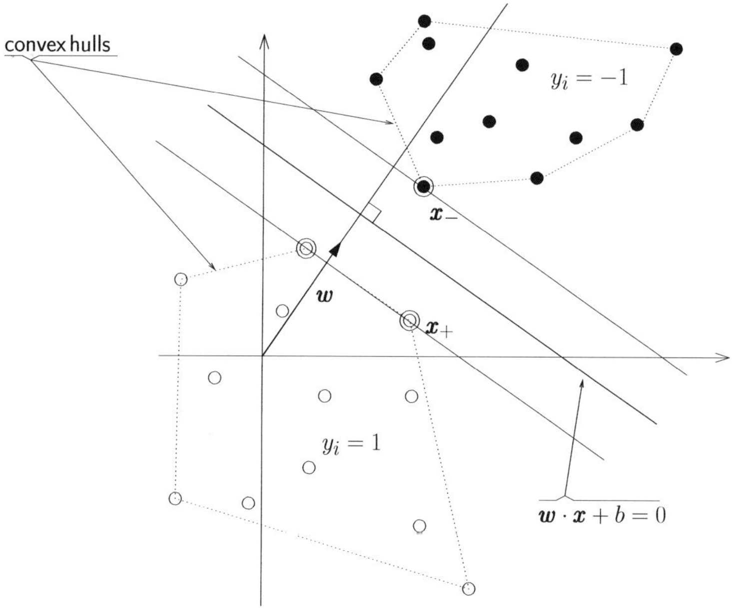


**Figure 12.3.** The basic geometry of a linear support vector machine exploited in Equations 12.56–12.60. The optimal separating plane can also be found by considering the convex hulls of the two classes; references to this geometrical approach can be found in Section 12.4.

A generalization of the above classification function is given by

$$\left(\sum_{i=1}^{N} \alpha_i \boldsymbol{x}_i \cdot \boldsymbol{x} + b\right) \underset{y=-1}{\overset{y=1}{\gtrless}} \theta \tag{12.54}$$

or, when the kernel trick is used (see Figure 12.4), by

$$\left(\sum_{i=1}^{N} \alpha_i K(\boldsymbol{x}_i, \boldsymbol{x}) + b\right) \underset{y=-1}{\overset{y=1}{\gtrless}} \theta \tag{12.55}$$

that reduces to the previous one for $\theta = 0$, but allows us to trade off false alarm and true positive detection rates and characterize the classifier by means of an ROC curve (see Section C.3). It is clear that there is some scaling freedom and, in the separable case, we may fix it by forcing to 1 the distances of the points of the two classes, letting them be $\boldsymbol{x}_+$ and $\boldsymbol{x}_-$, closest to the plane. From

$$\boldsymbol{w} \cdot \boldsymbol{x}_+ + b = d_+ \tag{12.56}$$

$$\boldsymbol{w} \cdot \boldsymbol{x}_- + b = -d_- \tag{12.57}$$

we go to a new classification function characterized by $\boldsymbol{w}' = s_w \boldsymbol{w}$ and $b' = s_b b$

$$\boldsymbol{w}' \cdot \boldsymbol{x}_+ + b' = 1 \tag{12.58}$$

$$\boldsymbol{w}' \cdot \boldsymbol{x}_- + b' = -1 \tag{12.59}$$

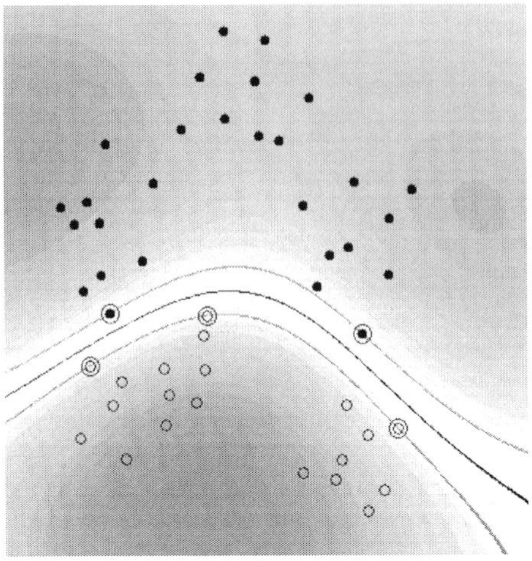

**Figure 12.4.** A Gaussian support vector classifier: the middle line identifies the zero-crossing line and the support vectors are highlighted. The darker the map, the greater the value of $\left|\sum_{i=1}^{N} \alpha_i K(\boldsymbol{x}_i, \boldsymbol{x}) + b\right|$ and the more certain the classification. Source: Scholkopf *et al.* (1996).

and the scaling factors $s_w$ and $s_b$ are found by substitution of the primed quantities in the above equations. The resulting hyperplane is said to be in canonical form and we will restrict our analysis it to, dropping the primes for convenience. The distance of $\boldsymbol{x}_+$ and $\boldsymbol{x}_-$ projected on the plane normal is then

$$\left(\frac{\boldsymbol{w}}{\|\boldsymbol{w}\|}\right) \cdot (\boldsymbol{x}_+ - \boldsymbol{x}_-) = \frac{2}{\|\boldsymbol{w}\|} \tag{12.60}$$

and is called the margin. In the linearly separable situation, the loss (or error) can be driven to 0 and the solution that minimizes the regularizing norm $\|\boldsymbol{w}\|^2/2$ maximizes the classification margin of Equation 12.60. The above reasoning carries over unchanged when the kernel trick (see Section 8.2) is employed in Equation 12.51. The solution provided by the representer theorem potentially involves a linear expansion with a number of terms equal to the number of training samples. In many cases this would lead to an unpractical solution due to excessive computational requirements. It turns out that some loss functions, the one in Equation 12.48 being a notable example, lead to a sparse solution: many of the $\alpha_i$ are 0, and the complexity of evaluation of the classification function is correspondingly reduced. The reasons behind the sparsity of the solution can be understood by considering the procedure for actually finding the coefficients. The minimization problem of Equation 12.50 can be reformulated as a constrained optimization problem

$$\hat{\boldsymbol{w}} = \operatorname*{argmin}_{\boldsymbol{w},b} \frac{1}{2}\|\boldsymbol{w}\|, \quad \text{subject to} \quad y_i(\boldsymbol{w} \cdot \boldsymbol{x}_i + b) \geq 1, \forall i \tag{12.61}$$

whose solution relies on the introduction of Lagrangian multipliers $\alpha_i \geq 0$ to incorporate the constraints into the Lagrangian $L$

$$L(\boldsymbol{w}, b, \boldsymbol{\alpha}) = \frac{1}{2}\|\boldsymbol{w}\| - \sum_{i=1}^{N} \alpha_i (y_i(\boldsymbol{w} \cdot \boldsymbol{x}_i + b) - 1). \tag{12.62}$$

The Lagrangian must be minimized with respect to the primal variables $\boldsymbol{w}$ and $b$ and maximized with respect to the so-called dual variable $\boldsymbol{\alpha}$, thereby finding a saddle point. If a constraint is violated, $y_i(\boldsymbol{w} \cdot \boldsymbol{x}_i + b) - 1 < 0$ and $L$ can be made arbitrarily large by increasing $\alpha_i$, while at the same time it should be minimized with respect to the primal variables. This requires that, if the problem is separable, the primal variables will be modified in order to meet the constraint. On the other side, when the constraint is met, $L$ will decrease by increasing $\alpha_i$, so that it will be maximized by the smallest possible value for $\alpha_i$, i.e. 0. These considerations can be formalized and are known as the Karush–Kuhn–Tucker (KKT) complementarity conditions. At a saddle point of $L$ its derivatives with respect to the primal variables must vanish

$$\frac{\partial}{\partial b} L(\boldsymbol{w}, b, \boldsymbol{\alpha}) = 0 \tag{12.63}$$

$$\frac{\partial}{\partial \boldsymbol{w}} L(\boldsymbol{w}, b, \boldsymbol{\alpha}) = 0 \tag{12.64}$$

leading to

$$0 = \sum_{i=1}^{N} \alpha_i y_i \tag{12.65}$$

$$\boldsymbol{w} = \sum_{i=1}^{N} \alpha_i y_i \boldsymbol{x}_i . \tag{12.66}$$

We then recover a solution of the form dictated by the representer theorem but with some additional information: the KKT conditions imply that the expansion is sparse and depends only on the points that meet the constraints exactly, the so-called support vectors (SVs). Classifiers based on the solution of the minimization problem of Equation 12.50 are then called support vector machines (SVMs). The fact that the support vectors condense all dataset information required for classification can be exploited to improve the handling of pattern variability with a limited number of additional examples as described by Intermezzo 12.2. Based on the above equations it is possible to get rid of the primal variables, obtaining the so-called dual optimization problem entirely expressed in terms of the dual variables $\alpha_i$

$$\hat{\boldsymbol{\alpha}} = \operatorname*{argmax}_{\boldsymbol{\alpha}} \left( \sum_{i=1}^{N} \alpha_i - \frac{1}{2} \sum_{i,j=1}^{N} \alpha_i \alpha_j y_i y_j (\boldsymbol{x}_i \cdot \boldsymbol{x}_j) \right) \tag{12.67}$$

subject to

$$\alpha_i \geq 0 \quad \forall i \quad \text{and} \quad \sum_{i=1}^{N} \alpha_i y_i = 0. \tag{12.68}$$

This is a standard quadratic programming problem that can be solved efficiently even for large datasets. The solution found can be given a physical interpretation substantiating our original idea of finding a stable separating surface. If we associate the separating hyperplane to a solid sheet and let each support vector exert a force

$$\boldsymbol{F}_i = \alpha_i \left( y_i \frac{\boldsymbol{w}}{\|\boldsymbol{w}\|} \right) \tag{12.69}$$

the constraint of Equation 12.65 implies that the sum of the forces acting on the plane is null, while Equation 12.66 implies that the torque is also null

$$\sum_i \left( \boldsymbol{x}_i \times \left( \alpha_i y_i \frac{\boldsymbol{w}}{\|\boldsymbol{w}\|} \right) \right) = \boldsymbol{w} \times \frac{\boldsymbol{w}}{\|\boldsymbol{w}\|} = 0 \tag{12.70}$$

so that the plane is mechanically stable.

**Intermezzo 12.2.** The virtual SV method

If we remove from the training set any point that is not an SV, the classifier will not change: in a sense, the SVs distill all the knowledge necessary to build the optimal classifier. This has been verified experimentally also by training an SVM solely on the SVs of another SVM based on a different kernel. The difference in performance with respect to training it on the whole database was found to be minor. We know that it is possible to enforce some invariances on the classifier by including in the training set some appropriate examples. For instance, if we want to make our eye detector (a classifier) invariant to small rotation and scale variations of eyes, we may generate from each given eye an additional set of templates obtained by rotating and scaling it. An apparent drawback of the strategy is that the size of the training set increases significantly leading to obvious computational difficulties.

The fact that SVs contain all the necessary classification information offers an easy way out: instead of generating the additional examples to enforce the required invariance for all training samples, we may proceed in the following way:

1. Build an SVM for the original dataset.
2. Generate the virtual examples necessary to enforce the required classification invariances only for the SVs, usually a limited fraction of the original database.
3. Build a new SVM using only the original SVs and the associated virtual examples.

If we limit ourselves to small transformations, the virtual SVs should be close to the SVs from which they are generated: this means that they will be close to the decision surface and are likely to become SVs for the final SVM computed in Step 3 above. On the other side, if it turns out that a significant fraction of the virtual examples become SVs, we can be confident that new valuable information has been provided to the classifier. If we start with a very large database, the information contained in the virtual SVs may already be present in the data, but not necessarily. Since applying the virtual SV method may provide increased performance at a comparable cost, it should be attempted when possible.

Let us now consider the non-separable case, trying to get a solution as close as possible to the separable one. We saw that if the constraints cannot be met, the Lagrangian formulation of the problem has no solution: the way out is to relax the constraints in a controlled way so that it will be possible to satisfy them while at the same time departing minimally from the original formulation. The approach can be formalized with the introduction of an additional set of parameters $\xi_i$, termed slack variables, transforming the original minimization problem into

$$\hat{\boldsymbol{w}} = \underset{\boldsymbol{w},b,\xi}{\operatorname{argmin}} \frac{1}{2}\|\boldsymbol{w}\| + \frac{C}{N}\sum_{i=1}^{N} \xi_i \quad \text{subject to } y_i(\boldsymbol{w} \cdot \boldsymbol{x}_i + b) \geq 1 - \xi_i,\ \xi_i > 0\ \forall i \tag{12.71}$$

where $C$ is a given constant. This results in the same dual maximization problem as before with the single important difference that the $\alpha_i$ are constrained within a finite size box

$$0 \leq \alpha_i \leq \frac{C}{N} \quad \forall\, i. \tag{12.72}$$

The KKT conditions for the dual problem in the non-separable case are:

$$\alpha_i = 0 \Rightarrow y_i f(\boldsymbol{x}_i) \geq 1 \quad \text{and} \quad \xi_i = 0;$$
$$0 < \alpha_i < \frac{C}{N} \Rightarrow y_i f(\boldsymbol{x}_i) = 1 \quad \text{and} \quad \xi_i = 0;$$
$$\alpha_i = \frac{C}{N} \Rightarrow y_i f(\boldsymbol{x}_i) \leq 1 \quad \text{and} \quad \xi_i \geq 0.$$

For any SV for which $0 < \alpha_i < C/N$, we have that $\xi_i = 0$ and the (original) constraint is met exactly so that we can recover the value of $b$ from averaging over

$$y_i \left( b + \sum_j y_j \alpha_j K(\boldsymbol{x}_i, \boldsymbol{x}_j) \right) = 1 \tag{12.73}$$

completing the solution. The SVs with maximal $\alpha$ are those corresponding to points that cannot be separated and for which the slack variables become active in changing the original constraint. Several algorithms exist for efficiently solving this kind of quadratic programming problem even when the number of points is large.

Successful use of SV machinery requires an appropriate choice of the kernel function, even if in many cases the differences in performance will not be so great, and, more importantly, an appropriate choice of the parameter $C$, the regularization constant that controls the soft margin, imposing an upper bound on the $\alpha_i$: these problems are considered in Section 12.3.3. The results obtained in a gender discrimination task are reported, as ROC curves, in Figure 12.5.

An important feature of SVMs is the possibility of deriving directly from the synthesized function an estimate of the leave-one-out error, or, to be more precise, to derive useful bounds on it. This is important for two reasons:

1. We do not incur any additional cost to compute it.

2. The leave-one-out error is an almost unbiased estimator of the expected risk (see Theorem C.2.1), the quantity that a classification system should minimize.

In the case of separable problems, the leave-one-out error is related to the number of SVs and the following theorem holds.

**Theorem 12.3.3.** *The expectation of the number of SVs obtained during training on a set of size N, divided by N, is an upper bound of the expected probability of test error of the SVM trained on training sets of size N − 1.*

One of the simplest bounds to compute is based on the value of the Lagrange multipliers $\boldsymbol{\alpha}$ and, for risk functionals without the constant threshold $b$, it is formalized by the following theorem.

**Theorem 12.3.4.** *If* $f(x) = \sum_{i=1}^{N} \alpha_i y_i K(x_i, x)$ *is the minimizer of the regularized risk functional of Equation 12.71, an upper bound on the leave-one-out error is*

$$R_{loo}(X \times Y) \leq \frac{1}{N} \sum_{i=1}^{N} \theta(y_i(f(x_i) - \alpha_i K(x_i, x_i))). \tag{12.74}$$

The theorem can be extended to the case of non-null $b$ but the result is overly conservative and the above value is often appropriate even in this case.

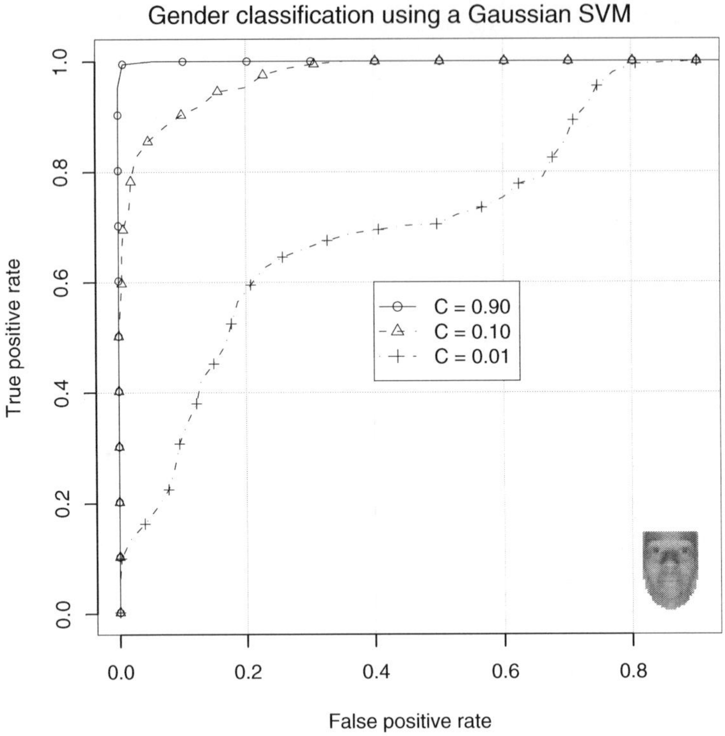


**Figure 12.5.** The plot shows the ROC curves for three different Gaussian SVMs using different $C$ values. The ROC is obtained by thresholding the output of the SVM at different values: the shape of the ROC with $C = 0.01$ shows that it is not the result of the optimal Neyman–Pearson classifier (see Section C.3). The SVMs with $C = 1$ attained perfect discrimination (not shown). The dataset included 800 synthetic images (400 males and 400 females) generated by the software FaceGen from Singular Inversions.

### 12.3.1. IMPROVING EFFICIENCY

A characterizing feature of SVMs is their ability to work efficiently with high-dimensional feature spaces, possibly with infinite dimension, by relying solely on kernel computations. When an SVM is trained we do not impose any constraint on the number of SVs. Actually, the automatic choice of an appropriate number is an attractive feature of SVMs. We have seen that the number of SVs chosen is inversely related to generalization and is influenced by SVM parameters. As discussed in Section 12.3.3, the parameters are usually chosen by cross-validation in order to optimize classification performance. A not uncommon situation is that, in order to achieve maximum performance, the number of SVs may be increased significantly, reaching a significant fraction of the available samples. As a result, the evaluation of the classification function becomes more and more expensive, scaling linearly with the number of SVs and becoming a practical concern in real-world applications.

There are two different aspects of the problem. One is related to the progressively smaller classification advantage with increasing number of SVs. The other one is related to the constraints imposed on the expansion coefficients $\alpha_i$ that are restricted to be in the interval

$[0, C/N]$: this may, for example, force the SVM to use multiple instances of a replicated data pattern in order to obtain the effect that could be reached with a single one but with a forbidden coefficient. These considerations suggest that we can attempt to approximate the solution found by the SVM using a reduced set of vectors, searching for an acceptable performance/computation trade-off. Note that the classification function actually computes the projection of a mapped input pattern $\Phi(\boldsymbol{x})$ on the normal vector of the SV hyperplane $\Psi$ expressed as a linear combination of mapped data points

$$\Psi = \sum_{i=1}^{N} \alpha_i \Phi(\boldsymbol{x}_i). \tag{12.75}$$

If we could find an exact preimage $\boldsymbol{x}_\Psi$ of $\Psi$ such $\Psi = \Phi(\boldsymbol{x}_\Psi)$ the classification step would require a single kernel computation. Unfortunately no exact preimage can generally be computed, and an approximation must be found. Since we know that building sequences of increasingly better approximations opens the way to efficient detection, we focus on the construction of a progressive approximation to $\Psi$. A basic requirement of the approximation is that it must rely exclusively on kernel evaluations. Let us seek an approximation of $\Psi$ in the form of $\Psi' = \beta\Phi(z)$: the additional freedom provided by $\beta$ does not impact on the classification function (the scaling can be compensated by rescaling and changing threshold $b$). The approximation can be found by minimizing the following squared error:

$$\|\Psi - \Psi'\|^2 = \sum_{i,j=1}^{N} \alpha_i\alpha_j K(\boldsymbol{x}_i, \boldsymbol{x}_j) + \beta^2 K(z, z) - 2\sum_{i=1}^{N} \alpha_i \beta K(\boldsymbol{x}_i, z) \tag{12.76}$$

or by minimizing the distance from the projection onto span($\Phi(z)$)

$$\left\| \frac{\langle \Psi, \Phi(z)\rangle}{\langle \Phi(z), \Phi(z)\rangle}\Phi(z) - \Psi \right\|^2 = \|\Psi\|^2 - \frac{\langle \Psi, \Phi(z)\rangle^2}{\langle \Phi(z), \Phi(z)\rangle} \tag{12.77}$$

so that we need to maximize

$$\frac{\langle \Psi, \Phi(z)\rangle^2}{\langle \Phi(z), \Phi(z)\rangle}. \tag{12.78}$$

Let us focus on Gaussian kernels $K(\boldsymbol{x}, z) = \exp[-\|\boldsymbol{x} - z\|^2/(2\sigma^2)]$ so that $\langle \Phi(z), \Phi(z)\rangle = 1$ and the maximization of Equation 12.78 reduces to the maximization of $\langle \Psi, \Phi(z)\rangle^2$ requiring that

$$0 = \nabla_z \langle \Psi, \Phi(z)\rangle^2 = 2\langle \Psi, \Phi(z)\rangle \nabla_z \langle \Psi, \Phi(z)\rangle = \sum_{i=1}^{N} \alpha_i \nabla_z K(\boldsymbol{x}_i, z). \tag{12.79}$$

For kernels depending only on the squared difference of the arguments as the Gaussian kernel we then get

$$0 = \sum_{i=1}^{N} \alpha_i K'(\|\boldsymbol{x}_i - z\|^2)(\boldsymbol{x}_i - z) \tag{12.80}$$

where $K'$ denotes the first derivative of $K$ with respect to the squared distance. We can then extract an expression of $z$ suitable for a fixed point iteration solution

$$z = \frac{\sum_{i=1}^{N} \alpha_i K'(\|\boldsymbol{x}_i - z\|^2)\boldsymbol{x}_i}{\sum_{i=1}^{N} \alpha_i K'(\|\boldsymbol{x}_i - z\|^2)} \tag{12.81}$$

leading, for Gaussian kernels, to

$$z_{t+1} = \frac{\sum_{i=1}^{N} \alpha_i \exp[-\|\boldsymbol{x}_i - z_t\|^2/(2\sigma^2)]\boldsymbol{x}_i}{\sum_{i=1}^{N} \alpha_i \exp[-\|\boldsymbol{x}_i - z_t\|^2/(2\sigma^2)]}. \tag{12.82}$$

Let us note that, by finding an approximate preimage of $\Psi$, we are trying to perform classification with a single Gaussian function centered at $z$, somehow recovering the RBF network idea. We now need to compute a sequence of approximations to $\Psi$

$$\Psi'_m = \sum_{i=1}^{m} \beta_{m,i}\Phi(z_i) \tag{12.83}$$

for $m = 1, \ldots, N_z$ where $N_z$ is fixed a priori as the maximum acceptable complexity of the classification function. Having obtained $z_1$ using Equation 12.82 (starting from an initial guess), we compute $\beta_{1,1}$

$$\beta_{1,1} = \langle \Psi, \Phi(z_1)\rangle. \tag{12.84}$$

We then proceed by finding the approximate preimage of $\Psi - \sum_{i=1}^{m-1} \beta_{m-1,i}\Phi(z_i)$. Given the set $\{\Phi(z_i)\}$, $i = 1, \ldots, m-1$, the computation of the optimal expansion coefficients $\beta_{m-1,i}$ relies on the following result.

**Theorem 12.3.5.** *The coefficients* $\boldsymbol{\beta} = (\beta_1, \ldots, \beta_m)$ *minimizing*

$$\left\| \sum_{i=1}^{N} \alpha_i \Phi(\boldsymbol{x}_i) - \sum_{i=1}^{m} \beta_i \Phi(z_i) \right\| \tag{12.85}$$

*for a set of independent vectors* $\{\Phi(z_i)\}$ *are given by*

$$\boldsymbol{\beta} = (K^{zz})^{-1} K^{zx} \boldsymbol{\alpha} \tag{12.86}$$

*where*

$$K_{ij}^{zx} = \langle \Phi(z_i), \Phi(\boldsymbol{x}_j)\rangle. \tag{12.87}$$

We now have all we need to construct a sequence of progressively better, and computationally more complex, approximations to the original SVM solution (see also Intermezzo 13.2).

### 12.3.2. MULTICLASS SVMS

So far we have addressed binary classification problems, to which many template matching tasks belong. The connection of SVMs to the optimal Bayes classifier is not obvious but recent results show that, whenever the RKHS is rich enough, the classification rule associated with an SVM, i.e. $\text{sign}[f(\boldsymbol{x})]$, approaches the Bayes decision rule $\text{sign}(p_1(\boldsymbol{x}) - 1/2)$ asymptotically, i.e. when the number of samples $N$ goes to $\infty$ and $\lambda$ assumes appropriate values. By rich enough we mean that the RKHS should include $\boldsymbol{p}_1(x)$ or, at least, $\boldsymbol{p}_1(x)$ should be close to the RKHS used. This is the technical reason why the performance of linear SVMs is often not good enough. As they simply divide the space into two half spaces, all situations where $\text{sign}[p_1(\boldsymbol{x}) - 1/2]$ divides the space into multiple regions cannot be modeled well. Among the kernels rich enough to approximate the Bayes classifier, we can mention the

**Table 12.1.** Different SVM kernels result in different numbers of SVs. The table lists the number of SVs used in a race discrimination task by a linear and a Gaussian SVM. Both SVMs achieved the same accuracy (0.9875%) but the linear one required fewer support vectors.

| | African | Asian | European | Indian |
|---|---|---|---|---|
| Linear SVM | 20 | 39 | 51 | 41 |
| Gaussian SVM | 21 | 42 | 67 | 57 |

Gaussian kernel. While the optimal Bayes classifier is based on an estimate of the probability $p_1(\boldsymbol{x})$, the SVM directly estimates the sign of $p_1(\boldsymbol{x}) - 1/2$ and not the probability itself.

Many pattern recognition tasks are best expressed as multicategory classification problems. A typical example is the problem of face recognition as opposed to the problem of face detection. While the latter is a binary classification problem, face versus not a face, the former requires the system to output the code of the person to which the face belongs. There are two strategies to solve the multiclass problem:

1. Mapping the multicategory problem into a set of binary problems.

2. Consider all classes jointly, adopting a truly multicategory approach.

The first general strategy can itself be divided into two distinct approaches that are usually called one versus the rest and pairwise classification. In the first case, in order to obtain an $m$-class classifier a set of $m$ binary classifiers $f_i$, $i = 1, \ldots, m$, is built, each of them discriminating class $i$ from the others. The pattern is then assigned to the class whose classifier responds more strongly to it, before the sign function is applied:

$$\hat{\jmath} = \operatorname*{argmax}_{j=1,\ldots,m} g^j(x) \quad \text{where} \quad g^j(x) = \left( \sum_{i=1}^{N} y_i \alpha_i^j K(x, x_i) + b^j \right). \tag{12.88}$$

A positive and a negative remark are in order. The values of $g^j(x)$ are useful for rejecting a pattern by not assigning it to any class when the difference between the two largest $g$ values is small. The negative consideration is that Equation 12.88 is of a heuristic nature: the classifiers are trained on different problems and their outputs are not necessarily on the same scale. The fact that SVMs do not estimate probabilities but directly the sign of the classification function further substantiates the possibility of incompatible value scales which may have a relevant impact when comparing $g^j$ values.

Another, more symmetric strategy is that of considering $m(m-1)/2$ binary classification problems by training a classifier for every possible pair of classes. When a pattern must be classified, all the binary classifiers express their vote and the pattern is assigned to the class that received the highest number of votes (see also Figure 12.6 and Table 12.1). While the number of classifiers to be trained is much higher than in the previous case, the fact that the single problems are much simpler (less data and less class overlap) may result in similar computational efforts. In many cases it is possible to structure a single classification as a path in a directed acyclic graph where the classifiers are embedded (see also Section 13.3). Strategies for selecting an optimal path can then be employed to avoid the evaluation of classifiers that cannot possibly influence the final result.

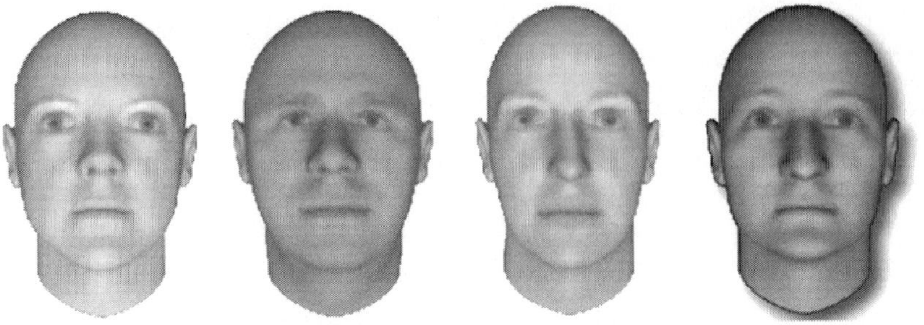

**Figure 12.6.** This shows four of the SVs chosen by a race classification multiclass Gaussian SVM based on multiple binary classification and subsequent voting. These SVs were used to discriminate Europeans from African, Asian, and Indian people: three of them represent European people, the shadowed one represents an Indian. The dataset set included 800 synthetic images generated with the software FaceGen from Singular Inversions.

### 12.3.3. BEST PRACTICE

Previous sections have shown that SVMs are a very flexible and robust tool for many pattern recognition tasks, including template matching. An important characteristic of SVMs is the limited number of choices that a practitioner has to perform in the design phase of an SVM classifier. Nevertheless, it is not unusual for an unfamiliar user to get less than optimal results when applying them, suggesting that a few practical hints may be useful. A common beginner approach is based on a three-step workflow:

1. Transform available data into the format required by the available SVM software.
2. Play with a few kernels and parameters.
3. Test.

This workflow is very simplistic and often leads to suboptimal results. Building on the theory presented and on a few simple considerations, the following six-step workflow generally provides good results:

1. Transform available data into the format required by the available SVM software.
2. Scale data.
3. Consider as a first testing kernel the RBF one $K(x, y) = e^{-\|\mathbf{x}-\mathbf{y}\|/2\sigma^2}$.
4. Use cross-validation to find the best values for parameter $C$ of the SVM and $\sigma$ of the Gaussian RBF kernel.
5. Use the values found in the previous step to train the classifier on the whole training set.
6. Test.

Let us now consider each single step. The input to an SVM is required to be a vector of real numbers. In some applications, the feature vectors include categorical values. A practical choice to code an $m$-category attribute is to expand it as a set of $m$ coordinates, representing the $i$th category with $(\delta_{1i}, \ldots, \delta_{mi})$ and not with the value $(i)$. Data scaling is important for two reasons:

1. To avoid dimensions (attributes) with more extended range dominating those with smaller ones.

2. To reduce numerical difficulties, particularly when linear or polynomial kernels are used.

A pragmatic choice is to normalize all coordinates in the range $[-1, 1]$. We have seen that many kernels can be used to build an SVM (they only need to be positive definite) and the suggestion to focus first on the RBF (Gaussian) kernel may seem limiting. There are, however, three important reasons that make it a very good starting choice:

1. If complete model selection is performed for the Gaussian kernel, i.e. the space $(C, \sigma)$ of its parameters is thoroughly explored, the linear kernel need not be considered as it is very well approximated by the Gaussian one for a proper choice of the model parameters.

2. It is characterized by a single parameter, hence less than a polynomial kernel.

3. It is numerically more stable than other kernels.

The choice of the model parameters $(C, \sigma)$ is extremely important and should be as exhaustive as possible (see Figure 12.7). A safe choice is to perform a grid search on the parameter space, evaluating the quality of each choice by means of cross-validation (see Appendix C.2). Exhaustive grid search is computationally expensive and can be improved in the following ways:

- use exponentially increasing intervals for the parameters instead of a uniform sampling;

- perform a coarse to fine search, progressively exploring parameter space in the neighborhood of the best position provided by the coarser grid;

- parallelize the computation.

Two important things on the use of cross-validation should be noted. Each cross-validation run uses only a fraction of the data, say $9n/10$ in the case of 10-fold cross-validation. At the end we will select for training the final classifier (with $n$ data) the best performing parameters (from classifiers with $9n/10$ data). It is important to note that they do not necessarily coincide, as smaller sets may require stronger regularization (such as a wider Gaussian kernel or a smaller $C$). Secondly, there could be a phase transition in the learning curve between $9n/10$ and $n$ where the generalization error as a function of the number of samples used changes significantly. Several free, efficient, software packages exist for the computation of SVMs.

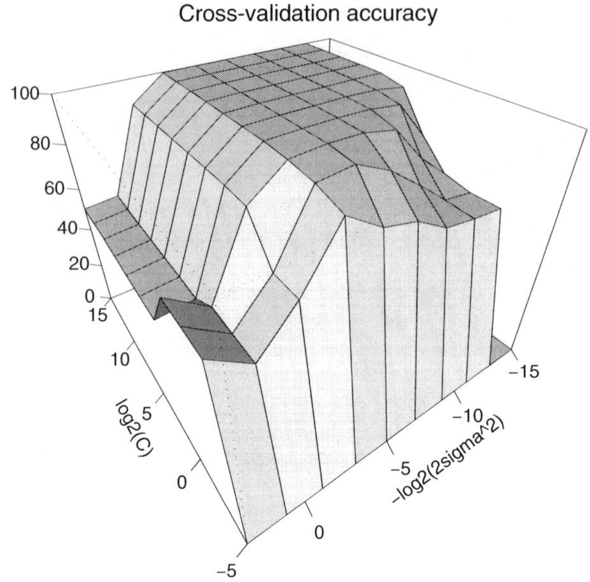


**Figure 12.7.** This shows the (cross-validation) accuracy of a Gaussian SVM for gender discrimination for different values of $C$ and of $(2\sigma^2)^{-1}$. These parameters have a significant impact on performance and an exhaustive search should be attempted whenever possible.

## 12.4. Bibliographical Remarks

A short, and dense, introduction to the concepts investigated in this chapter is the paper by Poggio and Smale (2003) while important technical results on regularization and learning can be found in Cucker and Smale (2002). A very detailed presentation of the regularization approach and its connection with SVMs is given in the paper by Evgeniou *et al.* (2000). The presentation of kernel methods is based strongly on material from the comprehensive book by Scholkopf and Smola (2002) and the more recent paper by Hofman *et al.* (2008).

RBFs are discussed in Poggio and Girosi (1990) and the gender classification example is from Brunelli and Poggio (1993). Regularization theory is not the only way to support vector machines, and a fairly recent account of the development from the original point of view can be found in the book by Vapnik (1998). A useful tutorial on SVMs is the paper by Burges (1998). An alternative approach to multicategory SVMs can be found in the paper by Lee *et al.* (2004) that also discusses the relation of SVMs to the Bayes classification rule. A geometrical perspective on SVMs is presented by Bennett and Bredensteiner (2000) and extended in Mavroforakis and Theodoridis (2006). The practical hints on successful use of SVMs are from Hsu *et al.* (2008). The results reported on SVM gender and race classification were obtained by using Chang and Lin (2001).

Applications of kernel methods to face analysis tasks are numerous. Gender detection is addressed by Moghaddam and Yang (2002) using SVMs while an early application of face detection can be found in Osuna *et al.* (1997) where an efficient training algorithm is introduced. The problem of multiview face detection and recognition is considered in several papers (Heisele *et al.* 2001; Li *et al.* 2001, 2004; Lu *et al.* 2003). An efficient architecture for face detection, based on the idea of a cascade classifier, is discussed in Romdhani *et al.* (2001). A more complex architecture based on a tree structure of SVMs exploiting a coarse to fine strategy is discussed in Sahbi and Geman (2006).

## References

Bennett K and Bredensteiner E 2000 Duality and geometry in SVM classifiers. *Proceedings of the 17th International Conference on Machine Learning (ICML'00)*, pp. 57–64.

Bottou L and Bengio Y 1994 Convergence properties of the k-means algorithm. *Proceedings of Advances in Neural Information Processing Systems,* vol. 7, pp. 585–592.

Brunelli R and Poggio T 1993 Caricatural effects in automated face perception. *Biological Cybernetics* **69**(3), 235–241.

Burges C 1998 A tutorial on support vector machines for pattern recognition. *Data Mining and Knowledge Discovery* **2**, 121–167.

Chang CC and Lin CJ 2001 *LIBSVM: A Library For Support Vector Machines*. Software available at http://www.csie.ntu.edu.tw/cjlin/libsvm.

Cucker F and Smale S 2002 Best choices for regularization parameters in learning theory: on the bias-variance problem. *Foundations of Computational Mathematics* **2**, 413–428.

Evgeniou T, Pontil M and Poggio T 2000 Regularization networks and support vector machines. *Advances in Computational Mathematics* **13**, 1–50.

Heisele B, Ho P and Poggio T 2001 Face recognition with support vector machines: global versus component-based approach. *Proceedings of the 8th International Conference on Computer Vision and Pattern Recognition (ICCV'01)*, vol. 2, pp. 688–694.

Hofman T, Scholkopf B and Smola A 2008 Kernel methods in machine learning. *Annals of Statistics* **36**, 1171–1220.

Hsu C, Chang C and Lin C 2008 A practical guide to support vector classification. Technical Report, Department of Computer Science, National Taiwan University.

Lee Y, Lin Y and Wahba G 2004 Multicategory support vector machines: theory and application to the classification of microarray data and satellite radiance data. *Journal of the American Statistical Association* **99**, 67–81.

Li S, Fu Q, Gu L, Scholkopf B, Cheng Y and Zhang H 2001 Kernel machine based learning for multi-view face detection and pose estimation. *Proceedings of the 8th International Conference on Computer Vision and Pattern Recognition (ICCV'01)*, vol. 2, pp. 674–679.

Li Y, Gong S, Sherrah J and Liddell H 2004 Support vector machine based multi-view face detection and recognition. *Image and Vision Computing* **22**, 413–427.

Lu J, Plataniotis K and Venetsanopoulos A 2003 Face recognition using kernel direct discriminant analysis algorithms. *IEEE Transactions on Neural Networks* **14**, 117–126.

Mavroforakis M and Theodoridis S 2006 A geometric approach to support vector machine (SVM) classification. *IEEE Transactions on Neural Networks* **17**, 671–682.

Moghaddam B and Yang MH 2002 Learning gender with support faces. *IEEE Transactions on Pattern Analysis and Machine Intelligence* **24**, 707–711.

Osuna E, Freund R and Girosi F 1997 Training support vector machines: an application to face detection. *Proceedings of the IEEE Conference on Computer Vision and Pattern Recognition (CVPR'97)*, pp. 130–136.

Poggio T and Girosi F 1990 Regularization algorithms for learning that are equivalent to multilayer networks. *Science* **247**, 978–982.

Poggio T and Smale S 2003 The mathematics of learning: dealing with data. *Notices of the AMS* **50**, 537–544.

Romdhani S, Torr P, Scholkopf B and Blake A 2001 Computationally efficient face detection. *Proceedings of the 8th International Conference on Computer Vision and Pattern Recognition (ICCV'01)*, pp. 695–700.

Sahbi H and Geman D 2006 A hierarchy of support vector machines for pattern detection. *Journal of Machine Learning Research* **7**, 2087–2123.

Scholkopf B and Smola A 2002 *Learning with Kernels*. MIT Press.

Scholkopf B, Burges C and Vapnik V 1996 Incorporating invariances in support vector machines. *Proceedings of the International Conference on Artificial Neural Networks*, Lecture Notes in Computer Science, vol. 1112, pp. 47–52. Springer.

Vapnik V 1998 *Statistical Learning Theory*. John Wiley & Sons, Inc.

# 13 FEATURE TEMPLATES

Your features! Lord warrant us! what features!

*As You Like It*
WILLIAM SHAKESPEARE

Many applications in image processing rely on robust detection of image features and accurate estimation of their parameters. Features may be too numerous to justify the process of deriving a new detector for each one. This chapter exploits the results presented in Chapter 8 to build a single, flexible, and efficient detection mechanism. The complementary aspect of detecting templates considered as a set of separate features will also be addressed and an efficient architecture presented.

## 13.1. Detecting Templates by Features

In spite of widespread use, there is not a universally accepted definition of what a feature is, and in the present chapter we will consider three pattern recognition tasks that illustrate different interpretations (and uses) of visual feature detection.

One of the main characteristics of the template matching approaches presented so far has been the fact that the template is considered as a single entity: in spite of its being built from many individual components (e.g. the pixels), nearly all of the techniques presented so far did not single out, at least explicitly, any subset of them. Even if there is no universal definition of what a feature is, a unifying aspect of the many available definitions is that a feature identifies a distinct(ive) part: feature-based approaches in computer vision de-emphasize the importance of the whole and focus more on a characterization of its constituent parts. The first example we consider is a rather extreme view of a visual feature and of its application to a template matching task: face detection. We will consider each pixel within a pattern as a distinct feature and we will use the resulting feature-based description to locate the associated templates. A couple of issues are immediately apparent:

1. The descriptive power of a pixel feature is very low and its value depends significantly on illumination.

2. We switch from the processing of a single, but complex, entity, the template, to a large number of (simpler) entities, the pixels: the impact on the efficiency and effectiveness of the template matching process is not obvious.

The first problem, descriptiveness and stability of pixel features, is usually addressed in computer vision by means of neighborhood operations. The original value of the pixel, representing image irradiance, is replaced by one (or more) values computed from its

*Template Matching Techniques in Computer Vision: Theory and Practice* Roberto Brunelli

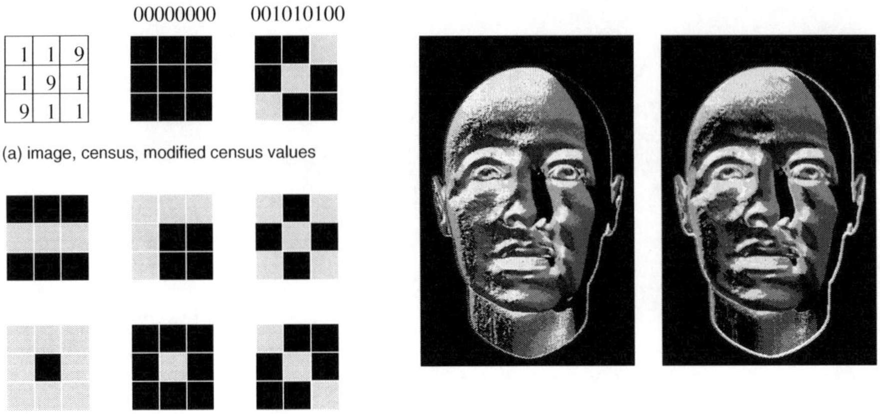


(a) image, census, modified census values

(b) sample modified census kernels

(c) standard (left) and modified (right) census transforms

**Figure 13.1.** The features identified by the modified census transform (a,b) provide an image description that is robust to illumination variations (c).

neighborhood, its feature value, an abstraction of original image information. This mapping may be constructed to highlight specific local image characteristics or structures (such as edges, corners, ridges), providing information that can be extracted in a reproducible way from multiple images. We will address the former type of mapping in Section 13.2 and we focus now on the latter. We have already seen in Section 5.2.1 that it is possible to obtain an image representation with reduced sensitivity to illumination by means of the census transform, a local (or neighborhood) transform. Besides being invariant to monotone intensity transformation, the value of the census transform provides a compact description of the $3 \times 3$ neighborhood of a pixel with a single 8-bit number. The descriptiveness of the transform can be enhanced by considering the central pixel as well, as reported in Equation 5.33, resulting in a 9-bit value (see also Figure 13.1). If we do not pack the information into a single value but build instead the corresponding binary image of the neighborhood, we see that the value of the modified census transform can be considered as the index of a dictionary of 512 local image structures, including edges, corners, junctions, and so on. Using a number of features equal to the number of pixels of the template clashes with the philosophy of feature-based approaches that aims at reducing the number of variables needed to describe data, striving for simpler models that can be handled efficiently. The task we are currently considering, that of face detection, requires a parsimonious number of features to improve computational efficiency, but at the same time also requires that the performance of the system be good enough for practical applications. The fact that the features we are considering are very simple is expected to have a positive impact on efficiency and a negative one on effectiveness. Is there an efficient and effective way to distill the most relevant pixel features in order to obtain a good face detection system? The answer is affirmative and it is based on two important facts:

1. The possibility of building an efficient template detection system as a cascade of fast rejectors, i.e. classifiers that reliably discard non-matching patterns, deferring (progressively more) difficult decisions to later stages.

2. The existence of an efficient way of boosting the performance of weak classifiers, with an error rate only marginally better than random guessing, into a strong classifier.

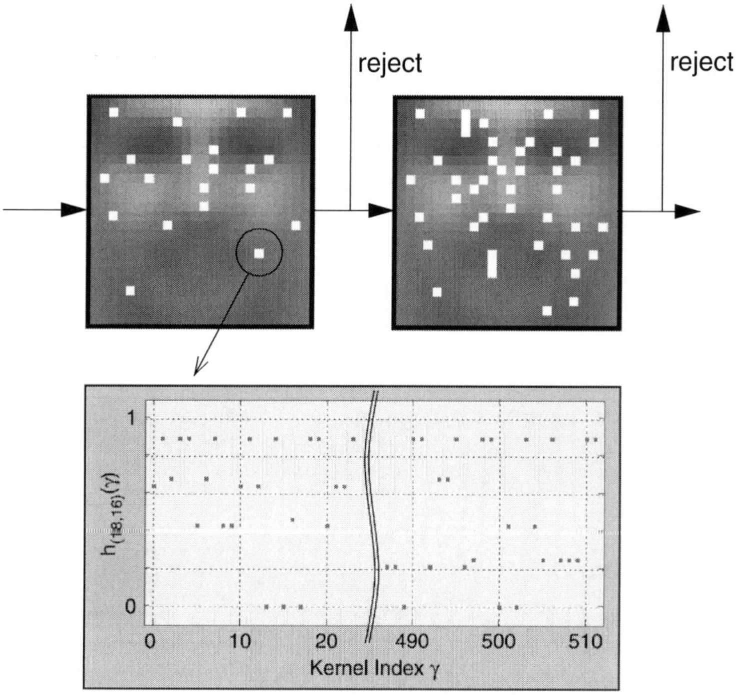


**Figure 13.2.** The architecture of a cascade template detection system based on boosting multiple pixel-level classifiers. Source: adapted from Küblbeck and Ernts (2006).

The architecture of a cascade template detection system is shown in Figure 13.2. The basic idea is to eliminate from further consideration, and as soon as possible, all patterns that cannot possibly match the template. The progressive reduction of patterns along the cascade allows the use of more complex rejectors and, finally, of a fully fledged detector that will have to decide upon a very small set of patterns. If we denote by $f_i$ the false positive rate of the rejector at the $i$th level of the cascade, the false positive rate $F$ of the whole chain of $K$ rejectors is

$$F = \prod_{i=1}^{K} f_i \tag{13.1}$$

and similarly for the detection rates $D$ and $d_i$

$$D = \prod_{i=1}^{K} d_i. \tag{13.2}$$

Achieving a final detection rate of 0.9 with a false positive rate of the order of $10^{-5}$with a 10-stage cascade requires that $d_i \approx 0.99$ (since $0.99^{10} \approx 0.9$), a not too difficult value to achieve considering that the requirements on the false positive rate are modest: $f_i \approx 0.3$ (since $0.3^{10} \approx 6 \times 10^{-6}$).

The next step is that of building the actual rejectors using the simple features provided by the modified census transform. As the rejectors must be very efficient, they should have a simple structure and use as few features as possible. One way to construct a rejector satisfying these constraints is to build it incrementally by aggregating weak but simple rejectors. A very effective algorithm for building this kind of integrated classifier is AdaBoost, a meta algorithm that builds a strong classifier as a weighted combination of weak, low-performance ones. The idea upon which AdaBoost itself is based is not too dissimilar from the idea of cascade classifiers: the final classifier is built by an iterative process that changes the weight given to the training samples at each iteration, increasing the importance of those that were not classified correctly. After $T$ iterations, the classifiers obtained at each iteration are combined into a single classifier $H_T$ by means of a weighted average (see Equation 13.7). Algorithm 13.1 is a description of AdaBoost for a binary classification task.

**Algorithm 13.1**: AdaBoost

**Data**: A set of training samples $(\boldsymbol{x}_i, y_i)$, $i = 1, \ldots, N$, where $\boldsymbol{x}_i \in X$, $y_i \in Y = \{-1, 1\}$, the maximum number of iterations $T$.

**Initialization**: $D_1(i) = N^{-1}, i = 1, \ldots, N$

**for** $t = 1, \ldots, T$ **do**

find the classifier $h_t : X \to \{-1, 1\}$ that minimizes the error with respect to the weight distribution $D_t$:

$$h_t = \operatorname*{argmin}_{h_j \in \mathfrak{H}} \epsilon_j \tag{13.3}$$

$$\epsilon_j = \sum_{i=1}^{N} \{D_t(i) : y_i \neq h_j(\boldsymbol{x}_i)\} \tag{13.4}$$

**if** $\epsilon_t \geq 0.5$ **then**

stop;

**else**

$$\alpha_t = \frac{1}{2} \ln \frac{1 - \epsilon_t}{\epsilon_t} \tag{13.5}$$

$$D_{t+1}(i) = \frac{D_t(i) e^{-\alpha_t y_i h_t(\boldsymbol{x}_i)}}{\sum_j D_t(j) e^{-\alpha_t y_j h_t(\boldsymbol{x}_j)}} \tag{13.6}$$

**end**

**end**

**return** the final classifier

$$H(\boldsymbol{x}) = \operatorname{sign}\left(\sum_{t=1}^{T} \alpha_t h_t(\boldsymbol{x})\right) \tag{13.7}$$

There are a few important remarks to make on AdaBoost. The update factor of the weight of a misclassified sample is greater than 1: this focuses classifiers of subsequent rounds on the difficult cases. The attentive reader will have noticed that Equation 13.7 is very similar to the equations reported in Chapter 12: the solution is a linear combination of functions, in this case classifiers. As a matter of fact, AdaBoost is a greedy optimization algorithm minimizing yet another (convex) loss function

$$C(\boldsymbol{x}) = e^{-yf(\boldsymbol{x})} \tag{13.8}$$

over a convex set of functions, in this case linear combinations of classifiers. The similarities with SVMs do not end here: it is possible to prove bounds on the generalization error of the classifier produced by AdaBoost

$$P(H_T(\boldsymbol{x}) \neq y(\boldsymbol{x})) \leq P_{\text{emp}}(H_T(\boldsymbol{x}) \neq y(\boldsymbol{x})) + O\left(\sqrt{\frac{T d_{VC}}{N}}\right) \tag{13.9}$$

where $N$ is the number of samples, $d_{VC}$ the VC dimension of the weak hypothesis space (see Intermezzo 13.1), and $P_{\text{emp}}(\cdot)$ represents the estimate derived from the training set.

**Intermezzo 13.1.** The VC dimension

The VC dimension is a measure of the complexity of a space of hypotheses. Formally, let $\{f_\alpha\}$ be a set of functions $f_\alpha : \mathbb{R}^n \to \mathbb{R}$. Given a set of $l$ points in $\mathbb{R}^N$, let us consider all the possible $2^l$ assignments of these points to two classes. If for each assignment there exists a decision function $f_{\alpha_0}$ that respects the division among the classes, the set of points is said to be shattered by the set of functions.

**Definition 13.1.1.** *The VC dimension of the set of functions $\{f_\alpha\}$ is the maximum number of points that it can shutter.*

The bound suggests that increasing the number of iterations $T$ may lead to overfit, due to the fact that many classifiers are combined. Furthermore, it is possible to define the margin $\rho_{AB}$ for a training sample as

$$\rho_{AB}(\boldsymbol{x}, y) = \left(\frac{\sum_t \alpha_t h_t(\boldsymbol{x})}{\sum_t \alpha_t}\right) \in [-1, 1] \tag{13.10}$$

which is positive only when the sample is correctly classified: the closer it is to 1, the higher the confidence in the classification, and to prove that

$$P(H_T(\boldsymbol{x}) \neq y(\boldsymbol{x})) \leq P_{\text{emp}}(\rho_{AB} \leq \theta) + O\left(\sqrt{\frac{d_{VC}}{N\theta^2}}\right) \tag{13.11}$$

for any $\theta > 0$ with high probability. An important characteristic of AdaBoost is that it strives to improve the margin even when the training error has reached zero: the above bound, which does not depend on the number of rounds, suggests that AdaBoost automatically controls overfitting by direct margin maximization: this is possible because the exponential loss function is not zero when the training error is zero, but decreases with increasing confidence in the classification, i.e. with increasing margin.

The last step in the pixel-level approach to face detection is the design of the weak classifier space $\mathfrak{H}$ from which AdaBoost extracts its weak classifiers. Following the original design that associates a feature to each pixel, we associate a classifier to each feature: it will be the task of AdaBoost to combine many pixel-level (weak) rejectors into a strong rejector.

A simple way to construct a pixel classifier is to look at the distribution, over face and non-face samples, of the values returned by the modified census transform at that position:

whenever a feature turns out to be more frequent over face samples, the pixel classifier will output a face label, otherwise it will output a non-face label. Such a classifier requires a lookup table (with 511 entries) for each pixel position within a face template, storing for each entry the label to be output when the corresponding feature is detected. Classification is then very fast: the census transform value is used to index the lookup table from which the output label is read. The requirements of AdaBoost of changing the sample weights can be easily accommodated as shown in the description of Algorithm 13.2, which provides the best pixel rejector for the weight distribution $D_t$ at round $t$ of the AdaBoost procedure.

---

**Algorithm 13.2**: Construction of the weak, pixel-level, face classifier

---

**Data**: A set of samples $(I_{C,i}, c_i)$, $i = 1, \ldots, N$, where $I_{C,i}$ denotes the census-transformed images (of homogeneous size) and $c_i \in \{-1, 1\}$ represents the corresponding class ($-1$: non-face, 1: face); the distribution of sample weights $D_t$ from AdaBoost; the set $S$ of pixels whose corresponding classifiers have already been incorporated by AdaBoost; and the maximum number $n_{\max}$ of different pixels to use.

**foreach** *pixel* **do**

generate the weighted frequency tables of the census-transformed images for the face/non-face classes:

$$b_t^{-1}(\boldsymbol{x}, l) = \sum_{i:c_i=-1, I_C(\boldsymbol{x})=l} D_t(i) \tag{13.12}$$

$$b_t^{1}(\boldsymbol{x}, l) = \sum_{i:c_i=1, I_C(\boldsymbol{x})=l} D_t(i) \tag{13.13}$$

compute the corresponding error $\epsilon_t(\boldsymbol{x})$

$$\epsilon_t(\boldsymbol{x}) = \sum_l \min\{b_t^{-1}(\boldsymbol{x}, l), b_t^{1}(\boldsymbol{x}, l)\} \tag{13.14}$$

**end**

select the best weak classifier

$$h_t(\boldsymbol{x}) = \text{sign}[b_t^{1}(\boldsymbol{x}_t, I_c(\boldsymbol{x})) - b_t^{-1}(\boldsymbol{x}_t, I_c(\boldsymbol{x}))]$$

where

$$\boldsymbol{x}_t = \begin{cases} \text{argmin}\,[\epsilon_t(\boldsymbol{x})] & \text{if } |S| < n_{\max} \\ \underset{\boldsymbol{x}\in S}{\text{argmin}}\,[\epsilon_t(\boldsymbol{x})] & \text{otherwise.} \end{cases} \tag{13.15}$$

---

A few concluding steps remain. The cascade structure requires progressively more sophisticated rejectors as they are facing more and more difficult decisions: this can be obtained by augmenting the number of weak classifiers combined into the strong one. This can be achieved by increasing the maximum number of AdaBoost rounds and the maximum number of pixel classifiers used (as described in Algorithm 13.2). The final step is that of optimizing the response of the strong classifier $H_s(x)$ at the $s$th cascade stage to maximize its detection rate at the expense of its rejection capability. This can be achieved by selecting a threshold $\theta_s$ for each stage of the cascade so that the classifier given by

$$H(x) \underset{\text{non-face}}{\overset{\text{face}}{\gtrless}} \theta_s \tag{13.16}$$

operates in the correct regime to satisfy the constraints of Equations 13.1–13.2. Note that changing the classifier of Equation 13.7 into that of Equation 13.16 does not necessarily preserve all the nice properties of the original AdaBoost classifier.

## 13.2. Parametric Feature Manifolds

Many computer vision applications rely on the robust detection of specific image features and of their parameters. One important example is that of step edge detection (see Chapter 7) for which we would also like to know orientation and contrast. Edges are not the only important feature, others being lines, corners, roof edges, etc. Many applications, especially in the field of industrial vision and vision-based quality control, need to locate task-specific features. Designing an effective feature detector is often a difficult task requiring careful modeling of the feature and the development of an efficient detection algorithm that, besides detecting it, also provides a reliable estimate of the main feature parameters. The development of a single detection algorithm that can be applied without significant modifications to any parameterized feature, providing accurate estimates of feature parameters, would clearly help. The framework of principal component analysis described in Chapter 8 provides a simple solution that can be implemented through the following steps:

1. Create an accurate generative model of the feature.
2. Sample the feature parameter space generating an example, i.e. a representative image, of each sample point.
3. Perform PCA on the resulting set of images.
4. Detect instances of the feature using the techniques described in Chapter 8.
5. Determine the associated feature parameters finding the most similar ones among the generated samples.

Let us consider all of the steps in detail. The development of an effective appearance-based feature detector requires accurate representative images. The description of the imaging process presented in Chapter 2 suggests that two important effects should be considered in building a generative feature model:

1. Image blurring due to out of focus imaging and diffraction effects.
2. Discrete sampling of scene irradiance.

The first effect can be modeled accurately by means of convolution of the signal $T^0$, corresponding to that of an idealized pinhole camera, with a Gaussian kernel with an appropriate choice of $\sigma$. As pointed out in Chapter 7, the same convolution is able to model the gradual nature of many edges found in natural images. The discrete sampling process can be easily simulated by integrating the image signal over an area corresponding to a single pixel and sampling the resulting image with a bidimensional Dirac comb (see the description of the sampling theorem in Section 2.4.1):

$$T(\boldsymbol{x}; \boldsymbol{q}) = \left\{ \left[ (T^0(x_1, x_2; \boldsymbol{q}) * N(\boldsymbol{x}; \sigma)) * \frac{1}{w_{x_1} w_{x_2}} \sqcap \left( \frac{x_1}{w_{x_1}}, \frac{x_2}{w_{x_2}} \right) \right] * \delta_{\boldsymbol{V}}(\boldsymbol{x}) \right\} \tag{13.17}$$

where $\boldsymbol{q} \in \mathbb{R}^{d_q}$ represents the feature parameters. We can obtain a discrete image by substituting $\boldsymbol{x}$ in Equation 13.17 with a vector of integer coordinates $\boldsymbol{i}$. Let us now focus on the modeling of a specific feature, namely corners, such as the one represented in Figure 13.3. The feature is characterized by five parameters:

- $A$ and $B$, representing the brightness values inside and outside the corner;
- $\theta_1$ and $\theta_2$, describing the orientation of the corner with respect to the positive $x$ semiaxis and the angular width of the corner;
- $\sigma$, jointly modeling the effects of blurring and the graduality of the transition from $A$ to $B$.

The number of parameters can be reduced to three if we normalize our feature templates to zero average and unit variance. The average intensity value $\mu_T$ and variance $\sigma_T^2$ can be expressed as a function of $A$ and $B$ as

$$\mu_T = \frac{1}{\pi r^2}[B(r\theta_2)^2 + (2\pi - r\theta_2)^2 A] \tag{13.18}$$

$$\sigma_T^2 = \frac{(r\theta_2)^2 B^2 + (2\pi - r\theta_2)^2 A^2}{\pi r^2} - \mu_T^2 \tag{13.19}$$

so that we may restrict ourselves to generate zero-average, unit-variance samples: the same normalization will be applied to the image windows analyzed by the detector and the actual values of $A$ and $B$ can be recovered by solving Equations 13.18–13.19. The way we sample the now three-dimensional parameter space is important: we would like to model accurately the distribution of patterns but, at the same time, we want to use as few samples as possible in order to speed up parameter determination that requires finding the sample most similar to the image patch we need to characterize. A simple, albeit suboptimal, solution is to fix the maximum number of samples $N_0$ we can afford to search, and sample each parameter with a density proportional to its variability:

1. Sample a small set of points $\{\boldsymbol{q}_i\}$, $i = 1, \ldots, k$, in parameter space.
2. Estimate the average rate of change of the template with respect to the $j$th parameter separately:

$$\delta_j(T) = \frac{1}{k}\sum_i \left\| T(q_{i,1}, \ldots, q_{i,j}, \ldots, q_{i,d_q}) - T\left(q_{i,1}, \ldots, q_{i,j} + \frac{\Delta_j}{m}, \ldots, q_{i,d_q}\right)\right\|_2 \tag{13.20}$$

where $\Delta_j$ is the range of the $j$th parameter and $m$ is a heuristically chosen maximum number of samples;

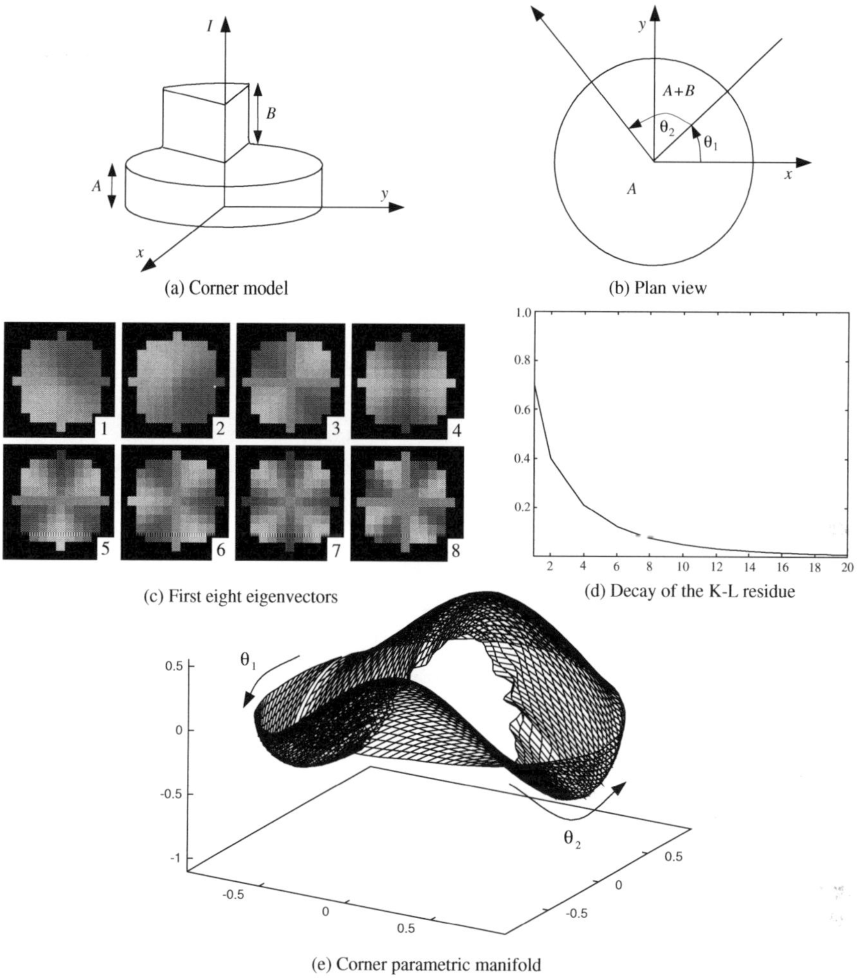


**Figure 13.3.** The model of a corner. The lowest picture shows the three-dimensional slice of the manifold spanned by the principal components of corner images when varying $(\theta_1, \theta_2)$. The normalized average approximation error (see Equation 8.9) from the Karhunen–Loeve transform (or PCA) is shown in (d). Source: Baker *et al.* (1998).

3. Sample the range of parameter $q_j$ with $n_j = n\delta_j(T)$ samples where

$$n = \left( \frac{N_0}{\prod_j \delta_j(T)} \right)^{1/d_q}. \tag{13.21}$$

Parameters that influence more the appearance of the feature are then sampled more frequently. The usefulness of principal components in template matching tasks is related to the possibility of finding a low-dimensional linear subspace where the patterns can be described. This can be easily verified for the specific feature considered: the first 10 components capture more than 90% of the complete variance. Note, however, that the distribution of the patterns within this linear subspace is not necessarily dense: they are distributed over a manifold that is well embedded in $\mathbb{R}^{10}$ but does not necessarily fill it completely. An important consequence, already remarked upon in Chapter 8, is that a pattern not belonging to the feature manifold but lying in the linear subspace where the manifold is embedded can be reconstructed well.

In order to classify an unknown pattern as a corner instance we need to know its distance from the manifold and not from the linear space. This can be achieved with a two-step process: we first check that the pattern lies in the correct subspace and, if it does, we proceed with the computation of its distance from the manifold. The first step can be performed by fully expanding the pattern onto the subspace and checking for the reconstruction residual, but faster solutions exist and will be addressed in the next section. Assuming that the manifold has been sampled densely enough, the distance of a pattern from the manifold is approximately given by the distance of the pattern from the nearest manifold sample, computed using the principal component expansion. A direct approach, even when the dimensionality is small, is computationally very expensive. A practical way to speed up the comparison by several orders of magnitude is to sample the manifold in a coarse to fine way as described in Algorithm 13.3 (see also Figure 13.4). The maximum number of levels depends on the resolution at which the manifold has been sampled: for a three-parameter manifold sampled with $10^5$ points setting $n_l = 4$ may result in a 100 times speed-up factor.

---

**Algorithm 13.3**: Coarse to fine manifold matching.

**Data**: A pattern $\boldsymbol{x}$ and a sampled manifold $\boldsymbol{M}(q_1, \ldots, q_d)$
**for** $l = 0, \ldots, n_l$ **do**

$$\hat{\boldsymbol{\imath}}_l = \operatorname*{argmin}_{|i_l - 2\hat{\imath}_{l-1}| \leq 1} |\boldsymbol{x} - \boldsymbol{M}(2^{n_l - l}(i_1, \ldots, i_d))|$$

$$\hat{\Delta}_l = \min_{|i_l - 2\hat{\imath}_{l-1}| \leq 1} |\boldsymbol{x} - \boldsymbol{M}(2^{n_l - l}(i_1, \ldots, i_d))|$$

**end**
**return**

$$(\hat{\Delta}_{n_l}, \hat{\boldsymbol{\imath}}_{n_l})$$

---

## 13.3. Multiclass Pattern Rejection

The idea of pattern rejection underlying the usage of the cascade classifier presented in Section 13.1 is a very general one and can be applied with success in conjunction with PCA. When the projection of a pattern onto the linear subspace $L$ determined by the first $k$ PCA components is a good approximation, the reconstruction error, corresponding to the projection onto the orthogonal subspace $L^\perp$, is small. This means that the projection of the pattern along any direction lying in the orthogonal subspace is small. This immediately suggests an efficient strategy for rejecting the pattern: if its projection onto $L^\perp$ is large, it can be safely rejected. This insight can be formalized and generalized to multiclass template detection with the introduction of composite rejectors.

Let $\{C_i\}, i = 1, \ldots, n_c$, be a set of disjoint classes and let $C(\boldsymbol{x})$ be the function returning the class to which $\boldsymbol{x}$ belongs. The idea of rejector is formalized by the following definition.

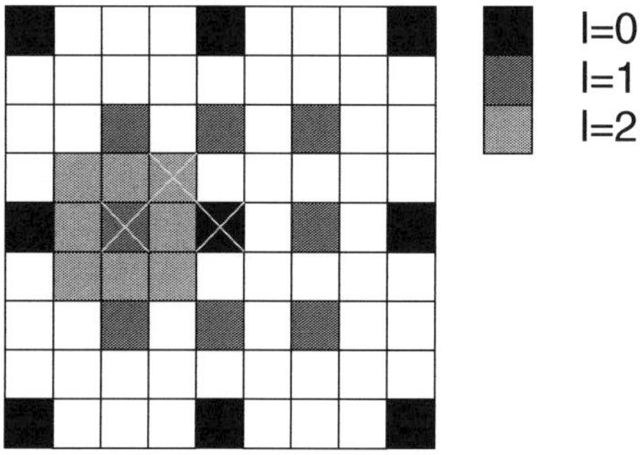


**Figure 13.4.** Coarse to fine sampling of a bidimensional manifold. The search for the manifold sample closest to a given pattern is performed using different sampling rates: when the manifold structure is smooth, significant speed-ups can be achieved without missing the correct match. The points marked with a cross represent the closest manifold sample at the different levels.

**Definition 13.3.1.** *A rejector is a function $\psi(\boldsymbol{x})$ that for any pattern $\boldsymbol{x}$ returns a set of labels $\psi(\boldsymbol{x})$ such that $\boldsymbol{x} \in C_i$ implies $i = C(\boldsymbol{x}) \in \psi(\boldsymbol{x})$, or, equivalently, $i \notin \psi(\boldsymbol{x})$ implies $\boldsymbol{x} \notin C_i$.*

The justification of the name rejector can be found in the closing part of the definition: if $i$ is not in the output of the rejector we can eliminate the class. However, $i \in \psi(\boldsymbol{x})$ does not imply that $\boldsymbol{x} \in C_i$. A composite rejector is a set of rejectors $\{\psi_r\}$ such that:

1. There is a rejector designed to be applied to the complete subset of classes.
2. For any rejector $\psi_r$ in the composite rejector there is at least a rejector $\psi_s$ designed to be applied to the classes identified by $\psi_r(\boldsymbol{x})$.

The second property is the more important as it is the key to successful application of the composite rejector: for any vector $\boldsymbol{x}$ we may apply a cascade of rejectors ending up in a single class label, i.e. the classification of the template, or in a set of class labels that cannot be further reduced.

The possibility of constructing an effective composite rejector depends on the structure of the underlying classes. In many practical cases, PCA provides a good approximation of the space spanned by the elements of class $C_i$

$$\|\boldsymbol{x} - (\boldsymbol{c}_i \oplus L_i)\| \leq \delta_i \quad \forall \boldsymbol{x} \in C_i \tag{13.22}$$

where $\boldsymbol{c}_i$ represents the average element of $C_i$ and $\oplus L_i$ represents a linear combination of vectors in the subspace $L_i$ given by the principal directions associated with class $C_i$. Let us now consider a unit vector $\hat{\boldsymbol{r}} \perp \bigoplus_{i=1}^{n_c} L_i$. As $\hat{\boldsymbol{r}}$ is of unitary norm, under the conditions of Equation 13.22 we have

$$\|\boldsymbol{x} - (\boldsymbol{c}_i \oplus L_i)\| \leq \delta_i \Rightarrow |\hat{\boldsymbol{r}} \cdot (\boldsymbol{x} - (\boldsymbol{c}_i \oplus L_i))| \leq \delta_i \tag{13.23}$$

while the fact that $\hat{\boldsymbol{r}}$ is perpendicular to all class subspaces $L_i$ ensures that

$$|\hat{\boldsymbol{r}} \cdot (\boldsymbol{x} - (\boldsymbol{c}_i \oplus L_i))| \leq \delta_i \Rightarrow |\hat{\boldsymbol{r}} \cdot \boldsymbol{x} - \hat{\boldsymbol{r}} \cdot \boldsymbol{c}_i| \leq \delta_i \quad \forall \boldsymbol{x} \in C_i. \tag{13.24}$$

The projections of vectors belonging to class $C_i$ are confined to a small interval of the line identified by $\hat{\boldsymbol{r}}$: whenever a vector projects outside the interval corresponding to class $C_i$ we

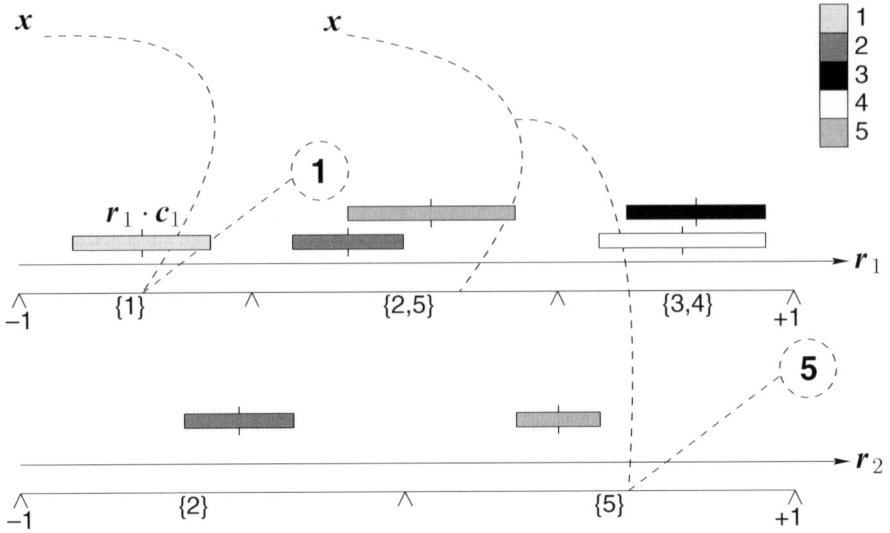


**Figure 13.5.** The action of a composite rejector. The pattern to be classified is projected in the direction of the first rejector. The range is partitioned into intervals compatible with different subsets of classes. Depending on the value of the projection, the set of candidate classes is restricted and additional rejectors may be activated progressively constraining the classification of $\boldsymbol{x}$.

may safely say that it cannot belong to class $C_i$ (see Figure 13.5). Since we want the rejector to be as effective as possible we must choose $\hat{\boldsymbol{r}}$ so that the projections of $\boldsymbol{c}_i$ on it are as spread out as possible: $\hat{\boldsymbol{r}}$ must be chosen as the first principal component of the set $\{\boldsymbol{c}_i\}$ projected on $\left(\bigoplus_{i=1}^{n_c} L_i\right)^{\perp}$. The procedure resembles Fisher linear discriminant analysis:

1. Working in the space orthogonal to the class subspaces minimizes within-class scatter.
2. Spreading out the projection of class centers maximizes between-class scatter.

Figure 13.5 presents a typical situation: some of the classes can be separated with the help of a single rejection vector while others overlap. For each set of overlapping classes we may repeat the above procedure using the corresponding linear subspaces: a specialized rejector is computed for each subset. A complete composite rejector can then be developed by iterating the steps described. Even if the task of a rejector is not to classify a pattern, the class structure may allow it to achieve classification with improved efficiency and better performance than dedicated approaches such as Fisher linear discriminant analysis. Rejection can also be exploited to improve the efficiency of expansion-based matching as described by Intermezzo 13.2.

## 13.4. Template Features

The features we considered in the previous sections were very local ones, little more than pixels in the case of face detection using the census transform and small image patches in the case of corner detection. Nothing prevents us from using larger features and, in fact, this is what we do when we split faces into eyes, nose, and mouth in the people recognition

**Intermezzo 13.2.** Rejection with progressive subspace expansion

The idea of rejection can be applied fruitfully in expansion-based matching. Instead of computing the full expansion on the principal subspace identified by PCA, and applying the classification step at the end, we compute a sequence of approximations

$$x_0 = c_i \tag{13.25}$$

$$x_l = x_{l-1} + (x \cdot e_l)e_l \tag{13.26}$$

$$\Delta_l = \|x_l - x\|_{L_1} \tag{13.27}$$

applying rejection after each step using precomputed bounds on $(x \cdot e_l)$ and $\Delta_l$.

In a similar way, the results presented in Section 12.3.1 can be used to implement a cascade SVM template detector. In order to build it, we must be able to control its detection performance at the different levels of the cascade. In an SVM this flexibility is offered by the $b$ parameter that effectively represents a decision threshold: by increasing (decreasing) it, we become more (less) conservative in accepting a pattern as belonging to the $+1$ $(-1)$ class. If we fix the overall performance of the detection system by selecting a point $(F_p, T_p)$ on its ROC curve (see Section C.3), we may propagate the corresponding performance requirements through the cascade. In the case of an $N_z + 1$ cascade classifier, where level $N_z + 1$ corresponds to using the complete SVM and level $m \leq N_z$ uses a reduced set of $m$ elements, we set a target false positive rate for level $m$ as $(f^{1/N_z})^m$ and we fix threshold $b_m$ at the minimum value for which the reduced classifier

$$\operatorname{sign}\left(\sum_{j=1}^{m} \beta_{m,j} K(x, z_j) + b_m\right) \tag{13.28}$$

is able to meet the constraint. The strategy can be quite effective and the literature reports gains of up to three orders of magnitude with respect to the use of the full SVM with negligible performance penalty (Romdhani *et al.* 2001).

system described in Chapter 14. The main reason for switching from a single face template to the use of multiple face regions represented by the corresponding templates is increased robustness. If one of the features changes, e.g. due to eyeglasses or beard, the matching of the remaining features would be unaffected and the performance of the biometric system would be only slighted degraded. On the contrary, if a single template is used for the whole face, changes in any of the facial features would result in a very low matching score, severely degrading system performance. A component-based representation of a face offers additional advantages, most notably an increased robustness to changes in pose (see Figure 13.6).

This suggests that a component-based representation can be useful for increasing the tolerance of face detection and recognition systems to pose changes. This insight is verified by several experiments reported in the literature and the reader can find the references in Section 13.5.

The workings of a multicomponent-based detection system differ from that of a single component one. Adopting a sliding window approach, for each position of the analysis window $W$ over the image we have multiple scores, and we get the candidate position of each component and the confidence of its localization

$$x_\alpha = \operatorname*{argmax}_{x \in W} \ p_\alpha(x) \tag{13.29}$$

$$\hat{p}_\alpha = \max_{x \in W} p_\alpha(x). \tag{13.30}$$

The resulting set

$$\{(x_\alpha, \hat{p}_\alpha)\}, \quad \alpha = 1, \ldots, n_c \tag{13.31}$$

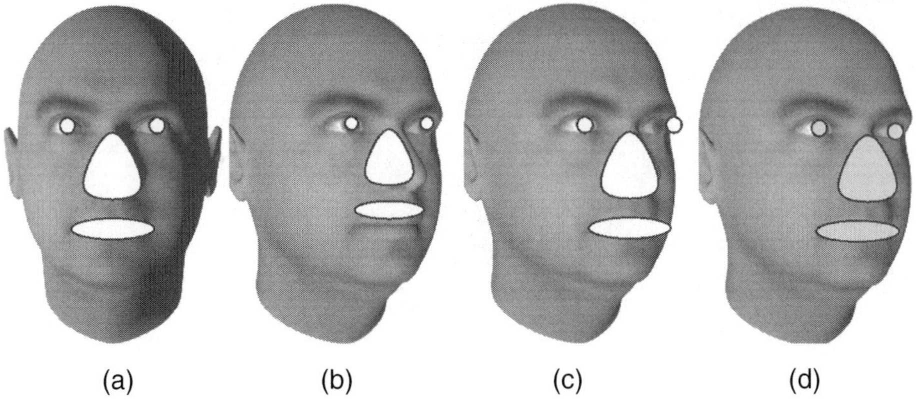


**Figure 13.6.** A rigid configuration of features is more sensitive to the geometrical deformations resulting from pose changes. The example in (b) shows the effect of rigid scaling of the configuration to match the eye distance while the one in (c) aligns the feature set on one of the eyes. The example in (d) shows that separate feature alignment provides a better result.

is called a constellation. The template is detected by evaluating the goodness of the constellation. The evaluation can be a very simple aggregation of the confidence of localization of each single component, e.g. summing all the values, or can be provided by a more sophisticated classifier, e.g. a previously trained SVM. A possible drawback of using a component-based face representation is that when working at medium to low resolution the matching scores are not as selective as one would like: the distribution of scores often presents local maxima where the corresponding feature is not present and the corresponding values may be higher than those at the correct feature location. The reduction in discriminative power is due to the small size of the feature templates and to the increased possibility of unrelated local image features matching them well. A solution to the problem is to exploit configural feature information. The structure of a face constrains the relative positions of the features: even if the localization of each single feature is uncertain, a reliable localization of the complete feature set may still be possible. The reason is that the location of each feature limits the regions where the other features may be located, providing a kind of prior probability distribution for them (Figure 13.7). If we consider the response of a template matcher for feature $\alpha$ as the probability $p_\alpha(\boldsymbol{x})$ of the template being located at the corresponding position, we may incorporate the constraints deriving from another feature $\beta$ in the following way:

$$p_\alpha(\boldsymbol{x}) \rightarrow p'_\alpha(\boldsymbol{x}) = p_\alpha(\boldsymbol{x}) p_\alpha(\boldsymbol{x}|\boldsymbol{x}_\beta). \tag{13.32}$$

The above equation, however, cannot be used directly as we do not yet know $\boldsymbol{x}_\beta$ and it requires knowledge of the conditional probability distribution $p_\alpha(\boldsymbol{x}|\boldsymbol{x}_\beta)$. The latter can be easily estimated from the training data by building a set of bidimensional histograms $H_{\beta,\alpha}$

$$H_{\beta,\alpha}(\boldsymbol{x}) = p_{\alpha,\beta}(\boldsymbol{x} = \boldsymbol{x}_\alpha - \boldsymbol{x}_\beta) \tag{13.33}$$

which count how many times feature $\alpha$ was found at displacement $\boldsymbol{x}$ from feature $\beta$. In many cases the number of samples is not enough to provide a smooth estimate of $p_{\alpha,\beta}$; it is then necessary to smooth $H_{\beta,\alpha}(\boldsymbol{x})$ with a small Gaussian kernel renormalizing the convolution

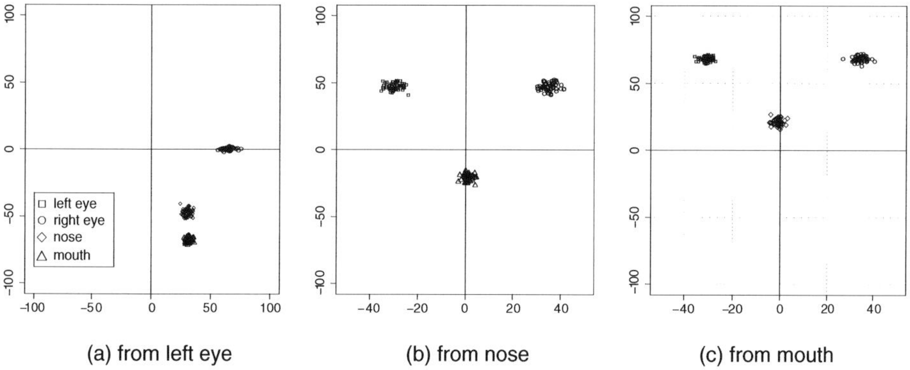


**Figure 13.7.** The spatial distribution of the main facial features assuming one of them as the coordinates origin: smoothing the resulting discrete distribution provides an estimate of $p_{\alpha,\beta}(\boldsymbol{x} = \boldsymbol{x}_\alpha - \boldsymbol{x}_\beta)$ (see Equation 13.33).

result to ensure that

$$\sum_{\boldsymbol{x}} H_{\beta,\alpha}(\boldsymbol{x}) = 1 \tag{13.34}$$

or

$$\max_{\boldsymbol{x},\beta} H_{\beta,\alpha}(\boldsymbol{x}) = 1. \tag{13.35}$$

We now need to address the facts that we do not know $\boldsymbol{x}_\beta$ and that we would like to condition each feature on all the other ones. The first problem can be solved by conditioning on the most plausible locations of feature $\beta$ as determined by the set of the $l$th most prominent local maxima of $p_\beta(\boldsymbol{x})$, $\{\hat{\boldsymbol{x}}_{\beta,(-i)}\}$, $i = 1, \ldots, l$, where the negative rank notation means that values are sorted by decreasing probability value:

$$p_\alpha(\boldsymbol{x}) \to p'_\alpha(\boldsymbol{x}) = p_\alpha(\boldsymbol{x}) \sum_{i=1}^{l} p_\alpha(\boldsymbol{x}|\hat{\boldsymbol{x}}_{\beta,(-i)}). \tag{13.36}$$

The second problem can be solved by assuming that each feature $\beta$ conditions feature $\alpha$ independently

$$p_\alpha(\boldsymbol{x}) \to p'_\alpha(\boldsymbol{x}) = p_\alpha(\boldsymbol{x}) \prod_{\beta\neq\alpha} \left( \sum_{i=1}^{l} p_\alpha(\boldsymbol{x}|\hat{\boldsymbol{x}}_{\beta,(-i)}) \right). \tag{13.37}$$

One reason of concern is the possibility that the true position of some features does not appear in the set of local maxima of the corresponding probability maps. The multiplicative conditioning operated by Equation 13.37 would (nearly) drive $p'_\alpha$ to zero even when the evidence for feature $\alpha$ is high. As we have already seen in many cases, an effective solution is that of bounding the influence of feature $\beta$. If we want to bound the overall effect of all features ensuring that $p'_\alpha(\boldsymbol{x})/p_\alpha(\boldsymbol{x}) > s$, we may introduce the following rescaling function:

$$f_s(x) = [(1 - s)x + s]^{1/n_f} \tag{13.38}$$

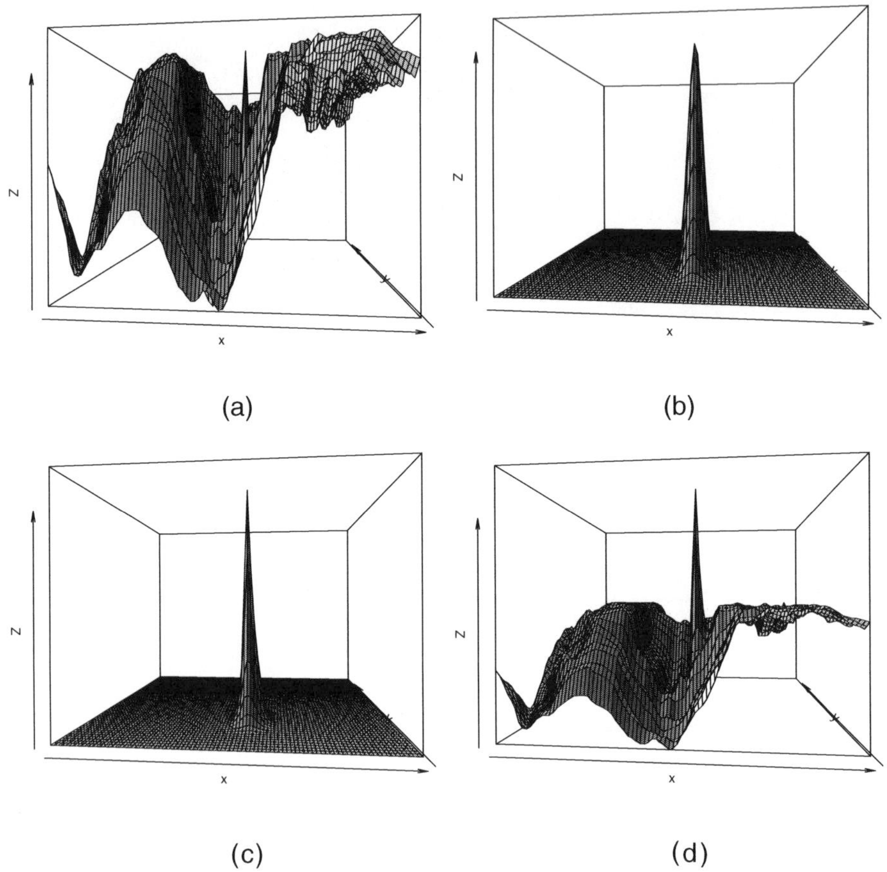


**Figure 13.8.** The results of an eye detector (a) can be biased by knowledge of the mouth position. The plot in (b) is the corresponding $p_{\alpha,\beta}(\boldsymbol{x} = \boldsymbol{x}_\alpha - \boldsymbol{x}_\beta)$ while the plots in (c) and (d) show the effect of using the normalization of Equation 13.34 and Equation 13.35. The latter seems to be preferable due to its softer impact. The reason is that the values normalized according to Equation 13.34 are too small to employ Equation 13.38. The plot in (c) was computed using $s = 0$ in Equation 13.38 while that in (d) was computed using $s = 0.5$.

where $n_f$ is the number of conditioning features, and use

$$p_\alpha(\boldsymbol{x}) \to p'_\alpha(\boldsymbol{x}) = p_\alpha(\boldsymbol{x}) \prod_{\beta \neq \alpha} f_s\left(\sum_{i=1}^{l} p_\alpha(\boldsymbol{x}|\hat{\boldsymbol{x}}_{\beta,(-i)})\right). \tag{13.39}$$

The constellation of Equation 13.31 can now be computed from the modified maps and leads to improved detection due to the incorporation of configural information (see Figure 13.8).

## 13.5. Bibliographical Remarks

A recent account of the cascade detection architecture and of its computation efficiency and robust performance can be found in Viola and Jones (2004). The pixelized version described in the chapter corresponds to the one presented in Küblbeck and Ernts (2006). The efficient architecture for face detection based on the idea of an SVM cascade classifier is discussed in Romdhani *et al.* (2001).

The idea of multiclass rejectors is described in Baker and Nayar (1996) while the approach of parametric feature detection has been explored thoroughly in the paper by Baker *et al.* (1998). A more recent paper considering the possibility of building maximal rejectors is the one by Elad *et al.* (2001).

The literature on the theoretical and practical aspects of AdaBoost is extensive. The description in the chapter is based on the papers by Freund and Schapire (1999) and Schapire *et al.* (1998). A general formulation of boosting algorithms as gradient descent is given by Mason *et al.* (1999).

The remarks on component-based face detection are from Bileschi and Heisele (2003). The idea of basing object recognition on parts and structure is considered extensively in Fergus *et al.* (2005) where the problem of learning parts and structure from data is also addressed.

## References


Baker S and Nayar S 1996 Algorithms for pattern rejection. *Proceedings of the 13th IAPR International Conference on Pattern Recognition (ICPR'96)*, vol. 2, pp. 869–874.

Baker S, Nayar S and Murase H 1998 Parametric feature detection. *International Journal of Computer Vision* **27**, 27–50.

Bileschi S and Heisele B 2003 Advances in component based face detection. *Proceedings of the IEEE International Workshop on Analysis and Modeling of Faces and Gestures (AMFG'03)*, pp. 149–156.

Elad M, Hel-Or Y and Keshet R 2001 Pattern detection using a maximal rejection classifier. *Proceedings of the 4th International Workshop on Visual Form*, Lecture Notes in Computer Science, vol. 2059, pp. 514–524. Springer.

Fergus R, Perona P and Zisserman A 2005 A sparse object category model for efficient learning and exhaustive recognition. *Proceedings of the IEEE Conference on Computer Vision and Pattern Recognition (CVPR'05)*, vol. 1, pp. 380–387.

Freund Y and Schapire R 1999 A short introduction to boosting. *Journal of the Japanese Society for Artificial Intelligence* **14**, 771–780.

Küblbeck C and Ernts A 2006 Face detection and tracking in video sequences using the modified census transformation. *Image and Vision Computing* **24**, 564–572.

Mason L, Baxter J, Bartlett P and Frean M 1999 Boosting algorithms as gradient descent. *Proceedings of Advances in Neural Information Processing Systems*, pp. 512–518.

Romdhani S, Torr P, Scholkopf B and Blake A 2001 Computationally efficient face detection. *Proceedings of the 8th International Conference on Computer Vision and Pattern Recognition (ICCV'01)*, pp. 695–700.

Schapire R, Freund Y, Bartlett P and Lee W 1998 Boosting the margin: a new explanation for the effectiveness of voting methods. *Annals of Statistics* **26**, 1651–1686.

Viola P and Jones M 2004 Robust real-time face detection. *International Journal of Computer Vision* **57**, 137–154.

# 14 BUILDING A MULTIBIOMETRIC SYSTEM

And these same crosses spoil me. Who are you?
Mine eyes are not o' the best: I'll tell you straight.

*King Lear*
WILLIAM SHAKESPEARE

Template matching techniques are a key ingredient in many computer vision systems, ranging from quality control to object recognition systems among which biometric identification systems today have a prominent position. Among biometric systems, those based on face recognition have been the subject of extensive research. This popularity is due to many factors, from the non-invasiveness of the technique, to the high expectations due to the widely held belief that human face recognition mechanisms perform flawlessly. Building a face recognition system from the ground up is a complex task and this chapter addresses all the required practical steps: preprocessing issues, feature scoring, integration of multiple features and modalities, and final classification stage.

## 14.1. Systems

In many real-world applications, template matching techniques are instrumental in solving complex tasks, which in turn may have to be integrated with additional subsystems in order to provide more advanced functionalities. A simple example is that of people identification. A biometric system determines the identity of a person and this information is exploited to let the user access services, such as on-line banking, or locations, such as restricted access areas. If we focus on a people identification system based on template matching techniques, we may identify interfacing issues at several levels:

- Template acquisition: there is a need to maximize the quality of the acquired templates while keeping user interaction as smooth and as unobtrusive as possible.
- Integration with system modules providing information useful for the solution of the same task, in this case people identification: the output of the template matching module should be designed to maximize the flexibility of integration with functionally similar systems, e.g. in order to create a more reliable and robust multibiometric system.
- Time and accuracy constraints imposed by the application.

This chapter describes one of the earliest multimodal biometric systems based on face and speech recognition developed in late 1992 and early 1993 and touches upon many of the techniques presented in the book. The identification system was one of the modules of the Electronic Librarian, itself one of the subsystems of MAIA (Modello Avanzato di Intelligenza Artificiale; Advanced Model of Artificial Intelligence).

## 14.2. The Electronic Librarian

The task of the Electronic Librarian was to identify a user by the face and speech, and to automate the operations of book lending and return by visual recognition of the books shown to the system by the user (see Intermezzo 14.1 and Figure 14.1). The system relied on template matching techniques for face recognition and on image processing for detecting the user and for book recognition. Even if the system was mainly a laboratory demonstrator of what could be achieved at the time, significant efforts went into devising a smooth user interaction and in the creation of a flexible system architecture.

**Intermezzo 14.1.** Meeting the Electronic Librarian

The interaction with the Electronic Librarian was easy and guided by the system that was able to speak, using a text-to-speech synthesizer, in two different languages: English and Italian. The following transcript reports a typical interaction:

**Electronic Librarian** Good morning. Please get closer to the camera above the monitor, looking straight into it and taking your glasses off, if any. A little bit to the right. Thanks! If you took your glasses off, please put them on. Now get close to the microphone and read loudly, without pausing, the digits as soon as they appear on the screen.

**User** [Reads digits loudly]

**EL** User identified [User name]. Please confirm.

**U** [Confirms by using the touchscreen interface]

**EL** Now show the book to the camera on your right.

**U** [Shows the book holding it with the right hand]

**EL** Book image acquired. Book identified [book code]. Please confirm.

**U** [Confirms]

**EL** Please select an operation.

**U** [Selects required operation, e.g. 'return', using the workstation touchscreen]

**EL** Operation registered. Please leave the booth. Have a nice day.

The Electronic Librarian was deployed in a small room dedicated to document printing and telefax (see Figure 14.2). People could enter and leave the room freely to get prints and send faxes, creating a significant amount of background noise. The lighting was complex and not (completely) under control: the ceiling of the room had a large skylight letting in unscreened sunlight and light from the corridor was an additional source of uncontrolled variation. The impact of audible noise was reduced by using a microphone with cardioid response, reducing pickup from the side and rear. An additional long fluorescent light was installed in the system on the stand of camera C2 (see Figure 14.2) devoted to the acquisition of face images. This long light provided soft illumination which did not cause any inconvenience to the user staring into the camera, in spite of looking straight at the camera (and the light). Experiments were

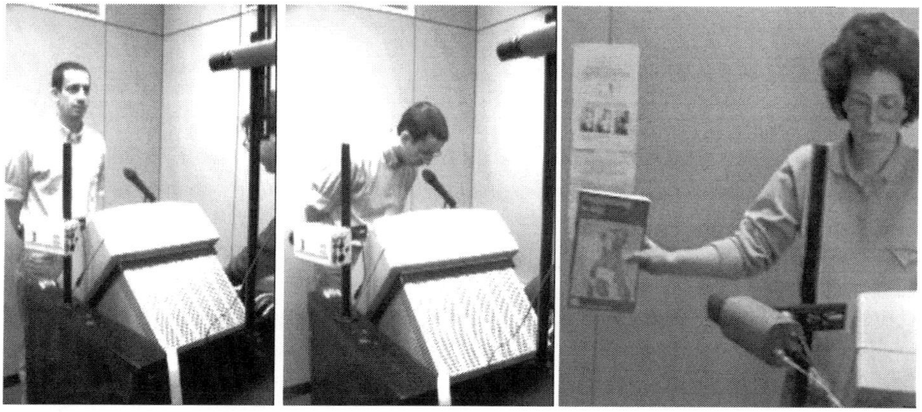

**Figure 14.1.** The three main phases of the interaction with the Electronic Librarian: acquisition of face images (left), spoken digits acquisition (middle), and book presentation (right, from the very first version of the Electronic Librarian).

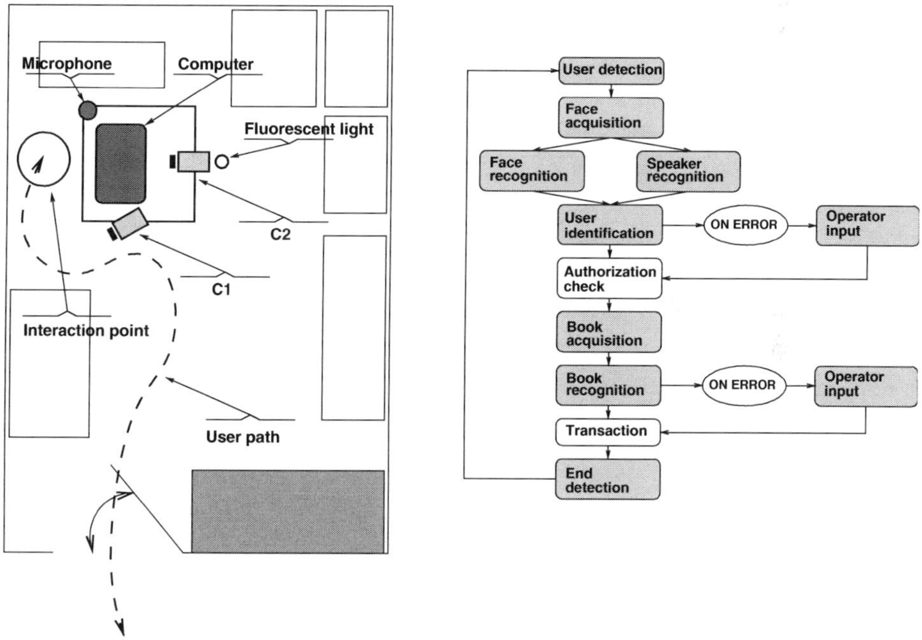


**Figure 14.2.** The layout of the environment where the Electronic Librarian was deployed (left) and the processing workflow of the system (right). The rectangles in the environment map correspond to printers and fax machines. The path of the user permitted a natural interaction and the use of cameras for multiple functions.

also performed using an infrared illuminator and removing the infrared filter from the camera (see Figure 2.13 for similar pictures), but the unavailability of extended infrared illuminators led to the adoption of a fluorescent, soft-light illuminator.

The booth area was monitored by camera C1 devoted to book acquisition, acting also as a visual gate. In order to speed up processing, the detection of a user entering the booth triggered the acquisition of a low-resolution background image $I_B$ from the camera dedicated to face acquisition. As the distance of the user from the camera was fixed (the user was supposed to stay close to the wall), focusing was fixed. Correct image exposure was checked by the system, verifying that the average image intensity $\bar{I}$ detected by C2 satisfied

$$I_{\text{inf}} < \bar{I} < I_{\text{sup}}.$$

The camera iris was computer controlled to satisfy the constraints. On being activated by the user entering the booth, the face acquisition system starts the acquisition of frames $I_t$ at the same resolution of $I_B$ verifying that:

- the changes with respect to $I_B$ are high enough to assume that a user is effectively in front of the camera;
- the user is located in the middle of the picture, in front of the camera, to minimize the chances of a non-frontal pose and self-shadowing due to the central fluorescent light;
- the user is standing still, so that no motion blur reduces image detail.

The required checks can be performed reliably and quickly using low-resolution images and simple image operations. While waiting for a still image, the system waits for the following condition to be satisfied:

$$a_1 < \frac{\|\theta(|I_t - I_B| - d_1)\|_{L_1}}{n_B} < a_2 \tag{14.1}$$

where $n_B$ is the number of pixels in $I_B$, $d_1$ is a threshold used to identify significant changes, $a_1$ and $a_2$ are two thresholds determining percentage area constraints, $\theta(\cdot)$ is the Heaviside function, and image operations are applied pixel by pixel, in a parallel fashion (see also Appendix A). When Equation 14.1 is satisfied, the system acquires a stable image and checks that the center of gravity of the difference from the background image $I_B$ is located approximately in the central band $(x_0, x_1)$ of the image:

$$\|\theta(|I_t - I_{t-1}|)\|_{L_1} < a_3 \tag{14.2}$$

$$x_0 < \bar{x} = \frac{1}{n_{tB}} \sum_{ij} i\, \theta(|I_t - I_B| - d_1)(i, j) < x_1 \tag{14.3}$$

where $a_3$ is an area threshold controlling the amount of change and $n_{tB} = \|\theta(|I_t - I_B| - d_1)\|_{L_1}$. If $\bar{x}$ does not satisfy Equation 14.3 the system tells the user, using text-to-speech synthesis, to move to the left or to the right in order to satisfy the constraint. When all constraints are satisfied, the system snaps three additional images: a low-resolution image $I_1$ ($64 \times 64$ pixels) and two high-resolution ($512 \times 512$ pixels) images $I_2$ and $I_3$ a few tenths of a second apart. The low-resolution image is used to perform an approximate localization of the user using image projections onto the horizontal and vertical axes

$$x_l = \min_x \left\{ x : \frac{1}{h} \sum_{j=1}^{h} \theta(|I_1 - I_B| - d_1)(x, j) > d_2 \right\} \tag{14.4}$$

$$x_r = \max_x \left\{ x : \frac{1}{h} \sum_{j} \theta(|I_1 - I_B| - d_1)(x, j) > d_2 \right\} \tag{14.5}$$

$$y_t = \min_y \left\{ y : \frac{1}{x_r - x_l + 1} \sum_{i=x_l}^{x_r} \theta(|I_1 - I_B| - d_1)(i, y) > d_2 \right\} \tag{14.6}$$

where $x_l$, $x_r$, and $y_t$ represent the left, right, and top bounds of the user in image $I_1$ and $h$ represents image height. Upon locating the user in the low-resolution image $I_1$, the location of the user in $I_2$ and $I_3$ can be easily obtained.

In order to perform a reliable pixel-by-pixel comparison of face images, it is necessary to align them, using some reference points that can be located reliably and with high accuracy in all images. As we restricted our system to frontal face images, the centers of the pupils are a natural choice. After having located them, the images can be rotated, scaled, and translated by means of an affine transform moving the center of the pupils to predefined locations in the image plane, e.g. (101, 100) and (155, 100), corresponding to an interpupilar distance of 55 pixels. The geometrical normalization phase can be split into four main steps:

1. Computation of an approximate symmetry axis $x_s$:

$$x_s = \operatorname{med}_i(P_V(|I_2 * K_V|)(i)) \tag{14.7}$$

   where $P_V(\cdot)(i)$ represents the sum of the image values in column $i$ and the convolution kernel $K_V = (-1, 0, 1)^T$. The left and right areas of the face are then determined, letting them overlap to account for the indeterminacy of $x_s$.

2. Localization of the two eyes using representative templates for which the position of the pupil is known. In order to limit the impact of uncontrolled illumination variations, the values of the image are normalized using a local contrast operator

$$N' = \frac{I_2}{I_2 * K_\sigma} \tag{14.8}$$

$$N = \begin{cases} N' & \text{if } N' \leq 1 \\ 2 - \dfrac{1}{N'} & \text{if } N' > 1 \end{cases} \tag{14.9}$$

   where $K_\sigma$ is a Gaussian kernel whose parameter is related to the expected inter-ocular distance $\Delta_{ee}$ expressed in pixels: $\sigma = 3(\Delta_{ee}/55)$. Image $N$ preserves all high-frequency details and removes the effect of a non-uniform ambient illumination. Eyes are then located, maximizing the correlation coefficient (with templates normalized according to Equation 14.8) using a hierarchical approach.

3. The eye positions found in the previous step are validated by scoring them with a coefficient $C_E$ that measures their symmetry with respect to $x_s$, their horizontal

alignment, and the compatibility of $\Delta_{ee}$ with the size of the templates used:

$$C_E = \frac{1}{2}(\rho_l + \rho_r) e^{-(2\sigma_l^2)^{-1}(l-1)^2} e^{-(2\sigma_\theta^2)^{-1}\theta^2} \frac{\min(\rho_l, \rho_r)}{\max(\rho_l, \rho_r)} > d_r \tag{14.10}$$

where $\rho_l$ and $\rho_r$ represent the correlation coefficients of the two eyes, $l$ is the ratio of the interocular distance to that of the eye templates (and $\sigma_l$ its expected deviation), and $\theta$ is the slope of the interocular axis with respect to the horizontal image axis. If Equation 14.10 is not satisfied, the system tries the localization on image $I_3$ as the failure may be due to the eyes being shut or not completely open. Should the localization fail again, the face recognition system declares forfeit and the multibiometric system will have to rely on a single modality: speech.

4. Upon successful eye localization, image $I_2$ (or $I_3$) undergoes an affine transform mapping the center of the pupils to standard positions (101, 100) and (155, 100).

The normalized image of the user is now ready for comparison to similar images representative of the users known to the system in order to determine identity. Due to geometrical standardization, the image regions containing eyes, nose, and mouth are characterized by (approximately) the same coordinates in every image: we do not need to localize them, either in the reference images or in $I_2$. Alternatively, we may automatically locate the nose and mouth in the reference images, storing the corresponding locations in the database, and exploit this information in the matching step.

The regions corresponding to the three features are then extracted from $I_2$ and compared in turn to the corresponding regions extracted from the database entries (normalized using Equation 14.8). In order to get a similarity score that is not too susceptible to noise we used the following similarity measure:

$$g(\boldsymbol{x}, \boldsymbol{y}) = 1 - \frac{\|\boldsymbol{x} - \boldsymbol{y}\|_{L_1}}{\|\boldsymbol{x}\|_{L_1} + \|\boldsymbol{y}\|_{L_1}} \tag{14.11}$$

where $\boldsymbol{x}$ and $\boldsymbol{y}$ are two vectors, and extended it to images (see Section 4.3). Patterns are normalized to common average intensity $\mu$ and standard deviation (or scale) $\sigma$ using robust location and scale estimators (e.g. the median and median absolute deviation discussed in Intermezzo 4.3). The values corresponding to database images can be computed off-line while the cost of computing them for $I_2$ is amortized by computing them once for all database comparisons. The matching $s$ of (a region) of image $B$ to a reference image $A$ is then given by

$$s = \max_{\boldsymbol{x} \in R_A} (g(A, B_{\boldsymbol{x}}))$$

where $R_A$ is a region of $A$ and $B_{\boldsymbol{x}}$ denotes the translation of $B$ at position $\boldsymbol{x}$. An advantage of $g(\cdot)$ over the normalized correlation coefficient is its reduced sensitivity to unusually high differences between corresponding pixels, such as those due to image specularities like iris highlights. Let us denote by $\{U_{k,m}\}$ the set of reference images for user $k$. We may then compare each facial feature from image $I_2$ to the corresponding regions of stored images, obtaining a list of similarity scores for each facial feature

$$\{s_{k,\alpha}\} = \left\{ \max_m (g(R_{\alpha,f}(I_2), R_\alpha(U_{k,m}))) \right\} \tag{14.12}$$

where region $R_{\alpha,f}$ contains $R_\alpha$ with a frame (border) whose size is related to the interocular distance and allows for imprecise alignment. The efficiency of the comparison process was increased by adopting a hierarchical approach (see Intermezzo 14.4).

**Intermezzo 14.2.** Speaker Identification

Acoustic features are extracted by spectral analysis of the speech signal digitized at 16 kHz every 10 ms. The analysis is performed on overlapping (Hamming) windows of 20 ms length. A set of 8 mel scaled cepstral coefficients (static features of the short-term power spectrum of sound, based on the cosine transform of a log power spectrum on the nonlinear mel frequency scale) and the corresponding first-order time derivatives (dynamic features) are computed and used to build two vectors for each analysis window. User $k$ is represented in the reference database by two sets of 64 reference vectors $\{\boldsymbol{\theta}_{k,(f,m)}\}_{m=1,\ldots,64}$ of the same type $f$ (static or dynamic), obtained by clustering available training data (see Intermezzo 12.1). As the system was tested several months after the acquisition of the training data, in different environmental conditions and using a different acquisition chain, the static reference vectors were adapted, shifting them towards a small collection of data derived from the new setup. After the removal of windows marked as background noise, the minimum squared Mahalanobis distance of each vector from the reference vectors (of the same type) of user $k$ is computed using the covariance matrix $\Sigma_f$ derived from the utterances of all users in the reference database and averaged over the available windows

$$d(\boldsymbol{x}_f, k) = \frac{1}{T}\sum_{t=1}^{T} \min_{m=1,\ldots,64} [(\boldsymbol{x}_{f,t} - \boldsymbol{\theta}_{k,(f,m)})^T \Sigma_f^{-1} (\boldsymbol{x}_{f,t} - \boldsymbol{\theta}_{k,(f,m)})] \tag{14.13}$$

where $f$ identifies the type of feature (static or dynamic). The lower the distance value, the higher the similarity. As the system integrating multimodal biometric scores followed the convention that the higher the value, the more similar the patterns, the sign of the distance values was reversed and the resulting two lists of scores delivered to the integration system.

## 14.3. Score Integration

The use of multiple cues, in our case face and voice, provides in a natural way the information necessary to build a reliable, high-performance system. Specialized subsystems can identify (or verify) each identification cue and the resulting outputs can be combined into a unique decision by some integration process. Multiple classifier systems can be subdivided into three broad categories depending on the level at which information is integrated:

- The abstract level: the output is a subset of the possible identification labels, without any additional qualifying information.
- The rank level: the output is a subset of the possible levels, sorted by decreasing confidence (which is not supplied).
- The measurement level: the output is a subset of labels qualified by a confidence measure.

The level at which the different classifiers of a composite system work constrains the ways that their responses can be merged. As the result of template matching is a similarity score, the most natural way is integration at the measurement level and this is the case that we will consider. The visual and acoustic identification systems described so far represent a multiple classifier system that can be further split to obtain a more modular system composed of five classifiers: two acoustic classifiers based on static and dynamic acoustic parameters (see Intermezzo 14.2), and three visual classifiers based on the different facial features (eyes, nose, and mouth).

A critical point in the design of an integration procedure at the measurement level is measurement normalization. The responses of the constituent classifiers often have different scales and possibly offsets, and a sensible combination of their output can be performed

**Intermezzo 14.3.** Chimeric identities

We have appreciated throughout the chapters that the availability of a large number of samples is key to the development of reliable classifiers. While reasonably large monomodal databases of biometric signatures appeared early in the research community, the availability of large multimodal corpora is not as great. A natural question then is whether it is useful to create virtual, chimeric identities by taking the Cartesian product of biometric signatures of different persons. Let us suppose that we have $N$ users and two biometric modalities for each of them $\{(\boldsymbol{x}_{1,i}, \boldsymbol{x}_{2,i})\}_{i=1,\ldots,N}$. We could be tempted to generate a new database with $N(N-1)$ entries in this way

$$\{(\boldsymbol{x}_{1,i}, \boldsymbol{x}_{2,j})\}, \quad i, j = 1, \ldots, N, \; j \neq i \tag{14.14}$$

to test multimodal integration strategies relying on the modality independence assumption by which the features associated to multiple biometric traits of an individual are considered to be independent. Unfortunately the approach is flawed even when the assumption holds. The reason is that the distribution of the points in feature space generated according to Equation 14.14 is very different from the real one: the same point $\boldsymbol{x}_{1,i}$ in the feature space of the first modality is associated to $N-1$ different identities only by means of the information derived from the second modality. The resulting distribution bias invalidates the strategy. Chimeric users can, however, be used if we restrict ourselves to the generation of a multimodal database of size $N$ from $k$ monomodal databases. We can then build $N^k$ different multimodal databases that can be used to compute statistics useful for characterizing the expected performance of multiple integration strategies.

only after having normalized the scores. In the particular case considered, the distributions of the scores from the different features are markedly unimodal and roughly symmetrical. A simple way to normalize them is to estimate, on a feature-by-feature basis, their average and variance so that their distributions can be translated and rescaled to zero average and unit variance. The values can then be forced into a standard interval, such as [0, 1], by means of an hyperbolic tangent mapping with a scale factor that prevents (excessive) tail saturation. The normalizing values can be fixed, derived from a training set, or estimated adaptively at each interaction with the system. The last strategy was adopted for the Electronic Librarian to cope with score variability due to different speech utterance length and varying environmental conditions. Score normalization should not be driven by extreme values in the set and robust estimators should be used. The reason is that the scores corresponding to correct matches are expected to be extremal, and therefore outliers with respect to those corresponding to wrong matches. These outliers bias scale estimators towards higher values. However, when the scale estimate is used to normalize the scores, an overestimated value results in score compression, minimizing the advantage of having an outstandingly good value for the correct match.

Some of them, such as those based on the median, have a low efficiency: their variance is relatively high with respect to the smallest achievable one, a fact that reduces their discrimination capacity. More efficient estimators exist, among which we may cite the hyperbolic tangent estimators proposed by Hampel that have been used in this system. Each list of scores $\{s_{i,\alpha}\}_{i=1,\ldots,n_p}$ from classifier $\alpha$, being $n_p$ the number of people in the database, is then normalized by the mapping

$$s'_{i,\alpha} = \frac{1}{2}\left[\tanh\left(0.001\frac{s_{i,\alpha} - \mu_{\text{tanh}}}{\sigma_{\text{tanh}}}\right)\right] \tag{14.15}$$

where $\mu_{\text{tanh}}$ and $\sigma_{\text{tanh}}$ are the average and standard deviation estimates of the scores as given by the Hampel estimators. An effective way to combine the normalized scores is by using a weighted geometric average

$$s_i = \left(\prod_j (s'_{i,\alpha})^{w_\alpha}\right)^{1/\sum_\alpha w_\alpha}. \tag{14.16}$$

Weights $w_\alpha$ represent an estimate of score dispersion in the right tail of the corresponding distributions

$$w_\alpha = \frac{s'_{(1),\alpha} - 0.5}{s'_{(2),\alpha} - 0.5} - 1.0 \tag{14.17}$$

where parenthesized indices $(i)$ denote the $i$th entry within the indexed set sorted by decreasing value. The reason for using a geometric average is that we can interpret normalized scores as probabilities and, in the case of independent features, the probability that a feature vector corresponds to a given person can be computed by taking the product of the probabilities of each single feature. Another way of looking at the geometric average is that of predicate conjunction using a continuous logic. The weights reflect the importance of the different predicates (features). Each feature is given an importance proportional to the separation of the two best scores. If the classification provided by a single feature is ambiguous it is then given low weight. A major advantage of the approach is that no detailed knowledge of the score distribution is needed; this eases the task of building a system that integrates many features.

**Intermezzo 14.4.** Hierarchical resolution matching

In order to speed up the computation, a hierarchical resolution approach can be employed. The image to be classified is first compared at low resolution to the complete database. For each person in the database, the most similar feature, among the set of available images, is chosen and the best matching location stored. The comparison is then continued at the upper resolution level limiting the search to the most promising candidates at the previous level. This results in a significant speed-up and in a major drawback: the score normalization adopted by the Electronic Librarian requires all the scores, but scores from different resolutions have different distributions. To overcome this difficulty, the scores from the previous level were reduced (scaled) according to the highest reduction factor obtained by comparing the newly computed scores to the corresponding scores from the lower resolution level. The performance of the system was unaffected and a significant speed-up was obtained.

The database includes data from 89 people and data from three out of four test sessions with the Electronic Librarian were used to evaluate the system (see Intermezzo 14.3 on the possibility of mixing multimodal cues to gather system statistics). The main performance measure of a biometric identification system is the percentage of persons correctly recognized. It can be further qualified by the average value $\bar{R}$ of the ratio

$$R_x = \frac{s'_x - s'_{(n_p)}}{\max_{i \neq x}(s'_i) - s'_{(n_p)}} \tag{14.18}$$

that measures the separation of the correct match $s'_x$ from the wrong ones. The value of $\bar{R}$ can be considered as an average margin, an indicator of the stability and generalization potential of the identification system (see Section 12.3).

This ratio is invariant to the scale and location parameters of the integrated score distribution and can be used to compare different integration strategies (weighted/unweighted geometric average, adaptive/fixed normalization). The weighted geometric average of the scores when adaptively normalized was found to provide the best results (see Table 14.1).

**Table 14.1.** The identification performance of the Electronic Librarian and the average separation ratio for each single feature and for their integration.

| Feature | Error rate (%) | $\bar{R}$ |
|---|---|---|
| **Voice** | **12** | **1.14** |
| Static | 23 | 1.08 |
| Dynamic | 29 | 1.08 |
| **Face** | **9** | **1.56** |
| Eyes | 20 | 1.25 |
| Nose | 23 | 1.25 |
| Mouth | 17 | 1.28 |
| **Integrated** | **2** | **1.65** |

## 14.4. Rejection

An important capability of a classifier is rejection of input patterns that cannot be classified with a sufficiently high degree of confidence. For a person verification system the ability to reject an impostor is critical. A simple rejection strategy can be based on the level of agreement of the classifiers in the identification of the best candidate. A simple measure of confidence is given by the integrated score itself: the higher the value, the higher the confidence of the identification. Another indicator is the difference of the two best scores: it is a measure of how sound the ranking of the best candidate is. The use of independent features (or feature sets) also provides valuable information in the form of the rankings of the identification labels across the outputs of the classifiers. If the pattern does not belong to any of the classes, its rank will vary significantly from classifier to classifier. On the contrary, if the pattern belongs to one of the known classes, rank agreement will be consistently high. The average rank and rank dispersion across the classifiers can then be used to quantify the agreement of the classifiers in the final identification. The rejector employed by the Electronic Librarian relied on a feature vector $\boldsymbol{f} \in \mathbb{R}^{18}$ comprising:

- the integrated score $s_{(1)}$ of the best candidate;
- the normalized ratio of the first to the second best integrated score

$$R = \frac{s_{(1)} - 0.5}{s_{(2)} - 0.5} \tag{14.19}$$

- the minimum and maximum ranks of the first and the second final best candidates (four entries);
- the rank standard deviation of the first and second final best candidates (two entries);
- the individual ranks of the first and second final best candidates (10 entries).

A set of positive examples $\{\boldsymbol{p}_i\}$ is derived from the data relative to people correctly identified by the system. A set of negative examples is given by the data corresponding to the best candidate when the system was unable to perform a correct identification. The resulting (small) set is augmented by the data of the best candidate obtained by removing the correct entry from the database, thereby simulating the interaction with a stranger. Final acceptance

**Table 14.2.** Percentage error per modality and per error type.

| Modality | False positives | False negatives | Familiar misrecognized |
|---|---|---|---|
| Face | 4.0 | 8.0 | 0.5 |
| Voice | 14.0 | 27.0 | 1.0 |
| Integrated | 0.5 | 1.5 | 0.0 |

or rejection of an identification associated to a vector $\boldsymbol{f}$ was done according to a linear rule

$$\sum_{i=1}^{18} w_i f_i + w_{19} > 0 \quad \text{accept} \tag{14.20}$$

$$\sum_{i=1}^{18} w_i f_i + w_{19} \leq 0 \quad \text{reject} \tag{14.21}$$

separating the two classes with a hyperplane whose normal is parallel to $\boldsymbol{w}$. The vector $\boldsymbol{w}$ characterizing the linear discriminant was determined minimizing the error

$$E = \alpha \sum_i \left(1 - \frac{1 - e^{-(\sum_{k=1}^{l} w_k p_{ik} + w_{l+1})}}{1 + e^{-(\sum_{k=1}^{l} w_k p_{ik} + w_{l+1})}}\right) + \beta \sum_j \left(1 - \frac{1 - e^{(\sum_{k=1}^{l} w_k n_{jk} + w_{l+1})}}{1 + e^{(\sum_{k=1}^{l} w_k n_{jk} + w_{l+1})}}\right) \tag{14.22}$$

where $\alpha$ and $\beta$ represent the weight of false negatives and false positives respectively, and $l = 18$ is the dimensionality of the input vectors. If the two classes are linearly separable it is possible to drive $E$ to zero. The ratio $\beta/\alpha$ determines the strictness of the system: the higher the value, the more difficult for a stranger to be accepted. For each value of the ratio $\beta/\alpha$ we obtain a different classifier and these can be integrated using the ROC hybrid approach described in Section C.3. The linear discriminant can be replaced with one of the more complex systems presented in Chapter 12. Different applications may require different ratios, depending on the required security level. The error contribution of points lying close to the discrimination plane is (approximately) $\alpha$ for false negatives and $\beta$ for false positives. The error contribution of points lying far away from the plane is bounded by $2\alpha$ and $2\beta$ so that they cannot excessively bias the computation of $\boldsymbol{w}$. The performance of the system required to operate at an equal error rate is reported in Table 14.2.

## 14.5. Bibliographical Remarks

The system described in the chapter was developed at the beginning of the 1990s and is described in a few papers. The face recognition system has evolved from the one presented by Brunelli and Poggio (1993), adopting one of the similarity estimators described in Chapter 4 from (Brunelli and Messelodi 1995). The integration of the speech modality was described in Brunelli and Falavigna (1995) where additional details can be found.

The scientific literature in the field of multimodal biometrics (or multi biometrics) is now vast. An easy introduction to multibiometric systems is the paper by Jain and Ross (2004) and a more comprehensive coverage can be found in the book by Jain *et al.* (2007). The problem of score normalization is still a research topic and is addressed in Jain *et al.* (2005). A recent approach based on the likelihood ratio test can be found in Nandakumar *et al.* (2008).

## References

Brunelli R and Falavigna D 1995 Person identification using multiple cues. *IEEE Transactions on Pattern Analysis and Machine Intelligence* **17**, 955–966.

Brunelli R and Messelodi S 1995 Robust estimation of correlation with applications to computer vision. *Pattern Recognition* **28**, 833–841.

Brunelli R and Poggio T 1993 Face recognition: features versus templates. *IEEE Transactions on Pattern Analysis and Machine Intelligence* **15**, 1042–1052.

Jain A and Ross A 2004 Multibiometric systems. *Communications of the ACM* **47**, 34–40.

Jain A, Flynn P and Ross A (eds.) 2007 *Handbook of Biometrics*. Springer.

Jain A, Nandakumar K and Ross A 2005 Score normalization in multimodal biometric systems. *Pattern Recognition* **38**, 2270–2285.

Nandakumar K, Yi C, Dass S and Jain A 2008 Likelihood ratio based biometric score fusion. *IEEE Transactions on Pattern Analysis and Machine Intelligence* **30**, 342–347.

# Appendices

# A AnImAl: A SOFTWARE ENVIRONMENT FOR FAST PROTOTYPING

There's many a beast then in a populous city. . . .

*Othello*
WILLIAM SHAKESPEARE

The process of developing a computer vision system for a specific task often requires the interactive exploration of several alternative approaches and variants, preliminary parameter tuning, and more. This chapter introduces AnImAl, an image processing package written for the R statistical software system. AnImAl, which relies on an algebraic formalization of the concept of image, supports interactive image processing by adding to images a self-documenting capability based on a history mechanism. The documentation facilities of the resulting interactive environment support a practical approach to reproducible research.

## A.1. AnImAl: An Image Algebra

The details of digital image formation presented in Chapter 2 together with the most common sensor architectures suggest that matrices provide an adequate representation for storing and processing digital images. This basic representation can be improved upon in order to better support both the characteristic processing patterns associated with image analysis and some general requirements typical of computational scientific research, to which computer vision belongs.

The typical question to which image processing must provide an answer is: what is where in the image? This implies that pixels (i.e. matrix entries), while useful atomic units for image representation, are not the best elements in terms of which the question can be answered. Bounded image regions of arbitrary shape provide a more useful descriptive level. A simple way to obtain such a representation is to associate to each region two matrices whose size corresponds to the minimum bounding rectangle of the region of interest: one of the matrices is used to store the characteristic function of the region, the other one to store the corresponding pixel values. In order to preserve the information on the position of the data within the original image, a way to store the position of the submatrices is required (e.g. the coordinates of the upper left corner of the bounding rectangle). A potential drawback of this approach is the resulting necessity of working on images of varying sizes.

Computer vision belongs to the general field of computational scientific research and inherits the corresponding working methodologies. We would like to comment on two

*Template Matching Techniques in Computer Vision: Theory and Practice* Roberto Brunelli

of them: investigation based on the iteration of hypothesize, test, and refinement cycles; and results reporting based on traditional scientific publications that can provide only outlines of the relevant computations. Both processes have a major, negative impact on the reproducibility of research. On the one hand, keeping track of multiple hypothesize, test, and refinement cycles imposes a significant burden on the researcher: continuous updating of documentation may be neglected, resulting in inaccurate reporting. On the other hand, the limitation of static documents, such as journal papers, prevents the distribution of the code used in the computational analysis, making verification difficult.

In the remainder of this appendix we show how it is possible to improve at the same time research practice and reproducibility in the field of image processing by careful formalization of basic concepts and choice of an appropriate development environment. The first step is a formalization of the concept of image as a structure supporting both common processing practices and accurate reporting requirements. The study of structure is one of the fields of mathematics and is associated with algebra. The concepts of elementary algebra, related to numbers and constituting common knowledge, are extended by modern and universal algebra to more abstract sets and structural relations. The generalizations of modern algebra facilitate the study of properties and relations that mathematical objects have in common despite their apparent differences. It turns out that images can be defined as elements of an image algebra whose characterization requires some concepts from the field of modern and universal algebra.

**Definition A.1.1.** *An* algebraic system, *or* algebra, *is a pair* $(A; F)$ *where* $A$ *is a non-empty set and* $F$ *is a (possibly empty or infinite) collection of operations: each operation* $f_i \in F$ *is a function*

$$f_i : A^{n(i)} \to A$$

*where* $n(i)$ *is a natural number, called* arity, *that depends on* $f_i$. *An operation like* $f_i$ *is said to be an* $n(i)$*-ary operation on* $A$. *A* finitary operation *is any* $n$*-ary operation,* $n \in \mathbb{N}$.

**Definition A.1.2.** *A* partial algebra *is an algebra where some of the operations may be defined only partially: an* $n$*-ary operation is not defined on the entire domain* $A^n$.

**Example A.1.3.** *Positive integers under binary subtraction* $(\mathbb{N}; -)$ *are a partial algebra because* $m - n$ *is undefined if* $m \leq n$.

**Definition A.1.4.** *A* heterogeneous algebra *(or* multisort algebra*) is a family of sets* $\{A_i\}_{i \in I}$ *together with a collection* $F$ *of (heterogeneous) functions: for each* $n$*-ary operation* $f \in F$ *there is an* $(n+1)$*-tuple* $(i_i, \ldots, i_n, i_{n+1}) \in I^{n+1}$ *such that*

$$f : A_{i_1} \times \cdots \times A_{i_n} \to A_{i_{n+1}}.$$

**Example A.1.5.** *Let us consider an additive group* $(A; +, -)$ *with a binary and a unary operation*

$$+ : A \times A \to A \quad \textit{and} \quad - : A \to A$$

*a field* $(K; \oplus, \ominus, \times)$*, and let* $\bullet : K \times A \to A$ *represent scalar multiplication. A vector space* $(A, K; +, -, \oplus, \ominus, \times, \bullet)$ *is a heterogeneous algebra with two carriers: vectors* $A$ *and scalars* $K$.

If $F$ is a set of operators it is possible to define an injective map $\omega$ from $F$ to the set of all names $\Sigma$. Let us now consider the set $M$ of all infinite $\mathbb{Z}$ matrices with entries in $\mathbb{Z}$ corresponding to the set of functions from $\mathbb{Z} \times \mathbb{Z}$ to $\mathbb{Z}$. We call $P$ (parameters) the set of all the $n$-tuples with integer or real values for each $n \in \mathbb{N}$ and we denote by $H$ the set of all ordered trees whose nodes belong to $\Sigma$ or $P$.

**Definition A.1.6.** *On the family of all the unions spanned by the sets*

$$M \times H, \{\mathbb{Z}^h\}_{h \in \mathbb{N}}, \{\mathbb{R}^h\}_{h \in \mathbb{N}}$$

*we can define a set $F$ of partial heterogeneous operators so that we obtain a partial multisort algebra. We call* image *each element of the set $M \times H$ and* AnImAl *the corresponding algebra (from An Image Algebra). The projection of an image onto $H$ is called* history.

It is often clear enough to use the term image also to refer to the matrix part only.

**Example A.1.7.** *Let*

$$\text{left}\colon M \times H \times Z \to M \times H$$

*be an operator of $F$ defined as*

$$\text{left}((m, h), z) = (\text{left}_M(m, z), \text{concat}(\omega(\text{left}), h, z))$$

*where* $\text{left}_M(m, z)\colon M \times Z \to M$ *is such that* $\text{left}_M(m, z)_{ij} = m_{i,j+z}$ *for each $i, j \in \mathbb{Z}$ and* concat *yields a new tree with name* $\omega(\text{left}) \in \Sigma$ *as new root, first subtree $h$, and second subtree the leaf $z$.*

**Definition A.1.8.** *An image $(m, h)$ is said to be* coherent *if*

$$\text{concat}(h) = (m, h)$$

*where* concat *is a function that substitutes each name $\omega(f) \in \Sigma$ within history $h$ with the operator $f$ exploiting the injectivity of $\omega$.*

## A.2. Image Representation and Processing Abstractions

The general image model outlined in the previous section cannot be implemented because an image would correspond to a boundless matrix. In many applications, however, it is possible to restrict computation to a limited region of the image plane (see Figure A.1). This region of interest is bounded by a rectangular region of the image plane. While the value of the pixels within the region can be arbitrary (within the assumed domain), pixels outside the given region share a constant value that is defined on an image-by-image basis. This implementation requirement is not restrictive as real images are bounded anyway. The resulting infinite matrix is then equivalent to a structure consisting of the starting point of the region of interest (focus) and of the constant value associated to the outside infinite area. In order to fully support the concept of image underlying AnImAl we need an additional entry in the structure: the tree representing the processing history. A support boolean flag, telling us whether updating history is required or not, completes the set of items required for a practical implementation.

The representation of images as boundless arrays offers several advantages:

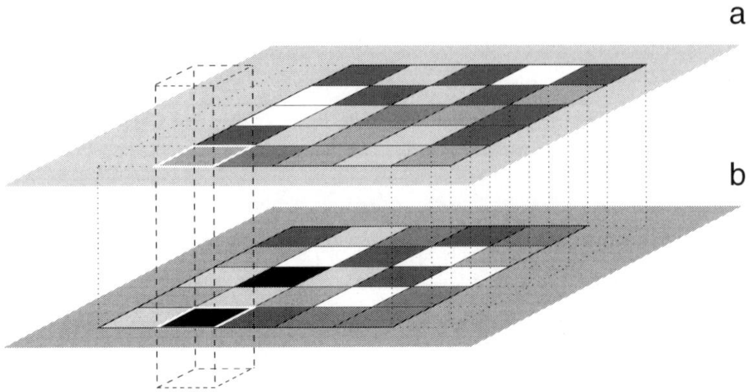


**Figure A.1.** A practical, albeit restricted, implementation of boundless images. The matrix component of an image is built from two components: a finite rectangular region represented by an array whose elements may assume arbitrary values within the computer's capability of representing elements of $\mathbb{Z}$; and its boundless complement described by a single integer value. Images have the same (infinite) size even if the size, and location, of their finite parts differ: image operations can then be performed homogeneously and in finite time.

- all points, including those on the boundary of the region of interest, have the same number of neighbors;
- translations are performed without loss of information and are simply realized by changing focus coordinates;
- operations on two images defined through binary commutative operations on integers are themselves commutative.

Each image is further characterized by its history, whose purpose is to trace back the elaborations performed on a single image: its usefulness in supporting interactive signal processing environments will be addressed in Section A.3.

A major application of AnImAl is the concise description and fast prototyping of algorithms for a specific class of image processing hardware: VLSI single instruction, multiple data (SIMD) machines, one of the possible parallel organizations for a set of processing elements (PEs). These machines are usually realized as array processors: synchronous parallel computers with multiple arithmetic logic units (the PEs) that can operate in parallel in lock step fashion. A set of masking schemes can be used to enable/disable the processing elements during the execution of vector (or, more appropriately, array) instructions. The topological structure of an SIMD array processor is mainly characterized by the data routing network used to interconnect the PEs. From a formal point of view, the communication network among the PEs can be specified by a set of data routing functions. If we identify the addresses of all PEs in an SIMD machine by the set $S$, each routing function is represented by a bijection from $S$ to $S$. When a routing function $f$ is applied via the interconnection network, PE($i$) copies its data into the data location of PE($f(i)$). An SIMD computer $\mathcal{C}$ can then be characterized by the following 4-tuple

$$\mathcal{C} = \langle N, F, I, M \rangle$$

where

- $N$ is the number of PEs in the system,
- $F$ is a set of data routing functions provided by the interconnection network,
- $I$ is the set of machine instructions for scalar, vector, data routing and network manipulation operations, and
- $M$ is the set of masking schemes.

The SIMD organization is a natural one for fine-grained parallel machines and for image processing tasks. Many early vision problems can be easily mapped onto an array of processors, each one performing the same operation using the data available in the local neighborhood. AnImAl can be used to simulate an SIMD machine by associating a processing element to every point of the plane $\mathbb{Z} \times \mathbb{Z}$. In order to realize the local memory of each PE (PEM) we can imagine an image overlain on the plane $\mathbb{Z} \times \mathbb{Z}$. Multiple local memories can be realized by means of a list of images, each pixel corresponding to a single local memory of the aligned PE. The interconnection network is realized through the shift operator: each PEM can be aligned with every other PEM in a rigid way for the whole image structure (see Figure A.1), supporting the use of near-neighbor meshes frequently used in image processing (4-connected, 6-connected, 8-connected) and more. Each PE can then operate on its data and on the data of virtually every other PE executing several classes of operations:

**arithmetical** (add, sub, mult, div),

**relational** ($>, \geq, <, \leq$, max, min),

**bit-level logical** (and, xor, not).

All these functions (with the exception of not) have a similar structure:

$$f(a_1, a_2) \quad \text{or} \quad f(a_1, a_2, a_3)$$

where

- $a_1$ must be an image,
- $a_2$ can be an image or an integer, and
- $a_3$ can be a masking image or a 4-tuple of integer values identifying a rectangular region in the plane.

The first argument corresponds to the data output of the corresponding PE while the second argument corresponds to the data obtained via the interconnection network and stored in the PEM. An integer second argument corresponds to data broadcasting from a centralized control unit. The modification of a single pixel via the usual array access operators corresponds to the distribution of data via the system bus. The third argument reflects the possibility of defining a mask to inhibit PE processing. Each operation has a default activation mask corresponding to the rectangular region bounding the image arguments: arbitrary activation masks can be specified using a masking image whose elements assume value 0

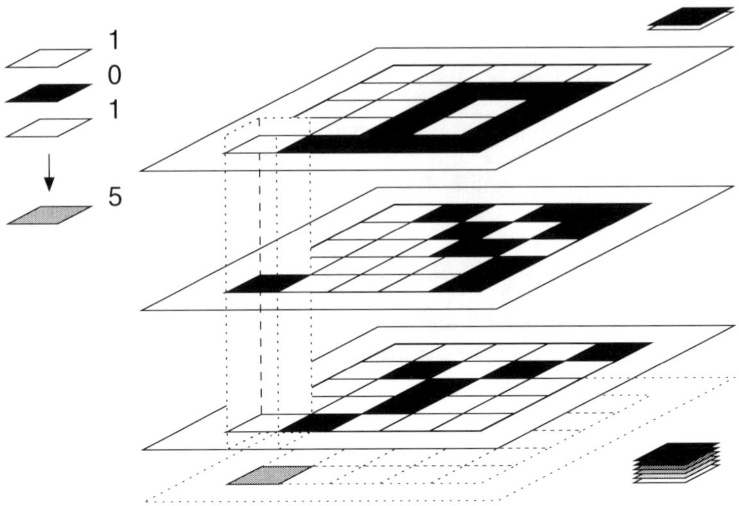


**Figure A.2.** The diagram shows how parallel computation in an SIMD architecture may have different facets. Multiple binary images can be packed into a single image with integer values and bit-level parallelism of the logical function provided by the PEs provides increased efficiency.

(1) for inhibited (active) processors. While the main computational parallelism exposed by array processors is at the level of the processing elements, the fact that each PE processes binary representations of integer values exposes an additional level of parallelism that can be exploited in the processing of binary images (see Figure A.2 and Figure A.5). Multiple binary images can be packed into a single integer image so that each PE effectively processes them at the same time when operating on the integer pixel values.

## A.3. The AnImAl Environment

The requirements of interactive image processing, particularly during the development of algorithms, are the same as those of the larger statistical community and of the even larger community of computational science. A good development environment should rely on a high-level programming language with extensive facilities for the manipulation of the most commonly used data structures, the possibility of saving the status of a processing session for later continuation, extensive community support, and means for code verification and high-quality documentation. Another important, often neglected, feature is the possibility of accessing the source code of the environment and of its libraries so that the appropriateness of the algorithms underlying the functions provided and the correctness of their implementation can be assessed. R, a language and environment for statistical computing and graphics, based on the system described in Chambers (1998), fulfills all these requirements and has been chosen as the programming environment for the implementation of the image algebra described in Section A.1. Several of the algorithms described in the present book have been implemented in the R environment and are provided as an accompanying set of computer files with extensive documentation and examples.

What we would like to address in the following paragraphs is the extent to which such a development environment can support investigations based on the iteration of hypothesize,

**Table A.1.** Co-occurrence of dark, midtone, and highlight pixels

| | Dark | Midtone | Highlight |
|---|---|---|---|
| Dark | 19 004 | 1 300 | 306 |
| Midtone | 1 096 | 11 722 | 1 001 |
| Highlight | 409 | 735 | 5 329 |

test, and refinement cycles, and the accurate reporting of results. The R environment supports selective saving of itself: objects, memory contents, and command history can be saved and imported at a later date or visually inspected. The history mechanism provided by AnImAl extends these facilities by adding self-documenting abilities to a specific data structure (the image) making it even easier to extract the details of every processing step in order to accurately document it. Accurate reporting of results, and of processing workflow, means that enough information should be provided to make the results reproducible: this is the essence of reproducible research. Traditional means of scientific dissemination, such as journal papers, are not up to the task: they merely cite the results supporting the claimed conclusions but do not (easily) lend themselves to independent verification. A viable solution in the case of the computational sciences is to adopt more flexible documentation tools that merge as far as possible data acquisition, algorithm description and implementation, and reporting of results and conclusions. A useful concept is that of the compendium: a dynamic document that includes both literate descriptions of algorithms and their actual implementation. The compendium is a dynamic entity: it can be automatically transformed by executing the algorithms it contains, obtaining the results commented upon by the literate part of the document. This approach has two significant advantages: enough information for the results to be reproducible by the community is provided; and results reported in the description are aligned to the actual processing workflow employed.

The R environment provides extensive support for the creation of compendiums. Code such as

```
sampleLuminance <- ia_averageImageChannels(sample,
                                           c(1,2,3),
                                           c(21,72,7))
dens            <- density(sampleLuminance@data)

tm_eps_on(file = "figures/sampleHisto.eps")

hist(sampleLuminance@data,
     xlab = "Intensity", ylab = "Probability",
     main = "Luminance histogram and density plot",
     probability = TRUE)
     lines(dens, lty = 2)

tm_eps_off()
```

can be freely inserted within the text, with the possibility of hiding it in the final document. The code can be executed, and its results, be they pictorial (see Figures A.3–A.5) or numeric (see Table A.1), automatically inserted in the right places.

Literate algorithm descriptions can be inserted as comments in the source code and automatically extracted and formatted to match high-quality scientific publication requirements. We report a simple example: the definition of an R function, based on AnImAl, for the computation of the boundaries of shapes in binary images. Insertion in the compendium of the directive

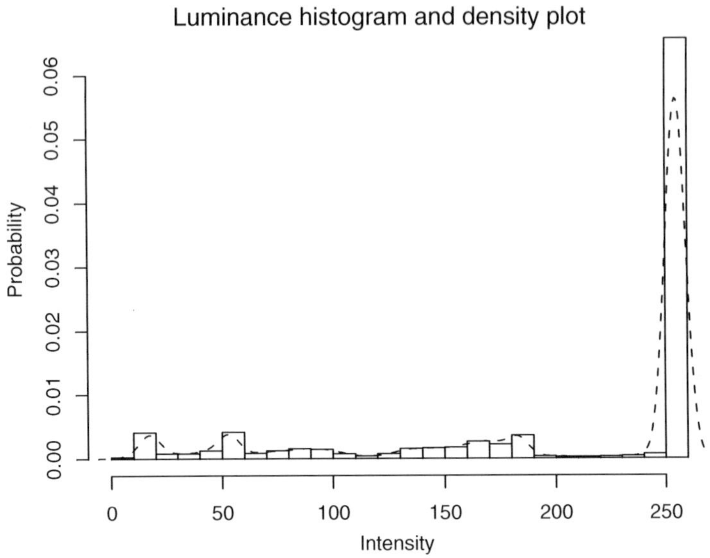


**Figure A.3.** The histogram of the luminance channel $L$ of a sample image (computed from the $R$, $G$, and $B$ color channels as $L = 0.2126R + 0.7152G + 0.0722B$) with overlain density information.

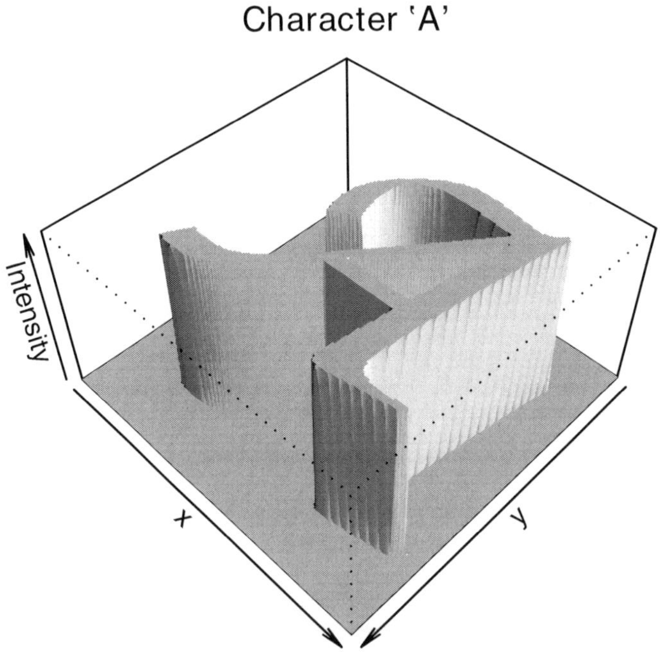


**Figure A.4.** The R environment provides extensive graphical support that can be applied to the visualization of AnImAl image data. This plot presents a perspective view of the character A obtained by mapping intensity information to surface height.

**Figure A.5.** An example of packed bit images (left) and the result of applying the operator `ia_edge` described in the text (right). Parallel processing occurs at two different levels: simulated, at the PEs which compute the same logical operation on their neighborhood; and real as each PE computes the same logical operation for all the bit planes at the same time. The resulting image can then be unpacked providing three separate images, each one containing the boundary of a single character shape.

```
\inputcode{../AnImAl/R/ia_edge.R}
```

generates the literate description of the function reported in the next paragraphs by suitably formatting code and comments included in the function definition file.

*A.3.0.0.1. Edge detection for a binary image*

This is an example of how the basic bit-level logical operators provided by AnImAl can be used to extend its functionalities. In this particular case we define a new function with name `ia_edge` and a single argument `img`:

```
ia_edge <- function (img) {
```

The first step is to check whether image history must be updated in order to produce a coherent image:

```
document <- img@document
```

We momentarily turn off self-documentation as we want to consider the current function as a whole and do not want its inner workings to be traced by the history mechanism:

```
img@document <- FALSE
```

In this particular case we fix the region of interest of the resulting image to that of the function argument. As subsequent operations may be enlarged we store the initial specification of the region

```
img_mask <- ia_mask(img)
```

A foreground pixel (value = 1) is considered to be an edge pixel if at least one of the pixels to its left, right, top, or bottom belongs to the image background (value = 0):

$$E_{i,j} \leftarrow \overline{(((I_{i,j-1} \wedge I_{i,j+1}) \wedge I_{i+1,j}) \wedge I_{i-1,j})} \wedge I_{i,j}.$$

This logical formula can be easily implemented using the bit-level logical operations provided by AnImAl:

```
res <- ia_get(ia_and(img,
                     ia_not(ia_and(ia_up(img),
                                   ia_and(ia_down(img),
                                          ia_and(ia_left(img),
                                                 ia_right(img)
                                                 ))))),
              img_mask)
```

If required by the configuration of the argument image `img` we update the history of the result image `res` to keep it coherent (see Definition A.1.8)

```
if(document) {
```

In this case, it is necessary to update the tracing flag

```
  img@document <- res@document <- document
```

and to get the new root of the history tree which is given by the current function invocation:

```
  resh         <- match.call()
```

We then need to expand the node corresponding to `img` with the corresponding history (see Example A.1.7)

```
  resh$img     <- img@history
```

so that complete history information can be stored in the resulting image:

```
  res@history  <- resh
}
```

before returning the final result:

```
res }
```

An example of the results provided by this function is reported in Figure A.5.

We have not commented so far on how data acquisition fits within the concept of a compendium. Unfortunately this stage is not reproducible unless data result from simulation experiments, a situation for which the reasoning already employed can be applied without any modification. The approach followed in this book is to leverage the capability of modern graphical rendering systems to automatically generate high-quality imagery on which algorithms are trained and compared, thereby extending the application of reproducible research ideas to the complete data flow. The entire Appendix B is devoted to the description of how synthetic, realistic images of complex objects can be generated. Data synthesis can then be regarded as an additional function and merging it with an active document does not require the introduction of any new concept or tool.

## A.4. Bibliographical Remarks

The image processing environment AnImAl is described in Brunelli and Modena (1989) where additional examples and details are presented. The definition of the more basic concepts present in the definition can be found in the *Encyclopedic Dictionary of Mathematics* (Mathematical Society of Japan 1993). The extension of the concepts of elementary algebra, leading to modern and universal algebra, is discussed by Birkhoff (1971) and Lipson (1981).

Several papers consider programming environments and languages for image processing. Among them, the closest to the one discussed are by Levialdi *et al.* (1981), and Kulpa (1981).

Image processing based on mathematical morphology, as discussed in Serra (1986) and in Haralick *et al.* (1987), is particularly convenient in the AnImAl environment.

Single instruction, multiple data machines are discussed in Flynn (1966) while Hwang and Briggs (1984) discuss the communication network of the processing elements as a set of data routing functions.

An introduction to the R environment can be found in Everitt and Hothorn (2006) while more detailed information can be found in the extensive documentation accompanying the software environment.

The approach to reproducible research is based on Gentleman and Lang (2007) and the idea of dynamic documents including algorithm description and implementation is considered also in Knuth (1992). Reproducible research is considered in many papers (e.g. Carey 2001; de Leeuw 1996; Mansmann *et al.* 2004; Rossini and Leisch 2003).

## References

Birkhoff G 1971 The role of algebra in computing. *SIAM–AMS Proceedings of Computers in Algebra and Number Theory*, vol. 4, pp. 1–48.

Brunelli R and Modena C 1989 ANIMAL: AN IMage ALgebra. *High Frequency* **LVIII**, 255–259.

Carey V 2001 Literate statistical programming: concepts and tools. *Chance* **14**, 46–50.

Chambers J 1998 *Programming with Data*. Springer.

de Leeuw J 1996 Reproducible research: the bottom line. Statistics Series 301, Department of Statistics, UCLA.

Everitt B and Hothorn T 2006 *A Handbook of Statistical Analyses Using R*. Chapman & Hall/CRC Press.

Flynn M 1966 Very high-speed computing systems. *Proceedings of the IEEE* **54**, 1901–1909.

Gentleman R and Lang D 2007 Statistical analyses and reproducible research. *Journal of Computational and Graphical Statistics* **16**, 1–23.

Haralick R, Sternberg S and Zhuang X 1987 Image analysis using mathematical morphology. *IEEE Transactions on Pattern Analysis and Machine Intelligence* **9**, 523–550.

Hwang K and Briggs F 1984 *Computer Architecture and Parallel Processing*. McGraw-Hill.

Knuth D 1992 Literate programming. CSLI Lecture Notes 27, Center for the Study of Language and Information, Stanford, California.

Kulpa Z 1981 PICASSO, PICASSO-SHOW and PAL–a development of a high-level software system for image processing. In *Languages and Architectures for Image Processing* (ed. MJB Duff and S Levialdi), pp. 13–24. Academic Press.

Levialdi S, Moggiolo-Schettini A, Napoli M, Tortora G and Uccella G 1981 On the design and implementation of PIXAL, a language for image processing. In *Languages and Architectures for Image Processing* (ed. MJB Duff and S Levialdi), pp. 89–98. Academic Press.

Lipson J 1981 *Elements of Algebra and Algebraic Computing*. Addison-Wesley Longman.

Mansmann U, Ruschhaupt M and Huber W 2004 Reproducible statistical analysis in microarray profiling studies. *Proceedings of the 7th International Conference on Applied Parallel Computing (PARA'04)*, Lecture Notes in Computer Science, vol. 3732, pp. 939–948. Springer.

Mathematical Society of Japan 1993 *Encyclopedic Dictionary of Mathematics* 2 edn. MIT Press.

Rossini A and Leisch F 2003 Literate statistical practice. UW Biostatistics Working Paper Series 194, University of Washington, Washington.

Serra J 1986 Introduction to mathematical morphology. *Computer Vision, Graphics and Image Processing* **35**, 283–305.

# B SYNTHETIC ORACLES FOR ALGORITHM DEVELOPMENT

This is as strange a maze as e'er men trod
And there is in this business more than nature
Was ever conduct of: some oracle
Must rectify our knowledge.

*The Tempest*
WILLIAM SHAKESPEARE

A key need in the development of algorithms in computer vision (as in many other fields) is the availability of large datasets for training and testing them. Ideally, datasets should cover the expected variability range of data and be supported by high-quality annotations describing what they represent so that the response of an algorithm can be compared to reality. Gathering large, high-quality datasets is, however, a time-consuming task. An alternative is available for computer vision research: computer graphics systems can be used to generate photorealistic images of complex environments together with supporting ground truth information. This appendix shows how these systems can be exploited to generate a flexible (and cheap) evaluation environment.

## B.1. Computer Graphics

The goal of computer vision is that of recovering as much information as possible on the scene structure from one or more images of it. Many of the techniques presented in the book are based on probabilistic or statistical models that must be estimated from suitably large datasets. In order to be useful, database images must often have associated information providing a detailed description of what image areas depict, camera parameters, light source configuration, object position and orientation, and so on. In many cases the information is provided by people and may contain errors and inaccuracies. As a result, the creation of large databases with the required supplemental, extensive, and accurate information is a lengthy and expensive task.

In some cases a practical solution is the generation of synthetic image databases using computer graphics techniques. Computer vision and computer graphics complement each other in the sense that the goal of the latter is the generation of images from an explicit description of a scene, while computer vision tries to guess at the scripts used by looking at the images. Tremendous efforts went into the development of computer graphics techniques for the generation of synthetic images that are indistinguishable from real ones and they were crowned by success.

*Template Matching Techniques in Computer Vision: Theory and Practice* Roberto Brunelli

It is not possible to provide a detailed description of the inner workings of a photorealistic computer graphics system in a few pages, but some notes are needed to explain how these systems can be used to support computer vision.

We saw in Chapter 2 that images derive from the interaction of electromagnetic waves with the physical world. No interesting things happen to photons traveling in empty space but the interaction of light with matter is complex and many phenomena arise. Photorealistic images generated by computer graphics systems need to simulate (or mimic) some of them. The difference between simulating and mimicking them is not so important for many applications of computer graphics imagery, especially those related to the movie industry, where the main goal is to generate images that satisfy the spectator, but it is of course important if we want to use them in computer vision applications. To help the reader appreciate the scope of optical phenomena investigated by computer graphics, we may mention:

- basic phenomena, including straight propagation, specular reflection, diffuse reflection (Lambertian surfaces), selective reflection, refraction, reflection and polarization (Fresnel's law), exponential absorption of light (Bouguer's law);
- complex phenomena, including non-Lambertian surfaces, anisotropic surfaces, multilayered surfaces, complex volumes, translucent materials, polarization;
- spectral effects, including spiky illumination, dispersion, interference, diffraction, Rayleigh scattering, fluorescence, and phosphorescence.

The most natural approach to the generation of realistic images is apparently that of simulating the photons emitted by the light sources, following them along their path during which they interact with matter. However, many of the generated photons would never make their way to the final image. The technique known as ray tracing follows the path of the photons backwards, from the image through the lens to the light interacting with the world surfaces (see Figure B.1). Let us assume, for the moment, that we have a convenient description of world objects and that we know how to compute the appearance of a point lit by a point (i.e. non-extended) light source to which we limit ourselves. The basic algorithm is simple:

1. Rays are emitted into the scene from a virtual eye through each pixel.
2. The first object intersected by the ray is computed:
   (a) if it is a diffusing surface, rays from the intersection to all light sources in the scene are cast and the lights that can be reached without intersecting other objects contribute to the shading of the pixel associated with the original ray: if no light can be reached, the intersection point will be black (shadowed);
   (b) if the surface is specular or semitransparent, a new ray will be emitted from the intersection according to Snell's law, and it will be followed as the original one. A limit on the depth of recursion of the process is imposed and the contributions of secondary rays are summed.

The basic algorithm has several shortcomings:

1. In the real world surfaces act as secondary illumination sources, modifying the illumination of all other surfaces.

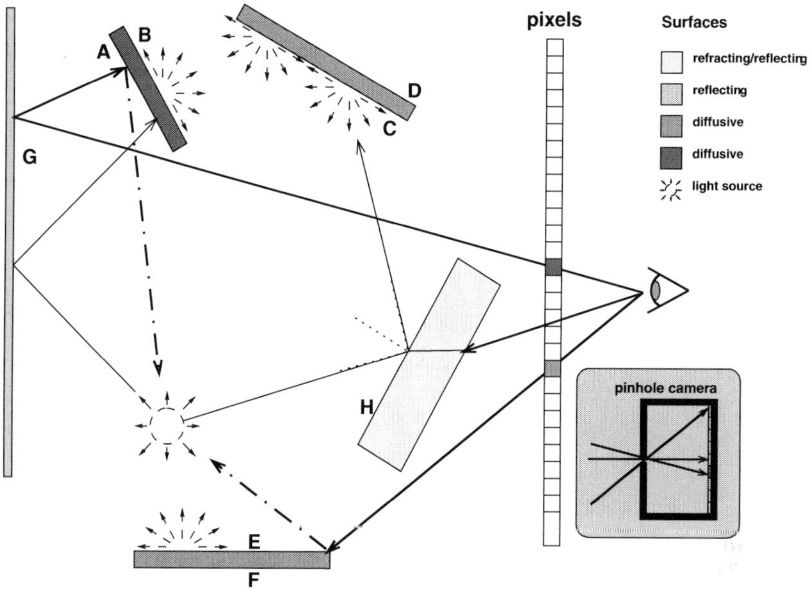


**Figure B.1.** The same world of Figure 2.1 from a backward ray tracing perspective. Backward rays are traced from the viewpoint through each image pixel into the world. The shaded inset shows the equivalence of this imaging model with that of the simple pinhole camera. Surface **B** is not directly illuminated and all its points are considered to be shadowed by the ray tracer. Diffuse illumination from surface **C** is taken care of by more sophisticated rendering algorithms such as radiosity and photon mapping.

2. Refracting surfaces focus light on surfaces, originating the so-called caustics. An example is that of light passing through a glass of wine: the light modifies its color and is refracted and focused on the table where the glass is located.
3. Images may exhibit jaggedness at image discontinuities.
4. Point light sources create unrealistic, sharp, shadowing.

While the last two can be accommodated within the framework of backward ray tracing by increasing the number of rays used, the first two require a different approach. Radiosity is a solution to the first problem based on ideas originally developed to describe heat transfer. It accounts for the effect of light that is reflected diffusively multiple times before hitting the (virtual) sensor. The idea underlying the solution is simple. All surfaces are divided into (small) patches: each of them will contribute to the illumination of all those that see it and it will contribute proportionally to its angular extension as seen from the patch receiving the contribution. Starting from an initial configuration, with no diffuse interreflections, the solution is found iteratively, by successive refinements. More formally

$$B_i = E_i + R_i \sum_{j=1}^{n} B_j f_{ij} \tag{B.1}$$

where $B_i$ is the radiosity of patch $i$, $E_i$ its emitted energy, $R_i$ the reflectivity, and $f_{ij}$ the constant and dimensionless form factor. The above equation is monochromatic and multiple equations must be used for color rendering. Images generated with ray tracing and radiosity have a very realistic appearance and can achieve photometric accuracy. A more recent and flexible solution overcoming the shortcomings of ray tracing is photon mapping, a forward ray tracing technique, capable of generating high-quality caustics, diffuse interreflections, realistic simulations of participating media resulting in complex volumetric effects, and also subsurface scattering effects. The basic idea is, once again, simple and is based on a two-pass algorithm:

1. Construction of the photon map. Photons are generated from the light sources and shot into the scene. When a photon hits a surface, the intersection and the direction of the photon are stored in the photon map. After hitting the surface, the fate of the photon is decided probabilistically. Depending on the characteristics of the material, a probability for the absorption, transmission/refraction, reflection is computed and one of the actions is chosen randomly based on the given probabilities. If the photon is absorbed, its path ends. Otherwise, the characteristics of the material dictate its path.

2. Rendering. This pass is similar to the original ray tracing algorithm, but the shading of each pixel additionally exploits the information contained in the photon map. Direct illumination effects are managed as in basic backward ray tracing. The information stored in the photon map on the photons associated with the ray intersection point is used to compute the contribution of soft, indirect illumination. Caustics need a high number of photons for a limited number of objects and their computation is usually based on a dedicated photon map.

While the ideas underlying radiosity and photon mapping are simple, their efficient and accurate implementation is complex. Very efficient implementations of related methods exist that are able to reach real-time performance.

An important point that was mentioned in the above paragraphs is that of computing the shading of a pixel using the characteristics of the material. This is a very important point because in a realistic image we want to simulate correctly not only light propagation, but also light interaction with the materials that our world is composed of. The appearance of a surface point is determined by solving the so-called rendering equation

$$L_o(\boldsymbol{x}, \hat{\boldsymbol{o}}, \lambda) = L_e(\boldsymbol{x}, \hat{\boldsymbol{o}}, \lambda) + \int_\Omega f_r(\boldsymbol{x}, \hat{\boldsymbol{\imath}}, \hat{\boldsymbol{o}}, \lambda) L_i(\boldsymbol{x}, \hat{\boldsymbol{\imath}}, \lambda)(\hat{\boldsymbol{\imath}} \cdot \hat{\boldsymbol{n}})\, d\hat{\imath} \tag{B.2}$$

where $\hat{\cdot}$ denotes the versor (direction) associated to a given vector, $L_o(\boldsymbol{x}, \hat{\boldsymbol{o}})$ is the amount of light leaving $\boldsymbol{x}$ in direction $\hat{\boldsymbol{o}}$, $\hat{\boldsymbol{\imath}}$ is the incoming direction, $\hat{\boldsymbol{n}}$ is the surface normal, $L_e$ and $L_i$ are respectively the emitted and inward light, the dot product $(\hat{\boldsymbol{\imath}} \cdot \hat{\boldsymbol{n}})$ accounts for the attenuation of light due to the foreshortening angle, and $f_r(\boldsymbol{x}, \hat{\boldsymbol{\imath}}, \hat{\boldsymbol{o}}, \lambda)$ is the bidirectional reflectance distribution function (BRDF) denoting the proportion of light from direction $\hat{\boldsymbol{\imath}}$ reflected in direction $\hat{\boldsymbol{o}}$ (see Figure B.2(a)). The BRDF is then a function of five parameters: four angles and light wavelength $\lambda$. In many practical cases the material can be considered isotropic and the BRDF is invariant to rotations about the surface normal, reducing the number of effective variables to four. In most situations a further simplification is introduced condensing wavelength dependency into a set of three measurements corresponding to trichromatic vision (the RGB channels). Radiosity and photon mapping are two ways of solving the rendering equation.

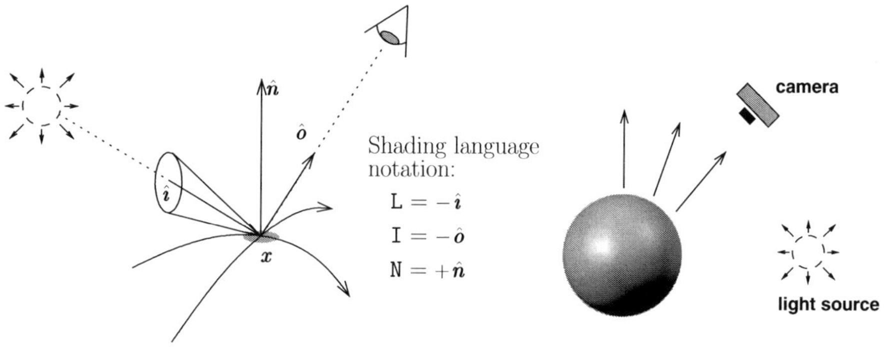


(a) BRDF geometry (b) image-based BRDF estimation

**Figure B.2.** BRDF geometry (a) and image-based estimation setup (b). Image-based estimation takes advantage of the fact that the sphere normals span all angles with respect to the light source, permitting the estimation of many BRDF values from a single snapshot.

The BRDF provides the radiometric characterization of a material and is used in computer graphics to generate photorealistic images and in computer vision to recognize objects from their radiometric signature. In order to provide a physically correct description, a BRDF must have the following two properties:

- obey the Helmholtz reciprocity law

$$f_r(\boldsymbol{x}, \hat{\boldsymbol{\imath}}, \hat{\boldsymbol{o}}, \lambda) = f_r(\boldsymbol{x}, \hat{\boldsymbol{o}}, \hat{\boldsymbol{\imath}}, \lambda); \tag{B.3}$$

- conserve energy

$$\int_\Omega f_r(\boldsymbol{x}, \hat{\boldsymbol{\imath}}, \hat{\boldsymbol{o}}, \lambda)\, d\hat{\boldsymbol{o}} \leq 1 \quad \forall \hat{\boldsymbol{\imath}}. \tag{B.4}$$

The BRDF of a material must usually be estimated and a simple way to do it is illustrated in Figure B.2(b).

**Intermezzo B.1.** The BRDF of human skin

The BRDF of human skin can be measured like that of any other material. The results reported in the literature show that it is somewhat unusual. At small incidence angles it is almost Lambertian, but it exhibits strong forward scattering at higher angles. The scattering does not correlate with specular direction and, as a consequence, cannot be described with the Phong model or with rough surface models of the Sparrow–Torrance type (surfaces are considered as composed of many microfacets, whose orientation distribution characterizes the optical properties of the surface). One of the reasons is that skin is actually a translucent material: light penetrates into the material, is scattered by interaction with the inner layers of the material, and finally emerges at a different point. In the case of skin only 6% of the reflectance is direct, the remaining part being due to subsurface scattering. The characteristics of the skin BRDF make it difficult to simulate it by traditional ray tracing techniques, requiring the use of the more advanced photon mapping technique for accurate rendition (Donner and Jensen 2006; Weyrich *et al.* 2006).

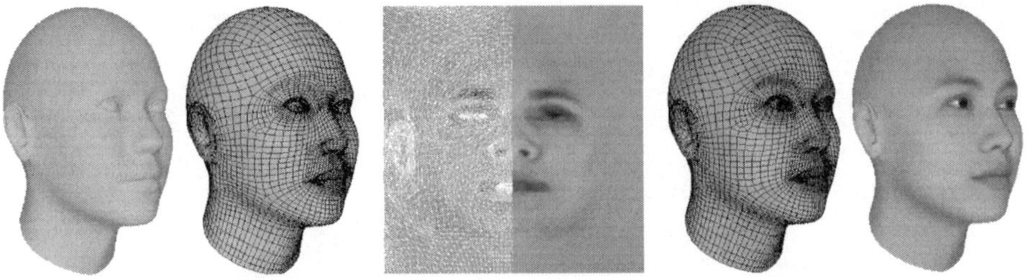

**Figure B.3.** The UV map processing for a head model.

## B.2. Describing Reality: Flexible Rendering Languages

The brief introduction to computer graphics presented in the previous section shows that the generation of realistic images requires a detailed description of the geometry of the world and its surfaces.

As far as geometry is concerned, several approaches have proven to be useful, ranging from surface modeling by means of a single geometric entity, the triangle, to the use of analytic surface expressions for planes, quadric, and spline surfaces. Complex surfaces and volumes can be described by means of constructive solid geometry which applies set operations like union and intersection to surfaces and volumes. The advantage of using analytic surface representations is clear when we consider that the rendering equation requires knowledge of the surface normal. When triangles are used, the surface is locally flat resulting in unrealistic flat shading. This artifact can be cured by generating the normals at each point by interpolation of those from nearby triangles. The resulting approximation is often adequate but not as precise as using the correct surface normals available from an analytic representation.

The next important step is that of linking the purely geometric description of the world to the corresponding BRDFs. Given the coordinates of a point belonging to an object we need to access all the information required for its accurate visual representation. The link is usually established by means of a system of surface coordinates: each point on the surface of an object is characterized by its three-dimensional coordinates in the world reference system and by its two-dimensional coordinates with respect to a surface-centric reference system. When the object moves, the three-dimensional coordinates of its points in the world reference system change, but the corresponding two-dimensional surface coordinates do not. The visual properties of the material are then naturally linked to the surface two-dimensional coordinates. While the choice of the surface coordinates is often a natural one, albeit not unique, for the basic geometric elements, the generation of a unique reference system for an object comprising multiple elements is not trivial. The way it is achieved, (semi)automatically or manually, is by means of a so-called UV map, unwrapping all surface elements onto a planar surface: each point on the object surface is then mapped onto the plane and its resulting UV coordinates are the associated rectangular coordinates (see Figure B.3). The UV coordinates of a point can then be used to index the corresponding rendering information.

The variety of BRDFs required for rendering realistic world scenes suggests the development of a shading language by which the different possibilities can be compactly described and made operational within a rendering system. Among the shading languages developed, we focus on the one associated with the RenderMan interface specification, developed by

Pixar Animation Studios to describe three-dimensional scenes for photorealistic rendering. The interface itself formalizes, among other things, how to describe cameras, viewing transformations, gamma correction, pixel filtering for proper sampling, output image formats, and geometric primitives. However, its shading language is the part of interest for two reasons:

1. It is key to the accurate rendering of materials.
2. It allows us to automatically generate image annotations.

When a geometric element is specified in the RenderMan description of a scene, a material can be associated to it. As different materials are characterized by specific parameters, the latter can also be specified in the description. When a point belonging to the surface must be shaded, the rendering system invokes a shading function, whose name corresponds to that of the material. The following code is an example of a description of a minimal world:

```
Projection "perspective" "fov" 35

WorldBegin
  LightSource "pointlight" 1 "intensity" 40 "from" [4 2 4]
  Translate    0 0 5

  Color        1 0 0
  Surface      "plastic" "roughness" 0.01
  Cylinder     1 0 1.5 360
WorldEnd
```

The information required for the computation of the BRDF is automatically computed by the renderer and passed to the shading function as a set of global (environment) variables:

- `E`, position of the eye;
- `P`, surface position;
- `dPdu,dPdv`, derivative of surface position along surface coordinates `u,v`;
- `N`, surface normal;
- `Cs, Os`, surface color and opacity;
- `L, Cl`, incoming light ray direction, color;
- `I`, incident ray direction (from the perspective of backward ray tracing).

All the available supplementary information is specified by means of the additional parameters. The evaluation of the shader results in:

- `Ci, Oi`, the incident ray color and opacity.

Writing of the shaders can take advantage of predefined functions:

- `color diffuse(vector N)`, computes the light scattered uniformly by the surface using Lambert's law

$$\sum_{j=1}^{n_l} C_j^l \max(0, N \cdot L_i) \tag{B.5}$$

with the summation taken over all $n_l$ light sources and a notation resembling that of shader variables is used. A surface fully characterized by the above law is called a Lambertian surface and is often used as a simplified surface model in computer vision;

- `color specular(vector N, I; float roughness)`, a non-physically correct rendition (Blinn–Phong model) of the highlights exhibited by glossy surfaces

$$\sum_{j=1}^{n_l} C_j^l \max\left(0, N \cdot \frac{(-I + L_j)}{\|-I + L_j\|}\right)^r \tag{B.6}$$

  where $r$ is a measure of surface roughness: when the viewing direction corresponds to the (opposite) of the reflection direction we get the maximum contribution, which decreases with surface roughness;

- `color ambient()`, returning the contribution of ambient light, a value computed by means of radiosity or photon mapping, or approximated by a low-level, constant, color.

A simple shader simulating a Lambertian surface is given by:

```
/* Compute the color of the surface using a simple Lambertian BRDF. */
color MaterialMatte (normal Nf;  color basecolor;  float Ka, Kd;)
{
    return basecolor * (Ka*ambient() + Kd*diffuse(Nf));
}
```

Basic differences between materials, such as white highlights of plastic and colored highlights of metals, can be easily simulated as the following two shaders show:

```
/* Compute the color of the surface using a simple plastic-like BRDF.
 * Typical values are Ka=1, Kd=0.8, Ks=0.5, roughness=0.1.
 */
color MaterialPlastic (normal Nf;  color basecolor;
                      float Ka, Kd, Ks, roughness;)
{
    extern vector I;
    return basecolor * (Ka*ambient() + Kd*diffuse(Nf))
          + Ks*specular(Nf,-normalize(I),roughness);
}

/* Compute the color of the surface using a simple metal-like BRDF.  To
 * give a metallic appearance, both diffuse and specular components are
 * scaled by the color of the metal.  It is recommended that Kd < 0.1,
 * Ks > 0.5, and roughness > 0.15 to give a believable metallic appearance.
 */
color MaterialRoughMetal (normal Nf;  color basecolor;
                         float Ka, Kd, Ks, roughness;)
{
    extern vector I;
    return basecolor * (Ka*ambient() + Kd*diffuse(Nf) +
                       Ks*specular(Nf,-normalize(I),roughness));
}
```

The use of a texture map providing information to color each point on a surface given its UV mapping is illustrated by the following shader:

```
surface
paintedplastic (float Ka = 1, Kd = .5, Ks = .5, roughness = .1;
               color specularcolor = 1;
               string texturename = ""; )
{
```

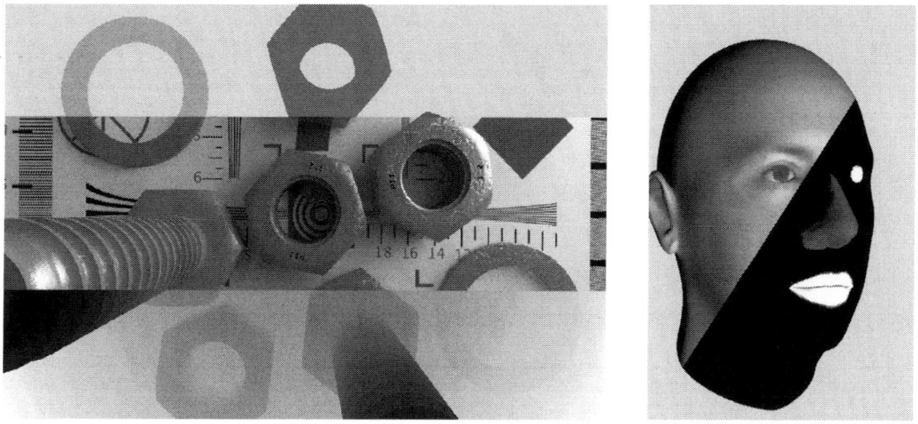

**Figure B.4.** The flexibility of the RenderMan shading language allows us to generate photorealistic images under varying conditions, with perfectly aligned description maps, providing us with depth information, object labels, surface coordinates, etc. The rows of the left image represent, in a top-down order, object labels, the original image, and its camera distance map. The right image shows how facial features can be shaded with custom colors automatically providing a feature map.

```
   Ci = Cs;
   if (texturename != "")
      Ci *= color texture (texturename);

   normal Nf = faceforward (normalize(N),I);
   Ci = Ci * (Ka*ambient() + Kd*diffuse(Nf)) +
        specularcolor * Ks * specular(Nf,-normalize(I),roughness);
   Oi = Os;  Ci *= Oi;
 }
```

The shader `paintedplastic` can then be used to project an unwrapped head texture onto the actual surface as in Figure B.3.

We remarked that the usefulness of shaders is not limited to realistic surface shading: we can shade a pixel so that its color represents the temperature of the surface, its distance from the observer, its surface coordinates, the material, or an object unique identification code. The advantages for computer vision investigations are immediately apparent: we may generate a set of physically correct images and, for each of them, with pixel-level resolution, we can automatically generate an accompanying set of thematic maps, providing us with all the ground truth information we need to train and test extensively computer vision algorithms and systems (see Figure B.4).

An example of a shader providing this kind of additional information is the following variation of `paintedplastic`:

```
surface paintedplastic(color Ka = .5;
                       color sheenColor = 1.;
                       float eta = 1./1.4, thickness = .5;
                       string texturename     = "";
                       string bodyPart        = "";
                       float  polyID          = 0.0;
                       float  hairContrast    = 1.0;
                       float  eyebrowContrast = 1.0;
                       float  mouthContrast   = 1.0)
{
  normal Nn = faceforward(normalize(N), I);
```

```
  vector Vf = -normalize(I);
  color  Ct;
  color  hairColor     = color(0.5,0.5,0.5);
  color  mouthColor    = color(1,0,0);
  color  eyebrowColor = color(1,1,1);
  color  skinColor     = color(1.0, 0.8,0.8);

  if (texturename != "") {
    Ct = color texture (texturename);
  } else if (bodyPart == "hair") {
    Ct = mix(skinColor, hairColor, hairContrast);
  } else if (bodyPart == "skin") {
    Ct = skinColor;
  } else if (bodyPart == "eyebrow") {
    Ct = mix(skinColor, eyebrowColor, eyebrowContrast);
  } else if (bodyPart == "mouth") {
    Ct = mix(skinColor, mouthColor, mouthContrast);
  }

  Oi = Os;
  Ci = Os * subsurfaceSkin(Vf, Nn, Ct, sheenColor, eta, thickness);

  Oi = color(1,1,1);
  Ci = color(u,v,clamp(distance(E,P)/maxdist, 0, 1));
}
```

This shader can be used to output annotation information, by using the appropriate selector in the file describing the synthesized scene:

```
Surface "paintedplastic" "bodyPart" ["mouth"] "polyID" 1421
Color [1.0 1.0 1.0]
Polygon "P" [  0.036682117730 -0.78101408481 -0.62228918075
                         0.039831120520 -0.82182705402 -0.59872508049
                        -0.003858879208 -0.84264004230 -0.60847401619 ]
        "N" [  0.159550771117 -0.24637593329 -0.95593124628
               0.378521084785 -0.67790764570 -0.63017672300
               0.028168585151 -0.73348188400 -0.67909789085 ]
```

Let us close the appendix with a remark on computation. In principle, a modern rendering system is able to simulate accurately all important aspects of a real imaging system, including motion blur, lens distortion, depth of field, and image noise. Unfortunately, the time required for some of them may be too long for a computer vision researcher. Some effects, however, can be simulated, with varying degrees of accuracy, by operating on the image generated by a rendering system. Typical examples include lens distortion and vignetting effects (see Figure B.5), noise, and even depth of field (see Figure B.6). Some sensor characteristics can also be readily simulated at a postprocessing level, such as the effect of using a Bayer mosaic layout (see Figure 2.14).

## B.3. Bibliographical Remarks

A classical reference textbook for computer graphics is Foley *et al.* (1995). Radiosity was introduced by Goral *et al.* (1984) who applied to computer graphics methods originally applied to model heat transfer. Photon mapping is a more recent addition to computer graphics due to Jensen (2001). A survey of optical phenomena that can be simulated with success is reported by Sun *et al.* (2001). A clear discussion on the constraints imposed by physics on shading models for rendering algorithms can be found in Lewis (1993). Human skin is an example of a hard-to-model material and recent references on realistic modeling of it are Marschner *et al.* (1999), Donner and Jensen (2006) and Weyrich *et al.* (2006).

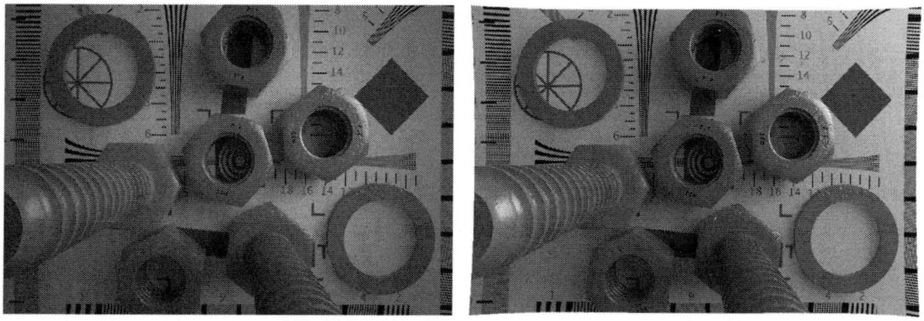

**Figure B.5.** Some optical effects like vignetting (left) and distortion (right) can be simulated using postprocessing techniques. These pictures are based on the data from a real photographic lens.

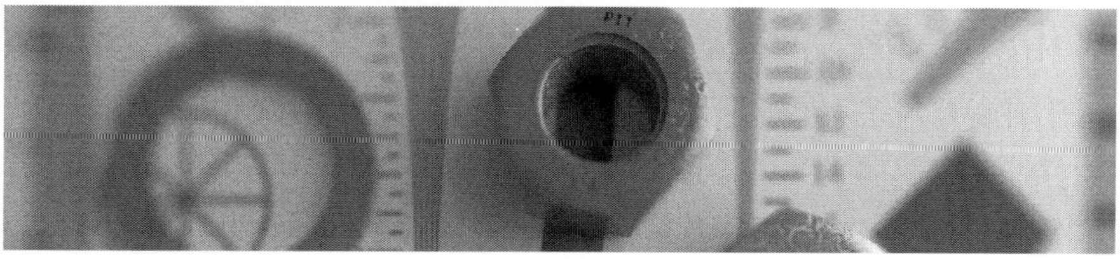

**Figure B.6.** Depth of field can be simulated with ray tracing or photon mapping techniques but getting high-quality results requires a long time as many rays must be traced. Postprocessing techniques based on automatically generated depth maps can be used to mimic the effect qualitatively well and, in some cases, also with good physical accuracy.

An example of using computer graphics for the evaluation of optical flow algorithms is McCane *et al.* (2001). The possibility of generating dynamic environments for testing and training tracking algorithms is discussed extensively in Bertamini *et al.* 2003, Santuari *et al.* 2003 and Qureshi and Terzopoulos (2006, 2008).

RenderMan and Pixar are registered trademarks of Pixar Animation Studios. The RenderMan specification provides a clear overview of the API structure and of basic shading possibilities. A very good guide to writing RenderMan shaders is the book by Apodaca and Gritz (1999). There are several free, good, RenderMan-compliant renderers, such as Pixie and Aqsis. Another free rendering system is PovRay: a variant of it including support for a subset of the RenderMan shading language has been used to generate some of the sample images in this book.

## References

Apodaca A and Gritz L 1999 *Advanced RenderMan: Creating CGI for Motion Pictures*. Morgan Kaufmann.

Bertamini F, Brunelli R, Lanz O, Roat A, Santuari A, Tobia F and Xu Q 2003 Olympus: an ambient intelligence architecture on the verge of reality. *International Conference on Image Analysis and Processing*, pp. 139–144.

Donner C and Jensen H 2006 A spectral BSSRDF for shading human skin. *Proceedings of the Eurographics Symposium on Rendering*, pp. 409–417.

Foley J, van Dam A, Feiner S and Hughes J 1995 *Computer Graphics: Principles and Practice in C*. Addison-Wesley Professional.

Goral C, Torrance K, Greenberg D and Battaile B 1984 Modeling the interaction of light between diffuse surfaces. *Computer Graphics* **18**, 213–222.

Jensen H 2001 *Realistic Image Synthesis Using Photon Mapping*. AK Peters.

Lewis R 1993 Making shaders more physically plausible. *Proceedings of the Eurographics Workshop on Rendering*, pp. 47–62.

Marschner S, Westin S, Lafortune E, Torrance K and Greenberg D 1999 Image-based BRDF measurement including human skin. *Proceedings of the Eurographics Workshop on Rendering*, pp. 139–152.

McCane B, Novins K, Crannitch D and Galvin B 2001 On benchmarking optical flow. *Computer Vision and Image Understanding* **84**, 126–143.

Qureshi F and Terzopoulos D 2006 Surveillance camera scheduling: a virtual vision approach. *Multimedia Systems* **12**, 269–283.

Qureshi F and Terzopoulos D 2008 Smart camera networks in virtual reality. *Proceedings of the IEEE* **96**, 1640–1656.

Santuari A, Lanz O and Brunelli R 2003 Synthetic movies for computer vision applications. *3rd IASTED International Conference: Visualization, Imaging, and Image Processing–VIIP 2003*, pp. 1–6.

Sun Y, Fracchia F, Drew M and Calvert T 2001 A spectrally based framework for realistic image synthesis. *The Visual Computer* **17**, 429–444.

Weyrich T, Matusik W, Pfister H, Bickel B, Donner C, Tu C, McAndless J, Lee J, Ngan A, Jensen H and Gross M 2006 Analysis of human faces using a measurement-based skin reflectance model. *ACM Transactions on Graphics* **25**, 1013–1024.

# C ON EVALUATION

> Sir, the event
> Is yet to name the winner: fare you well.
>
> *Cymbeline*
> WILLIAM SHAKESPEARE

Evaluation of algorithms and systems is a complex task. This chapter addresses four related questions that are important from a practical and methodological point of view: what is a good response of a template matching system, how can we exploit data to train and at the same time evaluate a classification system, how can we describe in a compact but informative way the performance of a classification system, and, finally, how can we compare multiple classification systems for the same task in order to assess the state of the art of a technology?

## C.1. A Note on Performance Evaluation

Choosing an error measure to quantify the performance of a recognition system is straightforward: the response of the system is correct if the label it provides corresponds to the true label. The situation for detection and localization systems is not as simple. When can we assume that a template has been correctly detected or localized? The fact that an appropriate error measure cannot but be task dependent already emerges when we compare detection to localization. In detection tasks, we may be interested in counting the number of appearances of a given template in an image or a set of images: we are interested in a correct count of the instances and not in their precise localization. In this case we may concede a correct detection even if the overlapping of the detected template with the actual image representation is minor. In localization tasks we may be interested in using the information on the location of the template to target it with a weapon, in the case of a military application, or to recognize or verify a person's identity in the case of a biometric system. In the former, a good localization is one that allows us to destroy the target, while in the latter a good localization is one that allows the biometric system to correctly assess the corresponding identity.

Let us investigate the issue in the case of face detection, grounding our measures of correct detection on the correspondence of the estimated eye positions $(\hat{\boldsymbol{x}}_l, \hat{\boldsymbol{x}}_r)$ with their true position $(\boldsymbol{x}_l, \boldsymbol{x}_r)$. A first step towards obtaining meaningful, comparable performance measures is the introduction of an error measure that is invariant to translation, scaling, and rotation. A simple measure is

$$d_{\text{eye}} = \frac{\max(d(\hat{\boldsymbol{x}}_l, \boldsymbol{x}_l), d(\hat{\boldsymbol{x}}_r, \boldsymbol{x}_r))}{d(\boldsymbol{x}_l, \boldsymbol{x}_r)} \tag{C.1}$$

and the corresponding criterion for correct localization, e.g. $d_{\text{eye}} < 0.25$. This error measure has two drawbacks: it is based on an arbitrary threshold and it does not differentiate between translation, scale, and rotation error. The cumulative error associated to eye positions described by Equation C.1 may be decomposed into four different errors that give a complete picture (in the sense that, given the estimated position of the eyes and the errors, the true position can be recovered):

$$\delta_{x_1} = \frac{(x_{l1} + x_{r1})}{2} - \frac{(\hat{x}_{l1} + \hat{x}_{r1})}{2} \tag{C.2}$$

$$\delta_{x_2} = \frac{(x_{l2} + x_{r2})}{2} - \frac{(\hat{x}_{l2} + \hat{x}_{r2})}{2} \tag{C.3}$$

$$\delta_s = \frac{d(\hat{\boldsymbol{x}}_l, \hat{\boldsymbol{x}}_r)}{d(\boldsymbol{x}_l, \boldsymbol{x}_r)} \tag{C.4}$$

$$\delta_\alpha = \arccos\left[\left(\frac{\boldsymbol{x}_l - \boldsymbol{x}_r}{\|\boldsymbol{x}_l - \boldsymbol{x}_r\|_{L_2}}\right) \cdot \left(\frac{\hat{\boldsymbol{x}}_l - \hat{\boldsymbol{x}}_r}{\|\hat{\boldsymbol{x}}_l - \hat{\boldsymbol{x}}_r\|_{L_2}}\right)\right]. \tag{C.5}$$

The errors are shown graphically in Figure C.1. A first step towards a more flexible error estimation procedure is the introduction of error weighting or scoring functions whose parameters may be adapted to different tasks

$$\psi(x; \gamma, \mu, \tau) = \begin{cases} e^{-\gamma^2 |(x-\mu)+\tau|^2} & x \le \mu - \tau \\ 1 & |x - \mu| \le \tau \\ e^{-\gamma^2 |(x-\mu)-\tau|^2} & \mu + \tau \le x \end{cases} \tag{C.6}$$

where $x$ is the error or criterion to be scored and errors are mapped into the range [0, 1]; the lower the score, the worse the error. A single face detection system can then be scored differently when considered as a detection or localization system by changing the parameters controlling the weighting functions. For instance, we could use $\boldsymbol{a}_d$

$$\boldsymbol{a}_d = (\gamma = 0.5, \mu = 0, \tau = 5) \tag{C.7}$$

to score detections and $\boldsymbol{a}_l$

$$\boldsymbol{a}_l = (\gamma = 1, \mu = 0, \tau = 2.5) \tag{C.8}$$

to score localization, using more peaked scoring functions. Having defined a scoring function allows us to estimate the detection and false alarm rates. The detection system provides a set of detected positions $\{l_i\}_{i=1}^n$ that must be compared to a known set of locations $\{L_j\}_{j=1}^m$. We associate to each true position the detected position with the highest score and we insert it into the set of good detection if the scoring is above a given acceptance threshold. If $r$ is the cardinality of the good detection set we define the detection rate as $r/m$ and the false alarm rate as $1 - r/n$.

While this procedure may be enough for some applications, it is still generalist, providing a flexible error measure that must be tuned to the different tasks. Let us now consider in detail a face verification application that includes as a preliminary step the localization of the face whose identity must be verified. In this case, what we would like is to score the localization system by the impact of its precision on the final verification error. The following remarks are in order:

- each type of error (translation, scale, and rotation) is expected to have a different impact on the final verification error;

- the impact of each error type depends on the verification system used.

Experimental support for both remarks is available in the literature. Let us assume that we want to optimize the localization stage of a biometric system whose final verification stage is fixed. The variability of the localization system may derive from the presence of several inner parameters (such as the different scales at which templates are used, or the sampling scheme of the image). The key step towards evaluating the influence of a localization subsystem on the complete verification system is the estimation of the final verification error $\Delta_v$

$$\Delta_v(\{\boldsymbol{x}_i\}) = \sum_i f(\boldsymbol{\delta}(\boldsymbol{x}_i); \boldsymbol{\theta}) \tag{C.9}$$

as a function of the detailed error information that can be associated to each localization $\boldsymbol{x}_i$

$$\boldsymbol{\delta}(\boldsymbol{x}_i) = (\delta_{x_1}(\boldsymbol{x}_i), \delta_{x_2}(\boldsymbol{x}_i), \delta_s(\boldsymbol{x}_i), \delta_\alpha(\boldsymbol{x}_i)). \tag{C.10}$$

The vector $\boldsymbol{\theta}$ represents all parametric information describing the post-localization processing of the verification system. The function $\Delta_v(\{\boldsymbol{x}_i\})$ can be estimated by subjecting a set of images that should be verified by the system to perturbations corresponding to the expected errors: each perturbed image is then associated to an error vector $\boldsymbol{\delta}$ and presented to the verification system and the resulting point $f(\boldsymbol{\delta}; \boldsymbol{\theta})$ used to estimate $f(\cdot)$. A simple way to estimate $f$ is that of using the average over the $k$ nearest neighbors of the function argument

$$f_{\mathrm{knn}}(\boldsymbol{\delta}) = \frac{1}{k} \sum_{j \in \mathrm{knn}(\boldsymbol{\delta})} C_j \tag{C.11}$$

where index $j$ runs over the available samples and $C_j$ represents the cost associated to the $j$th classification, for instance

$$C_j = \begin{cases} 0 & \text{client accepted} \\ 1 & \text{client rejected.} \end{cases} \tag{C.12}$$

The resulting estimated function $f_{\mathrm{knn}}$ can be used to gauge different localization systems by the induced verification error (given the verification system for which it was computed). If the localization system is trained from data, $f_{\mathrm{knn}}$ can be given a more active role, biasing the training of the localization systems towards errors that have a reduced impact on the final face verification task.

## C.2. Training a Classifier

We have seen in Chapter 3 that a useful way to look at template matching is that of classification: to synthesize a function of sensory input that can be used to assess the presence of a specific template (recognition) or of a generic instance of a given template class (detection). We have seen that synthesizing the classifier requires the estimation of one or more parameters, such as the covariance matrix of noise or a thresholding value. In the general case we have a set of data and the possibility of building multiple classifiers by estimating their operating parameters. The goal is to synthesize the classifier whose performance on the complete set of possible inputs is maximal. We then face a classifier (model) selection

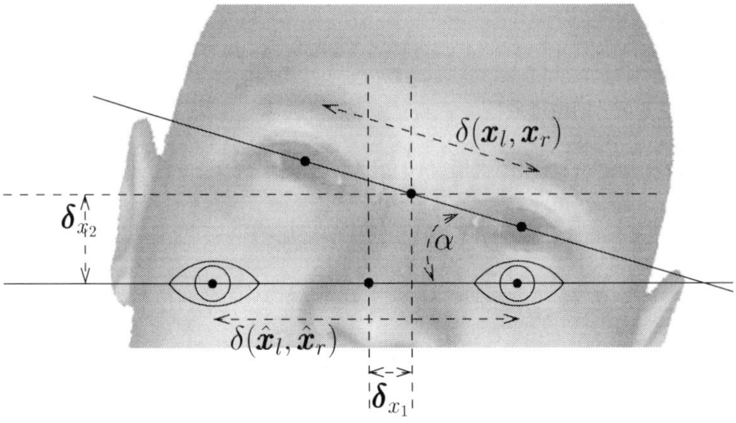


**Figure C.1.** The decomposition of localization error into multiple components.

task that includes the estimation of the corresponding optimal operating parameters, and a performance estimation task, as we are interested in the performance of the chosen classifier on the universe of possible inputs. A naive, but incorrect approach is to use all available data for training and testing the classifier. If we use all available data to build our classifier, the estimate of performance we get from applying it to the same dataset may be severely biased: the system has already seen the data (during the training) and it may rely on its memory to recognize them. The memory of the patterns need not be explicit or complete. Let us think about the simple problem of estimating the mean and the variance of a set of points. The value of the mean represents our target function while the variance gauges its error in representing our data. If we further assume that our points are Gaussian distributed, it is a basic result that the expected value of the arithmetic mean corresponds to the average of the Gaussian and the variance of the estimated value decreases with the number of samples used to compute the average. However, the sample variance is not an unbiased estimator of the variance of the distribution: as we already used the data to estimate the mean, the value of the sample variance underestimates the true value. However, the bias becomes negligibly small as we consider more and more data: it is asymptotically unbiased.

Let us now formalize the problem and see how a limited set of samples can be used to select the most appropriate model and to estimate its performance. Let $\mathcal{X}$ be the space of possible inputs (without label), $\mathcal{L}$ the set of labels, $\mathcal{S} = \mathcal{X} \times \mathcal{L}$ the space of labeled samples, and $D = \{s_1, \ldots, s_N\}$, where $s_i = (x_i, l_i) \in \mathcal{S}$, our dataset. A classifier is a function $\mathfrak{C} : \mathcal{X} \to \mathcal{L}$, while an inducer is an operator $\mathfrak{I} : D \to \mathfrak{C}$ that maps a dataset into a classifier. The accuracy $\epsilon$ of a classifier is the probability $p(\mathfrak{C}(\boldsymbol{x}) = l, (\boldsymbol{x}, l) \in S)$ that its label attribution is correct. The problem is to find a low-bias and low-variance estimate $\hat{\epsilon}(\mathfrak{C})$ of $\epsilon$. There are three main different approaches to accuracy estimation and model selection:

- hold-out;
- bootstrap;
- $k$-fold cross-validation.

The hold-out method is the simplest: a subset $D_h$ of $n_h$ points is extracted from the complete dataset and used as the testing set while the remaining set $D_t = D \setminus D_h$ of $N - n_h$ points is provided to the inducer to train the classifier. Stratified hold-out selects the samples from the different classes so that their distributions in the training and testing set match that of the complete dataset. The induced classifier $J(D_t; \cdot)$ is applied in turn to each sample in the testing set, resulting in a sequence of Bernoulli trials: correct or incorrect prediction. The accuracy is estimated as

$$\hat{\epsilon}_h = \frac{1}{n_h} \sum_{\boldsymbol{x}_i \in D_h} \delta[J(D_t; \boldsymbol{x}_i), l_i] \tag{C.13}$$

where $\delta(i, j) = 1$ when $i = j$ and 0 otherwise. Being the sum of $n$ Bernoulli trials, $\hat{\epsilon}_h$ follows a binomial distribution that can be approximated for reasonably large $n_h$ by a Gaussian distribution $N(\epsilon, \epsilon(1-\epsilon)/n_h)$, from which an estimate of the variance of $\epsilon$ follows. The hold-out method may be applied multiple times by changing the split, a procedure called random sampling. The accuracy of the classifier and its variance can be estimated from the results over the multiple samples.

A different approach that by its very definition preserves the original distribution of data is provided by the bootstrap. The accuracy and its variance are estimated from the results of the classifier over a sequence of bootstrap samples, each of them obtained by random sampling with replacement $N$ instances from the original dataset. Sampling with replacement has two important consequences:

1. The distribution of data is preserved in each bootstrap sample.
2. The probability of any given data point not being selected after $N$ samples is $(1 - 1/N)^N \approx e^{-1} \approx 0.368$, so that the expected number of distinct instances in each bootstrap sample is $0.632N$.

The instances within the bootstrap subset are used for testing, the remaining ones being given to the inducer. The accuracy $\epsilon_{\text{boot}}$ is then estimated as

$$\epsilon_{\text{boot}} = 0.632\epsilon_b + 0.368\epsilon_r \tag{C.14}$$

where $\epsilon_r$ is the re-substitution accuracy, i.e. the accuracy of the classifier on the training set given to the inducer, and $\epsilon_b$ is the accuracy on the bootstrap subset. Multiple bootstrap subsets $D_{b,i}$ must be generated and the corresponding values are used to estimate the accuracy by averaging the results

$$\bar{\epsilon}_{\text{boot}} = \frac{1}{n_\epsilon} \sum_{i=1}^{n_\epsilon} \epsilon_{\text{boot}}(D_{b,i}) \tag{C.15}$$

and its variance. For some classifiers, e.g. perfect memorizers like single neighbor, nearest neighbor ones, the re-substitution accuracy is 100% resulting in optimistic estimates. A refinement of this .632 rule, the .632+ rule, has been proposed modifying the weights in Equation C.14 to accommodate the degree of overfitting $R$ of the classifier. The latter is estimated by the performance $\hat{\gamma}$ of the classifier on a zero-information synthetic dataset obtained by randomly associating points and labels from the original dataset:

$$\{(\boldsymbol{x}_i, l_i)\} \to \{(\boldsymbol{x}_i, l_j)\}_{i,j} \tag{C.16}$$

leading to

$$\hat{R} = \frac{\epsilon_n - \epsilon_r}{\hat{\gamma} - \epsilon_r} \tag{C.17}$$

and the new rule is given by

$$\epsilon^{+}_{\text{boot}} = \hat{w}\epsilon_b + (1 - \hat{w})\epsilon_r \tag{C.18}$$

where

$$\hat{w} = \frac{0.632}{1 - 0.328\hat{R}}. \tag{C.19}$$

The third method, $k$-fold cross-validation, is based on the subdivision of the dataset into $k$ mutually exclusive subsets of (approximately) equal size: each one of them is used in turn for testing while the remaining $k - 1$ groups are given to the inducer to estimate the parameters of the classifier. If we denote by $D_{\{i\}}$ the set that includes instance $i$, then

$$\hat{\epsilon}_k = \frac{1}{N} \sum_i \delta[J(D \setminus D_{\{i\}}; \boldsymbol{x}_i), l_i]. \tag{C.20}$$

As in the hold-out method, the resulting value depends on the split and stratification is recommended. Complete cross-validation would require averaging over all $\binom{N}{N/k}$ possible choices of the $N/k$ testing instances out of $N$ and is too expensive except for the case $k = 1$ which is also known as leave-one-out (LOO). Besides being the only complete cross-validation used in practice, leave-one-out error (or accuracy) estimation is important because of the following theorem:

**Theorem C.2.1 (Leave-one-out error is almost unbiased).** *Let $P$ be a distribution over $\mathcal{S} = \mathcal{X} \times \mathcal{L}$, and $D_N$ and $D_{N-1}$ two samples (datasets) of size $N$ and $N - 1$ respectively drawn independently and identically distributed from $P$. Let $\epsilon[J(D_N)]$ be the expected accuracy of an estimator derived from sample $D_N$. Then, for any learning algorithm, the leave-one-out estimate is almost unbiased*

$$E_{D_{N-1}}\{\epsilon[J(D_{N-1})]\} = E_{D_N}\{\epsilon_{\text{LOO}}[J(D_N)]\}. \tag{C.21}$$

An important concept is that of the stability of an inducer:

**Definition C.2.2.** *An inducer is said to be stable for a given dataset and a set of perturbations if it induces classifiers making the same predictions when given the perturbed datasets.*

The following proposition addresses the computation of variance of the cross-validation estimates for stable inducers:

**Proposition C.2.3.** *If the inducer is stable under the perturbations caused by deleting instances for the folds in $k$-fold cross-validation, the cross-validation estimate will be unbiased and the variance of the estimated accuracy will be approximately $\epsilon(1 - \epsilon)/N$ where $N$ is the number of instances in the dataset.*

The proposition can be proved by observing that as the $k$ classifiers make the same predictions under the hypothesis of stability, the situation is equivalent to the hold-out case with a testing set equal to the complete dataset. Let us further note that the expression of variance does not depend on $k$. The smaller the perturbation, the more stable the inducer is expected to be: the stability of $k$-fold cross-validation is expected to increase with $k$, suggesting that in order to get a good estimate of the variance of the accuracy estimate, large values are to be preferred. However, when stability is reached, usually with $k \in [10, 20]$, there is no point in going further.

The results reported in the literature suggest that bootstrap, at least when the basic .632 rule is used, may be severely biased, while 10-fold cross-validation turns out to be a good choice in many situations.

If we want to select the optimal model and get an estimate of its accuracy, the procedures previously described must be modified to accommodate a three-way split:

1. Training data for the different models.
2. Validation data to select the best model.
3. Testing data to estimate the performance of the best model retrained with the training and validation data.

The reason for the three-way split is that the choice of the model is yet another parameter of the complete system and it may bias performance estimation.

## C.3. Analyzing the Performance of a Classifier

We saw in the previous section how available data should be used to obtain a reliable estimate of the accuracy of a classifier, supplemented by variance information, quantifying the confidence in it. There are cases, however, and biometric identification is one of them, where scoring a classifier by a single number is not informative enough. More specifically, when the number of samples in each class varies significantly or the possible classification errors have uneven relevance, average accuracy is not a good qualifier of the performance of a classifier. In such cases we characterize a classifier $J$ not by a single number but with a matrix, called the confusion (or matching) matrix

$$M_{hr} = P(h = J(\boldsymbol{x}, r), r) \tag{C.22}$$

whose rows/columns are associated respectively to the hypothesized/real class labels. For an identification system the labels can correspond to the different persons plus a reject label, providing a detailed view of the results of the classifier, while for a verification system two labels for *positive match* (i.e. it verifies) and *negative match* (it does not verify) are enough. Let us focus on the latter scenario and let $\{p, n\}$ represent the true (positive and negative) classes and $\{\pi, \eta\}$ represent the corresponding predicted classes (see Figure C.2): the off-diagonal entries represent errors. Several metrics can be built from the four elements of the two-class confusion matrix:

$$\text{precision: } \frac{T_p}{T_p + F_p} \tag{C.23}$$

$$\text{accuracy: } \frac{T_p + T_n}{P + N} \tag{C.24}$$

$$\text{recall: } \frac{T_p}{P} \tag{C.25}$$

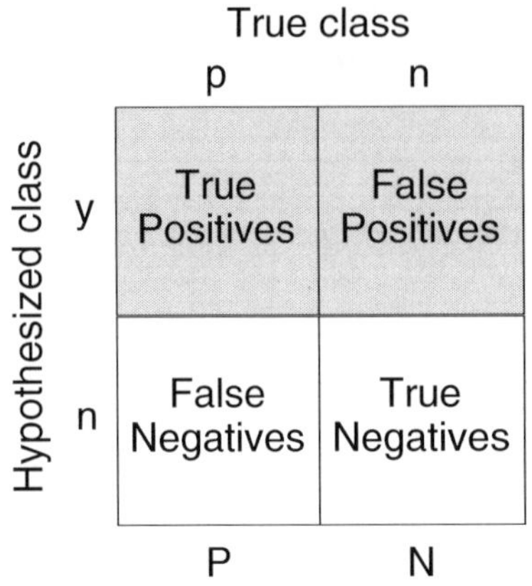


**Figure C.2.** The confusion matrix for a two-class classification problem. Source: Fawcett (2006).

where $P$ ($N$) stands for the number of positives (negatives) in the test set and $T_x$ ($F_x$) denotes the number of correct (wrong) label assignments for class $x$. Note that errors falling within $F_p$ are also known in the statistical literature as type I (or $\alpha$) errors, while those falling within $F_n$ are known as type II (or $\beta$) errors. Two quantities are routinely used to quantify the performance of a classifier: the true positive rate $T_p$ (hit rate or recall) and the false positive rate $F_p$ (false alarm rate). A classifier that produces a single label per input can be characterized by a single point in a two-dimensional ROC curve whose horizontal/vertical axis reports respectively $F_p/T_p$: the ideal classifier is the one located in the upper left corner (0, 1). Many applications are characterized by a large number of negative instances, e.g. non-face patterns when searching for a face in a movie. In these cases, the smaller the value of $F_p$, the better: the left area of the ROC graph is more appealing. A random classifier would lie on the line $y = x$, the exact position being determined by the frequency with which it guesses the positive class. A classifier lying below the line $y = x$ would be worse than random: reversing its output would yield a better than random classifier located above the line. While a classifier producing a single label per instance generates a single representative point in ROC space, any classifier generating scores can generate multiple binary classifiers, one for each threshold applied to the scores it produces. Let us assume that positive instances are characterized by high scores: we can generate a discrete classifier by binarizing the scores with a threshold $\theta$. When $\theta \to +\infty$ nearly all instances will fall in the negative class and $F_p \to 0$; when $\theta \to -\infty$ nearly all instances will fall in the positive class and $T_p \to 1$. Moving $\theta$ from $+\infty$ to $-\infty$ will trace a curve in ROC space from left to right (see Intermezzo C.1 for an efficient way of computing the curve). Most classifiers considered in this book are scoring classifiers and can be characterized by a complete ROC curve.

An important feature of ROC diagrams is that they are not affected by class skewness: the values used to compute the coordinates are derived from a single column of the confusion matrix and are not affected by the relative cardinalities as would happen for quantities such as precision and recall that are computed from values belonging to both columns. The definitions of $T_p$ and $F_p$ make ROC diagrams invariant also to error costs. These invariances do not detract from the usefulness of the diagrams: different error costs and class distributions change the region of interest within the graph without altering the graph itself.

**Intermezzo C.1.** Efficient generation of an ROC curve

Let us assume that we have the set of testing scores $\{s(\boldsymbol{x}_i)\}$ from a classifier $J_{\theta_0}$. As

$$J_{\theta_0}(s(\boldsymbol{x}_i)) = \mathrm{p} \quad \text{when} \quad s(\boldsymbol{x}_i) > \theta_0 \Rightarrow J_\theta(s(\boldsymbol{x}_i)) = \mathrm{p}, \quad \forall \theta < \theta_0 \tag{C.26}$$

we can exploit the monotonicity by sorting all test instances by decreasing scores, and checking each of them in turn, progressively updating $T_\mathrm{p}$ and $F_\mathrm{p}$ based on the true data label, without even comparing the scores with the threshold. Whenever the score of the instance considered differs from that of the previous one, a new point $(F_\mathrm{p}, T_\mathrm{p})$ of the ROC curve is generated. If the score does not change, we keep updating $T_\mathrm{p}$ and $F_\mathrm{p}$ but we do not generate any new points: the tract of the curve corresponding to the sequence of equal values will then be a segment.

Chapter 4 addressed the important issue of robustness of statistical estimators, but there are other levels at which the concept of robustness plays a critical role. An important level at which robustness is required is that of the operating conditions of a classifier. In the previous section we focused on how to estimate the accuracy of a classifier, a figure of merit that depends on class distribution. The same statistical estimation approach can be extended to different figures of merit such as the cost of a classifier, which depends on the class distribution and the weights given to the different classification errors. Choosing the best classifier according to this type of score entails a commitment to a specific operating condition, the one specified by the class distribution and weights upon which the discriminating score is computed. It is not uncommon for the knowledge of class distribution to be imprecise and error costs may vary during the operation cycle of the classifier. This variability may be aggravated by the time delay between the construction of the classifier and its application. The ROC curve provides the basis for an analysis of the sensitivity of a classifier to variable operating conditions and also for the synthesis of hybrid classifiers that are robust and optimal under changes in the operating conditions. Let us assume as classifier score its expected cost $\hat{C}$, a quantity that can be expressed in terms of its coordinate $(F_\mathrm{p}, T_\mathrm{p})$ in ROC space:

$$\hat{C} = p(\mathrm{p})(1 - T_\mathrm{p})C_{\eta\mathrm{p}} + p(\mathrm{n})F_\mathrm{p}C_{\pi\mathrm{n}} \tag{C.27}$$

where $C_{\eta\mathrm{p}}$ and $C_{\pi\mathrm{n}}$ represent the costs for the corresponding classification errors. Two classifiers represented by $(F_{\mathrm{p}1}, T_{\mathrm{p}1})$ and $(F_{\mathrm{p}2}, T_{\mathrm{p}2})$ have the same cost $m_{\mathrm{ec}}$ if

$$m_{\mathrm{ec}} = \frac{T_{\mathrm{p}2} - T_{\mathrm{p}1}}{T_{\mathrm{p}2} - T_{\mathrm{p}1}} = \frac{C_{\pi\mathrm{n}}p(\mathrm{n})}{C_{\eta\mathrm{p}}p(\mathrm{p})} \tag{C.28}$$

so that all points lying on a line whose slope is $m$ have the same performance, therefore identifying an iso-performance line. Inspection of Equation C.27 shows that the cost can be computed from the intercept of the line with the ordinate axis ($F_p = 0$): the higher the intercept, the lower the cost. Let us now consider the convex hull of a set of classifiers (see Figure C.3). The reason for considering the convex hull of a set of available classifiers is that it is possible to synthesize the classifier corresponding to any point within the convex hull relying on the following result:

**Proposition C.3.1.** *For any set of cost and class distributions, there is a point on the ROC convex hull (ROCCH) with minimum expected cost.*

In fact, among the set of parallel lines identified by the operating conditions described by Equations C.27–C.28, there will be one passing through one of the vertexes of the convex

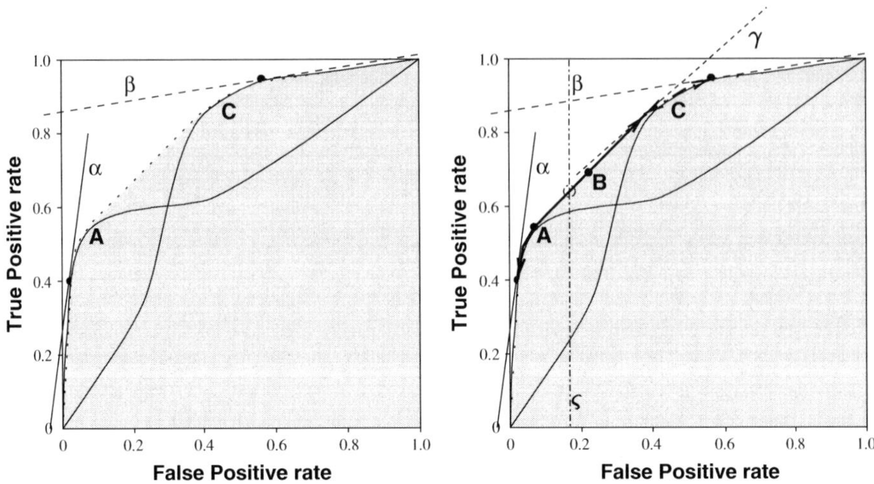


**Figure C.3.** Convex hull of ROC curves and ROCCH hybrid classifiers: the shaded area represents the AUC. The plot on the right illustrates two important facts related to ROC analysis. A vertical line at $F_p = \varsigma$ corresponds to a Neyman–Pearson constraint. Given the two classifiers $A$ and $B$ we can interpolate between them obtaining a classifier that satisfies the constraint. The two sets of lines $\{\alpha, \gamma\}$ and $\{\gamma, \beta\}$ identify two different operating ranges. The classifier considered is sensitive to operating condition variations in $\{\alpha, \gamma\}$ but not as much to variations in $\{\gamma, \beta\}$. Source: Fawcett (2006).

hull whose intercept will be maximal: this line identifies the minimum expected cost classifier available as the one corresponding to the convex hull vertex. This means that, given a set of classifiers, characterized by single points or curves in ROC space, it is possible to select for each operating condition, as determined by the cost, an optimal classifier. The concept of ROC convex hull is useful also under other optimality conditions. As we have seen in Chapter 3, an important criterion in classification is the Neyman–Pearson criterion, which specifies a maximum allowable $F_p$ rate. This constraint can be visualized in ROC space by a vertical line at $F_\mathrm{p}$. The optimal classifier is given by the intersection $\boldsymbol{a}_x$ of the constraint line with the convex hull. The corresponding classifier can be built using the following result that allows us to build a novel classifier, the ROC convex hull hybrid, for any point in ROC space on the segment joining any two available classifiers:

**Proposition C.3.2 (ROC convex hull hybrid).** *Given two classifiers $J_1$ and $J_2$ represented within ROC space by the points $\boldsymbol{a}_1 = (F_{\mathrm{p}1}, T_{\mathrm{p}1})$ and $\boldsymbol{a}_2 = (F_{\mathrm{p}2}, T_{\mathrm{p}2})$, it is possible to generate a classifier for each point $\boldsymbol{a}_x$ on the segment joining $\boldsymbol{a}_1$ and $\boldsymbol{a}_2$ with a randomized decision rule that samples $J_1$ with probability*

$$p(J_1) = \frac{\|\boldsymbol{a}_2 - \boldsymbol{a}_x\|}{\|\boldsymbol{a}_2 - \boldsymbol{a}_1\|}. \tag{C.29}$$

This can be extended to the general case of iso-performance lines by considering their intersection with the ROCCH. An immediate consequence is that, for some operating conditions, the ROCCH hybrid classifier may beat the constituent classifiers. The slope of the ROC curve corresponds to the likelihood ratio of the decision criterion $\theta$: it need not be monotonic, but it actually is when the decision criterion function is the likelihood ratio (otherwise the curve would not be concave and we could find an ROC hybrid improving on the Neyman–Pearson classifier). An ROC curve with non-increasing slope is said to be a proper ROC curve. If we consider two iso-performance lines corresponding to the extrema of a range of operating conditions, they will identify a portion of the convex hull: the length of the arc is a measure of the sensitivity of the choice of the optimal classifier to the operating conditions. The ROC curve can also be used to compute a single scalar value characterizing the quality of a classifier. The area under the curve (AUC) gives the probability that the classifier will score, and hence rank, a randomly given positive instance higher than a randomly chosen one. This value is equivalent to the Wilcoxon rank test statistic $W$

$$W = \frac{1}{PN} \sum_{i:l_i=\mathrm{p}} \sum_{j:l_j=\mathrm{n}} w(s(\boldsymbol{x}_i), s(\boldsymbol{x}_j)) \tag{C.30}$$

where, assuming no ties,

$$w(s(\boldsymbol{x}_i), s(\boldsymbol{x}_j)) = 1 \quad \text{if } s(\boldsymbol{x}_i) > s(\boldsymbol{x}_j) \tag{C.31}$$

(see Intermezzo C.2). The Wilcoxon statistic provides a non-parametric estimate of the separability of positive from negative instances based on the rank of their scores: the closer the area to 1, the better the classifier.

**Intermezzo C.2.** AUC equivalence to the Wilcoxon statistic

The equivalence can be most easily appreciated in the continuous case. Let $P_{\mathrm{p}}^{s}(\theta)$ be the probability that $s(\boldsymbol{x}) = \theta$ when $\boldsymbol{x} \in \mathrm{p}$, $p_{\mathrm{p}}^{s}(\theta)$ the corresponding density, and similarly for class n. The Wilcoxon statistic can be written as

$$W = \int p_{\mathrm{p}}^{s}(\theta) P_{\mathrm{n}}^{s}(\theta)\, d\theta. \tag{C.32}$$

As

$$F_{\mathrm{p}} = 1 - P_{\mathrm{n}}^{s}(\theta) \tag{C.33}$$

$$T_{\mathrm{p}} = \int_{\theta}^{\infty} p_{\mathrm{p}}^{s}(\theta)\, d\theta \tag{C.34}$$

$$\frac{dT_{\mathrm{p}}}{d\theta} = p_{\mathrm{p}}^{s}(\theta) \tag{C.35}$$

we can rewrite Equation C.32 as

$$W = \int (1 - F_{\mathrm{p}})\, dT_{\mathrm{p}} = 1 - \int F_{\mathrm{p}}\, dT_{\mathrm{p}} \tag{C.36}$$

where the last integral on the right represents the area above the curve, thereby proving the result.

The AUC can be used to characterize classifiers represented by a single point in ROC space. Besides the straightforward possibility of building a complete curve by connecting the point to the two extremal ones (0, 0) and (1, 1), it is possible to compute an explicit lower bound $A_-$ and upper bound $A_+$ for proper ROC curves passing through a given point $(F, T)$:

$$A_- = \frac{1}{2}(1 + T - F) \tag{C.37}$$

$$A_+ = \begin{cases} 1 - 2T(1 - F) & \text{if } F < 0.5 < T \\ \dfrac{(1 - F)}{2T} & \text{if } F < T < 0.5 \\ 1 - \dfrac{(1 - T)}{2(1 - F)} & \text{if } 0.5 < F < T \end{cases} \tag{C.38}$$

and we can use $\hat{A} = (A_- + A_+)/2$ to quantify the discriminability of a single point classifier.

It is possible to map an ROC curve into spaces characterized by different evaluation metrics. An example is the possibility of mapping an ROC curve, and its convex hull, into the precision recall (PR) space, provided that the recall is non-null, using Equations C.23–C.25. The projection of the ROCCH into a PR space identifies the so-called achievable PR curve, which cannot be obtained in PR space by linear interpolation. The following theorem is of interest:

**Theorem C.3.3.** *For a fixed number of positive and negative examples, one curve dominates a second curve, i.e. has higher $T_p$ for any $F_p$, in ROC space if and only if the first dominates the second in PR space.*

This result tells us that there is no real advantage for characterizing the performance in PR space, in spite of the preference gathered by PR space especially in the case of highly skewed databases.

## C.4. Evaluating a Technology

In the previous sections we investigated how a classification system can be scored, addressing some of the subtleties arising in the definition of performance metrics and in estimating the real accuracy of the system when a limited set of data is available. When we move our focus from the evaluation of a single classification system, or the comparison of a limited set of systems, to the comparative evaluation of a large number of systems addressing a given task, in an attempt to get a critical, overall view of the state of the art of technology, additional issues surface. A first remark is that an extensive comparison of algorithms cannot (usually) be performed by a single group: the required effort would be huge and very often critical details of the algorithms to be compared have not been reported in the literature. This means that in order to compare the different solutions, the best implementation of each system should be used, usually from the people that developed it. The evaluation should then be administered by independent groups, providing a fair comparison of the submitted systems. Another key issue is that the testing set used for the comparison should not be available to the developer of the systems prior to the evaluation. As we remarked in Section C.2, this would permit tuning of the systems, biasing the estimation of their performance. Another important feature of a well-designed evaluation is the difficulty of the tests it employs. If the tests to be performed are too difficult or too easy, the performances of the systems will be too grouped, being statistically indistinguishable (especially in the case of easy evaluations) or meaningless (merging with noise if the tests were too difficult). A related problem is the cardinality of the dataset used in the evaluation. The scoring of systems is of a statistical nature and the reliability of the resulting figures depends significantly on the size of the datasets. The performance of a system is usually associated to a measure of its variability: in order to reliably discriminate between two systems, the uncertainty in the estimate of their performance should be as small as possible, requiring the largest available datasets.

A current major application of template matching techniques, such as those presented in this book, is biometrics: the development of systems that recognize a person by his biometric signature such as a face image, a voice recording, a three-dimensional head scan, a fingerprint, a retinal or iris image. Each biometric signature associated to a person exhibits characteristic variability associated to the conditions (controlled illumination, outdoor environments) and timing (aging) of the acquisitions. The selectivity of the signature may depend on the demographics of the populations (composition and size). Furthermore, as we remarked in Section C.1, different tasks may be performed on (approximately) the same input data but requiring different evaluation metrics. An important case in biometrics is that of verification and identification. In order to provide reliable and fair evaluation of the compared systems, the evaluation procedures should be fully automated, based on a clearly defined protocol and data format. Furthermore, the size of the database required for significant statistical evaluation requires, on the one hand, that the algorithms themselves be reasonably efficient to compute their response on all available data and, on the other hand, that repeated applications of the algorithms for different experiments be avoided whenever possible, the evaluation relying on previously computed data. The community working on the development of face recognition systems has become increasingly aware of the issues involved in a large-scale evaluation of technologies and the Face Recognition Vendor Test 2002 (FRVT 2002) represents a remarkable achievement in the field. Let us now see in detail how this test was organized.

Each system must be able to compare a biometric signature, composed of multiple images for an individual, reporting a scalar similarity score. Images from the same person (should) have a large similarity score. This is not restrictive as distances can be turned into similarity scores by simply changing their sign. Each algorithm compares the images in a query set $Q$ to the images in a target set $T$. The result is a (rectangular) similarity matrix $S$ whose entry $S_{ij}$ reports the score of the $i$th target element with the $j$th element in the query set. If we consider as target and query set the set of all available signatures we obtain a square matrix that contains (nearly) all the information needed to perform all possible experiments by restricting the set of scores used. This is important because there is no need to compute multiple times the same data once the complete matrix $S$ has been computed. Each experiment is then based on the use of virtual image sets, as represented by subsets of $S$ entries. The gallery $G$ is a subset of $T$ that contains a single biometric signature per person and represents the data enrolled in a biometric system. The probe set $P_G$ is a subset of $Q$ such that all of its images have a match in $G$. The impostor set $P_I$ is a subset of $Q$ whose images have no match in the gallery and represents the set of persons trying to defeat the system: the results of all comparisons can be extracted from $S$. The performance metrics employed by FRVT 2002 are centered on the formalization of the so-called watch list task, considered as a generalization of both identification and verification tasks. A watch list performance experiment requires a watch list gallery $G$, a probe set $P_G$ with people to be identified, and a probe set $P_I$ with people that should not be identified. The detection and identification rate is defined for a given threshold $\theta$ and maximum rank $r$ in the score list as

$$P_{\mathrm{DI}}(\theta, r) = \frac{|\{j \in P_G : \mathrm{rank}_{i\in G}(s_{ij}) \leq r,\ s_{ij} > \theta,\ \mathrm{id}(j) = \mathrm{id}(i)\}|}{|P_G|} \qquad \text{(C.39)}$$

while the false alarm rate is computed as

$$P_{\mathrm{F}} = \frac{|\{j \in P_I : \max_{i \in G}(s_{ij}) \geq \theta\}|}{|P_I|}. \tag{C.40}$$

Identification is a special case of the watch list case in a closed universe. As in this case the false alarm rate is not defined (all people are known to the system), the identification rate completely characterizes the performance of the system:

$$P_{\mathrm{I}}(r) = \frac{|\{j \in P_G : \operatorname{rank}(s_{ij}) \leq r\}|}{|G|}. \tag{C.41}$$

Verification is characterized by the comparison of a probe to a single biometric signature gallery, corresponding to the claimed identity: the resulting similarity score is compared to a threshold to verify the claim, following the Neyman–Pearson paradigm. By varying the threshold we obtain the ROC curve described in the previous section:

$$\left(\frac{|\{s_{ij} : s_{ij} \geq \theta,\ i \in G,\ j \in P_I\}|}{|P_I||G|}, \frac{\{s_{ij} : s_{ij} \geq \theta,\ j \in P_G, \operatorname{id}(j) = \operatorname{id}(i)\}}{|P_G|}\right). \tag{C.42}$$

Random sampling of the space of possible galleries and probes permits the computation of confidence ranges for the performance values. In order to increase the flexibility of the scoring procedure, the possibility of normalizing the set of scores resulting from the comparison of a probe to a gallery was introduced in FRVT 2002. Given the vector $\boldsymbol{s}_p$ representing the matches of probe $p$ to all images in the gallery, a score normalization function is a function $f : \mathbb{R}^N \to \mathbb{R}^N$. Two possibilities were considered:

$$\boldsymbol{t} = f(\boldsymbol{s}) \tag{C.43}$$

$$\boldsymbol{t} = f(\boldsymbol{s}, S_{GG}) \tag{C.44}$$

where $S_{GG}$ is a matrix of similarities between all gallery pairs. Note that, in the case of verification, the use of a normalization function transforms the original 1–1 problem into a 1-to-many problem, with a consequent increase in the required computation.

The large databases used by FRVT 2002, which included a set of 121 589 images from 37 437 different subjects, supported the performance of several experiments based on multiple image subsets. Each experiment assessed the performance of face recognition algorithms under different operating conditions characterized by the way images in the gallery and probe sets were acquired, their demographics, and the parameters used for the watch list task:

- same day indoor images, expression/illumination change;
- indoor different day (and more than 18 month lapse);
- outdoor same/different day;
- impact of sex, age, and their interaction;
- watch list performance as a function of gallery size and rank.

A detailed analysis of the results would take up too much space, but among the findings we would like to report that:

- the use of a normalizing function improved verification but not identification;
- the verification performance of face matching systems is comparable to that of fingerprint systems at a false acceptance rate of 0.01, being worse at lower rates and better at higher ones;
- identification performance decreases logarithmically with gallery size.

FRVT 2002 was followed by a similar test, FRVT 2006, focusing on the possibility of reliably estimating false reject rates as low as 0.01 at a false acceptance rate as low as 0.001. This required the acquisition of large image databases at high resolution but with a somewhat limited number of different subjects. The new tests included experiments on the large database of FRVT 2002 in order to measure the advancement of the technology: results showed a four-fold reduction of the false rejection rate at a false acceptance rate of 0.01. A comparison to iris identification also showed that at a false acceptance ratio of 0.01, high-resolution still image face recognition has comparable performance. The Face Recognition Vendor Test is thus an interesting example of sustained technology evaluation from many perspectives: fairness, completeness, incremental testing, and more.

## C.5. Bibliographical Remarks

The section on error measures for detection and localization systems is based on the papers Popovici *et al.* (2004) and Rodriguez *et al.* (2006).

The literature on cross-validation and the bootstrap is vast, but the corresponding section is mainly based on the paper by Kohavi (1995), a lucid account of the main features of the two methods supported by significant experimental evidence. The .632+ bootstrap rule was introduced by Efron and Tibshirani (1997).

The analysis of the ROC curve as a means of representing the performance of a classifier is mainly based on the paper by Fawcett (2006). The proofs of some of the results can be found in the paper by Provost and Fawcett (2001). The paper by Barreno *et al.* (2007) extends these results to obtain the optimal decision rule in a more general setting. The probabilistic interpretation of the AUC is discussed at length in Hanley and McNeil (1982) while the computation of explicit upper and lower bounds for proper ROC curves passing through a given point is presented by Mueller and Zhang (2006). The AUC ignores probability values and only considers ranking information: an attempt to extend it to incorporate probabilistic information and to associate it to a modified probabilistic ROC curve is reported by Ferri *et al.* (2005).

The large-scale evaluation of face recognition technology performed initially under the DARPA FERET evaluation program and subsumed by the Face Recognition Vendor Test series is described in several papers (Grother *et al.* 2003; Phillips *et al.* 2006, 2005, 2003, 2000a,b, 2007).

## References

Barreno M, Cardenas A and Tygar J 2007 Optimal ROC curve for a combination of classifiers. *Proceedings of Advances in Neural Information Processing Systems*, vol. 20, pp. 57–64.

Efron B and Tibshirani R 1997 Improvements on cross-validation: The .632+ bootstrap method. *Journal of the American Statistical Association* **92**, 548–560.

Fawcett T 2006 An introduction to ROC analysis. *Pattern Recognition Letters* **27**, 861–874.

Ferri C, Flach P, Hernandez-Orallo J and Senad A 2005 Modifying ROC curves to incorporate predicted probabilities. *Proceedings of the ICML 2005 Workshop on ROC Analysis in Machine Learning.*

Grother P, Micheals R and Phillips P 2003 Face Recognition Vendor Test 2002 Performance Metrics. *Proceedings of the 4th International Conference on Audio- and Video-Based Biometric Person Authentication*, Lecture Notes in Computer Science, vol. 2688, pp. 937–945. Springer.

Hanley J and McNeil B 1982 The meaning and use of the area under a receiver operating characteristic (ROC) curve. *Radiology* **143**, 29–36.

Kohavi R 1995 A study of cross-validation and bootstrap for accuracy estimation and model. *Proceedings of the International Joint Conference on Artificial Intelligence*, pp. 1137–1145.

Mueller S and Zhang J 2006 Upper and lower bounds of area under ROC curves and index of discriminability of classifier performance. *Proceedings of the ICML 2006 Workshop on ROC Analysis in Machine Learning*, pp. 41–46.

Phillips P, Flynn P, Scruggs T, Bowyer K and Worek W 2006 Preliminary face recognition grand challenge results. *Proceedings of the 7th International Conference on Automatic Face and Gesture Recognition (FG'06)*, pp. 15–24.

Phillips P, Flynn P, Scruggs T, Bowyer K, Jin C, Hoffman K, Marques J, Jaesik M and Worek W 2005 Overview of the face recognition grand challenge. *Proceedings of the IEEE Conference on Computer Vision and Pattern Recognition (CVPR'05)*, vol. 1, pp. 947–954.

Phillips P, Grother P, Micheals R, Blackburne D, Tabassi E and Bone M 2003 Face Recognition Vendor Test 2002 Evaluation Report. Technical Report NISTIR 6965, National Institute of Standards and Technology.

Phillips P, Hyeonjoon M, Rizvi S and Rauss P 2000a The FERET evaluation methodology for face-recognition algorithms. *IEEE Transactions on Pattern Analysis and Machine Intelligence* **22**, 1090–1104.

Phillips P, Martin A, Wilson C and Przybocki M 2000b An introduction to evaluating biometric systems. *IEEE Computer* **33**, 56–63.

Phillips P, Scruggs W, O'Toole A, Flynn P, Bowyer K, Schott C and Sharpe M 2007 FRVT 2006 and ICE 2006 large-scale results. Technical Report NISTIR 7408, National Institute of Standards and Technology.

Popovici V, Thiran J, Rodriguez Y and Marcel S 2004 On performance evaluation of face detection and localization algorithms. *Proceedings of the 17th IAPR International Conference on Pattern Recognition (ICPR'04)*, vol. 1, pp. 313–317.

Provost F and Fawcett T 2001 Robust classification for imprecise environments. *Machine Learning* **42**, 203–231.

Rodriguez Y, Carinaux F, Bengio S and Mariethoz J 2006 Measuring the performance of face localization systems. *Image and Vision Computing* **224**, 882–893.

# INDEX



*Template Matching Techniques in Computer Vision: Theory and Practice* Roberto Brunelli

Printed and bound by CPI Group (UK) Ltd, Croydon, CR0 4YY